Soft and Biological Matter

"Soft and Biological Matter" is a series of authoritative books covering established and emergent areas in the realm of soft matter science, including biological systems spanning from the molecular to the mesoscale. It aims to serve a broad interdisciplinary community of students and researchers in physics, chemistry, biophysics and materials science.

Pure research monographs in the series as well as those of more pedagogical nature, will emphasize topics in fundamental physics, synthesis and design, characterization and new prospective applications of soft and biological matter systems. The series will encompass experimental, theoretical and computational approaches.

Both authored and edited volumes will be considered.

More information about this series at http://www.springer.com/series/10783

Igor Muševič

Liquid Crystal Colloids

Springer

Igor Muševič
Jožef Stefan Institute
University of Ljubljana
Ljubljana
Slovenia

ISSN 2213-1736 ISSN 2213-1744 (electronic)
Soft and Biological Matter
ISBN 978-3-319-54914-9 ISBN 978-3-319-54916-3 (eBook)
DOI 10.1007/978-3-319-54916-3

Library of Congress Control Number: 2017933441

Printed on acid-free paper

This Springer imprint is published by Springer Nature
The registered company is Springer International Publishing AG
The registered company address is: Gewerbestrasse 11, 6330 Cham, Switzerland

For Maja, Sašo and Nataša

Preface

I started the work on liquid crystal colloids around the year 2000 as a continuation of our experiments on forces between surfaces in liquid crystals. We were using the atomic force microscope (AFM) as a tool to measure structural forces in the nematic and smectic liquid crystals. In those experiments, liquid crystal was confined to a very narrow gap, between 1 and 100 nanometres, typically between a perfectly smooth crystalline surface of mica and the micrometre diameter glass microsphere attached to the cantilever of the AFM. We were looking for the fluctuation forces, also called the Casimir force, of the nematic director field confined to the thin gap between the mica and the microsphere. We never observed the Casimir force, simply because it was to tricky to separate it from the much stronger mean-field forces, which were caused by the gradient of the order parameter in that gap. These mean-field forces between the two surfaces at the nanometre separation were indeed strong, and I was questioning myself whether similar forces between the surfaces of the particles in the nematic liquid crystal exist at a much larger separation, of the order of a micron or so. At the same time, two articles came along my desk: one was the *Science* paper by Poulin, Stark, Lubensky and Wietz, showing beautiful images of the attraction of water droplets in nematics. The other was a paper on photonic crystals by Yablonovitch and Gmitter, published in *Physical Review Letters* in 1989. When reading these papers, I asked myself whether it is possible to build photonic crystals by putting small colloidal particles into nematic liquid crystals, combining the essence the above-mentioned both papers. This was around the year 2000.

At that time, the experiments with colloidal particles in nematics were practically a kind of science fiction, since no tool was available for grabbing individual micrometre diameter particles inside the measuring cell and moving them to an arbitrary position. However, we were lucky that in that period laser tweezers started to emerge as very powerful non-contact tweezers, which use light to trap and manipulate the particles even when they are floating in the nematic liquid crystal sandwiched between two rather thick glass plates.

As mentioned before, we started the experiments with laser tweezers by good luck, using low refractive glass beads, putting them in the nematic liquid crystal and trying to see whether the tweezers are able to grab such a particle and move it. It came as a big surprise to the three of us, Miha Škarabot, Igor Poberaj and myself, when we saw that glass beads could nicely be trapped in the nematic liquid crystal, although the refractive index of glass is lower than the refractive indices of both nematic liquid crystal used in the experiments. I explained this anomalous trapping by noting that polarisation of the laser tweezers is an important parameter and that the second important phenomenon is the action of the strong electric field of the laser tweezers on the nematic liquid crystal itself. We could soon explain this anomalous trapping on a general grounds by considering elastic deformation of nematic liquid crystal by light and the polarisation of light, which selectively grabs the nematic director along the direction of larger polarizibility and moves the particle together with the field.

After the work on this anomalous trapping of colloids by optical tweezers was published, there was a lot of excitement and new ideas on what physics experiments could be done with the laser tweezers. This result opened the exciting pathway to the assembly of the first 2D nematic colloidal crystal in 2006. Another milestone occurred in the same year 2006, when Miha Ravnik proposed to use local quench of the nematic liquid crystal colloids from the isotropic phase to the nematic phase and entangle two colloidal particles. This was soon realised in the experiment by using the absorption of the laser tweezers in the nematic liquid crystal colloids to melt the nematic and then quench it. The result was colloidal entanglement, where colloidal particles could be entangled by topological defect loops. The entanglement later provided the most complex colloidal binding in the nematic liquid crystals, i.e. knotting and linking of colloids, which was observed by Uroš Tkalec and myself in 2010. Linking and knotting of the nematic liquid crystal opened the doors to the experimental topology in liquid crystals and what followed were several years of very creative and exciting exploration of this complex phenomenon by many different groups all over the world.

Along the work on nematic colloids, I began in 2007 with Matjaž Humar a parallel work on an entirely new line of research with the aim of using the nematic dispersions for microphotonics. There was only a slight hint at that time that liquid crystals could provide a beautiful setting for the realisation of tiny photonic elements, such as optical microcavities, microlasers and photonic microfibers. Indeed, all this was found latter in the dispersions of nematic or cholesteric liquid crystals in other immiscible fluids, such as water. This led me to consider whether one could use the structural forces for binding and entangling colloidal particles in the nematics together with the photonic properties of nematic dispersions. The aim is to create a soft matter analogue to the solid state microphotonic circuits, where instead of silicon, a soft matter is used to self-assemble into a topological photonic soft matter. At present, we have clear proof that such technology based on soft matter self-assembly is indeed possible.

This work is written as a kind of retrospective of all the mentioned studies and it follows, more or less chronologically, the experimental work performed in the last decade, starting from the year 2004. It turns out that telling the story in a chronological order is the best way to present the development of ideas from simple beginnings to very complex issues. Most of the work in this book was performed in my laboratory at the J. Stefan Institute; however, the book would be far from complete without including the work of other researchers from very different laboratories around the world. I did my best to include all relevant work in this field and I hope that I have not missed any important references.

Ljubljana, Slovenia
December 2016

Igor Muševič

Acknowledgements

I owe many thanks to a number of colleagues and friends with whom I have been working in all these years. In the first place, I would like to mention Miha Škarabot, with whom I started the experiments on nematic colloids around the year 2000. Several years before, Klemen Kočevar and myself performed the first Atomic Force Microscopy experiments of measuring forces between the particles in the nematic liquid crystals. At the same time, Giovanni Carbone collaborated with me and developed a new force measuring technique for detecting presmectic forces in the nematics. I also owe thanks to Igor Poberaj and Dušan Babič for enabling our work with their laser tweezers, which were later on used in so many fascinating experiments. One of the most demanding and exciting experiments was performed by Uroš Tkalec, who noticed strange knotting and linking of colloidal particles in chiral nematic cells. I still vividly remember many hours of discussion we had on this phenomenon, and how long it took for us to understand what was really happening in knotting and linking colloidal experiments. I owe many thanks to Ulyana Ognysta, Andriy Nych, Christian Bahr, Giorgio Mirri and Surajit Dhara. This work would not be possible without collaboration with my colleagues from the physics theory of liquid crystals, first Slobodan Žumer, then Miha Ravnik and Simon Čopar, with whom I spent countless hours of vivid discussions, analysis of the experiments, comparisons of experimental findings and theoretical predictions, etc. Without this synergy between the experiments and the theory, much less would be achieved in this field. I owe special thanks to Matjaž Humar, who was brave enough to start an entirely new line of research on the microphotonics of liquid crystals. A lot of this work was done by my Ph.D. students and post-docs, who visited my lab, and I would like to thank Venkata Subba Rao Jampani, Marjetka Conradi, Anna Ryzhkova, Maryam Nikkhou, Karthik Peddireddy, Gregor Posnjak, Maruša Mur and Uroš Jagodič.

Igor Muševič

Contents

Chapter 1
Introduction

Abstract Basics of the orientational order, phase transitions and Landau-de Gennes theory are presented. We briefly discuss optical properties of liquid crystals, which is necessary for understanding optical experiments in these materials. Basic introduction to topological defects in nematic liquid crystals is given, including winding number, topological charge, Euler characteristic, genus of colloids, and elastic energy of topological defects.

1.1 Molecular Order in Nematic Liquid Crystals

A nematic liquid crystal is a state of the matter which is intermediate between the isotropic liquid and the fully ordered solid crystal [1]. A nematic liquid-crystalline phase is usually formed of elongated, rod-like organic molecules, which are spontaneously ordered with their long axes parallel to each other, as illustrated in Fig. 1.1. The direction of this average molecular ordering is called the director. The centres of gravity of molecules show no long-range ordering and the matter is therefore a fluid.

The nematic phase is usually obtained by changing the temperature and that is why it is also called a thermotropic nematic liquid crystal. It is therefore stable within a certain temperature range. If heated above a certain temperature, which is called the clearing point (T_c), the orientational order of long molecular axes is lost and the nematic liquid crystal transforms into an isotropic fluid, as illustrated in Fig. 1.1. In this isotropic phase, there is no orientational order of long molecular axes and the centres of gravity of molecules are randomly distributed in space. On the other hand, if a nematic liquid crystal phase is cooled down, it usually transforms into a more solid-like phase; usually, this is the smectic-A phase, which still displays orientational order of liquid crystal molecules, but its positional disorder transforms into a 1-dimensional (1D) solid. This means that in the smectic-A phase, the molecules form layers, which are parallel to each other, as illustrated in Fig. 1.1. Within each layer, which is liquid-like, the molecules are oriented along the normal to the layers. Upon further cooling, the smectic A phase may transform into other partially positionally ordered, i.e. layered phases, such as the smectic C, smectic B, smectic H, etc.

I. Muševič, *Liquid Crystal Colloids*, Soft and Biological Matter,
DOI 10.1007/978-3-319-54916-3_1

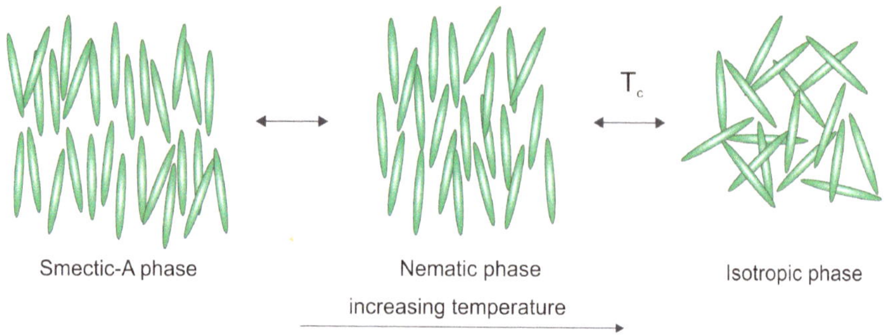

Fig. 1.1 Different phases of a thermotropic liquid crystal. The smectic-A, nematic and isotropic phase of a liquid crystal

The degree of orientational ordering of the rod-like molecules in the nematic liquid-crystalline phase is determined by the nematic order parameter S. Like any other order parameter within Landau theory of phase transitions (LdG) [2–4], this order parameter is equal to zero, $S = 0$, in the isotropic phase, and obtains a finite value in the nematic phase. It is given in a form of a second order Legendre polynomial:

$$S = \frac{1}{2} \cdot \langle (3cos^2\theta - 1) \rangle = \int f(\theta)\frac{1}{2}(3cos^2\theta - 1)d\Omega \qquad (1.1)$$

Here, θ is the angle between long axis of a selected molecule within the molecular ensemble and the average direction of all molecules forming the nematic phase, which is called a director **n**, see Fig. 1.2. The brackets $\langle\rangle$ denote the time -average of the angles θ over the ensemble of molecules. $f(\theta)$ is the orientational distribution function of molecular axes in space, and $d\Omega$ is the differential of the solid angle. It is clear from this expression that the nematic order parameter of a fully disordered isotropic phase is $S = 0$, whereas the maximum possible order parameter is $S = +1$. This occurs when all the molecules are perfectly aligned along the director. The values of the order parameter S are therefore formally restricted to the interval $[-1/2, 1]$, with $+1$ corresponding to fully aligned molecules along the director. The negative value of $S = -1/2$ formally corresponds to a state where all the molecules are aligned perpendicularly to the director. This state is different from the fully aligned state with $S = +1$ and is physically realised in liquid crystals with disc-like molecules or in special cases when the rod-like molecules are orientationally disordered and form pancake-like probability distribution.

Let us remind that the order parameter S, which specifies the local degree of ordering, is a scalar quantity and therefor cannot not describe the orientational collective ordering of the nematic molecules in space. To this aim, one has to combine the director field **n** with the degree of local order S, which naturally leads to a second-rank tensor Q, also called the Q-tensor, which describes both the degree of local order and its "direction". This tensor is expressed as:

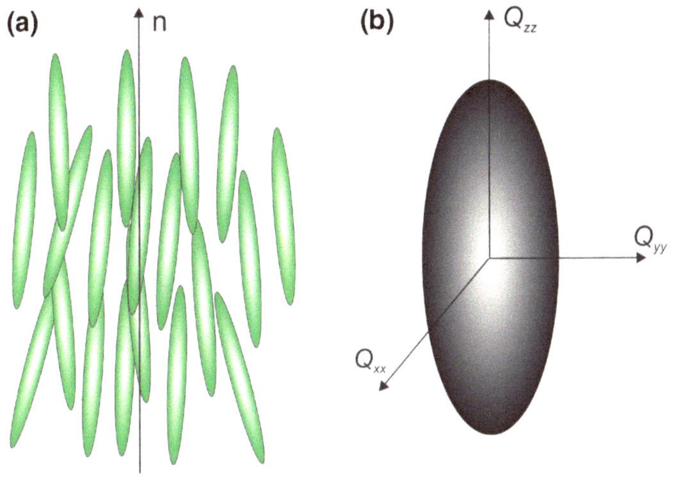

Fig. 1.2 **a** Snapshot of molecular ordering in the nematic phase. Note the uniaxial symmetry of this phase. **b** The corresponding order parameter tensor has the shape of a rotational ellipsoid with the largest eigenvalue along the director

$$Q_{ij} = \frac{S}{2}(3n_i n_j - \delta_{ij}) \tag{1.2}$$

Here, $n_i, i = x, y, z$ are the Cartesian components of the director **n**, The Q tensor is traceless *uniaxial* tensor, with its largest eigenvalue S along the director, which is the C_∞ symmetry axis of the nematic phase, allowing arbitrary rotation of this phase around the director. The other two eigenvalues are equal and smaller, $-\frac{S}{2}$ and their corresponding eigenvectors are directed perpendicular to the director.

In some special cases, such as in external electric or magnetic field, and confinement, the local ordering is not uniaxial, but is forced into biaxial form [5]. In this case the two degenerate eigenvalues become different and the biaxial order parameter tensor is given by:

$$Q_{ij} = \frac{S}{2}(3n_i n_j - \delta_{ij}) + \frac{P}{2}(e_i^{(1)} e_j^{(1)} - e_i^{(2)} e_j^{(2)}) \tag{1.3}$$

In the above expression, $e^{(2)} = \mathbf{n} \times e^{(1)}$ is the secondary director and $P = \langle sin^2\theta cos(2\Phi)\rangle$. In this biaxial case, the order parameter tensor is a real, symmetric and traceless tensor, which has now three distinct eigenvalues. The first eigenvalue is, like before, equal to S, which is the nematic degree of order, and the corresponding eigenvector is along the director. Obviously, the nematic director represents the direction corresponding to the eigenvalue S and it is clear that the two orientations of the vectors **n** and −**n** are equivalent. The director field is, loosely speaking, the "headless" vector, which specifies the direction of the largest tensor eigenvalue. The other two eigenvalues are $-\frac{1}{2}(S+P)$ and $-\frac{1}{2}(S-P)$. Their two corresponding eigenvectors specify the orientation and the biaxiality P of the order parameter tensor.

The biaxiality P therefore describes the lost uniaxiality and the broken continuous rotational symmetry around the director. The possible values of the biaxiality P are in the interval $[-3/2, +3/2]$. A special value $P = 0$ corresponds to the uniaxial nematic order and $|P| = \frac{3}{2}$ corresponds to complete ordering along the secondary director $\mathbf{e}^{(1)}$. Whereas the known nematic phases are uniaxial in bulk, local biaxiality can develop in confined nematics and especially in the cores of topological defects. In this case, the tensorial ordering field displays fully complex behaviour.

Because the order parameter tensor describes the uniaxial ordering of the molecules in the nematic phase, this tensor is also proportional to the material properties of this phase. Namely, the material properties, or better, the corresponding tensors (such as dielectric and magnetic) must be of the same symmetry and proportional to the tensor, describing the molecular ordering, which are the sources of material properties. This means that the dielectric, magnetic and optical properties of the nematic liquid crystals are uniaxial. This unaxiality can be visualised in 3D as a closed surface in the shape of a rotational ellipsoid, as illustrated in Fig. 1.2

The uniaxial material properties of the nematic phase are responsible for practical applications of these materials. As an example, one can see that an external electric field applied to a nematic liquid crystal will force the molecules to align along the field direction. This will induce rotation of liquid crystal molecules, which will find the minimum free energy, when the director is aligned with the electric field. This collective molecular rotation will change the optical properties of the nematic phase, which will be discussed in the next subsection.

1.2 Landau-de Gennes Theory of the Nematic-Isotropic Phase Transition

Phase transitions in thermotropic liquid crystals can be described within the Landau-de Gennes (LdG) theory, which is built around the order parameter of any specific phase transition, which is of interest to us [1]. More generally, the order parameter is a spatially inhomogeneous tensorial field and the free energy is a functional of this field. LdG theory is a mean-field theory that considers the free energy density expansion in terms of the equilibrium order parameter and neglects the time fluctuations of the order parameter field. In practice, the temperature region, where the fluctuation effects are relevant for the liquid crystal physics, is in most cases very small, i.e. of the order of milli Kelvin.

The free energy functional of the isotropic-nematic (I-N) phase transition is constructed as a Taylor expansion of the local free-energy density f in the order parameter tensor, and takes into consideration also spatial derivatives of this field [5]. If the nematic liquid crystal is confined by surfaces, additional surface terms that specify the energy costs related to the molecular orientation at the surface have to be considered in the free energy expansion. The LdG free energy density expansion

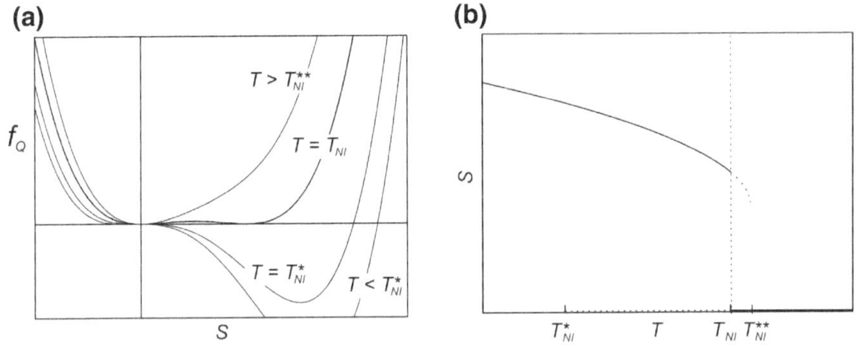

Fig. 1.3 **a** Free energy density f_Q as a function of the order parameter S, calculated for different temperatures around the isotropic-nematic phase transition. The local minima determine the stability of the phase with a given value of S. **b** Temperature dependence of the equilibrium value of the order parameter S. Because this is the first-order phase transition, the order parameter "jumps" from zero to a finite value at the phase transition. Image courtesy of M. Ravnik [5]

for the I-N phase transition in bulk, therefor not considering the surface effects, is (Fig. 1.3):

$$f_Q = \frac{1}{2}A \cdot Q_{ij}Q_{ji} + \frac{1}{3}B \cdot Q_{ij}Q_{jk}Q_{ki}Q_{ji} + \frac{1}{4}C \cdot (Q_{ij}Q_{ji})^2 \qquad (1.4)$$

All terms in this expansion have to be invariant to all symmetry operations, which are allowed in the nematic phase, or the symmetry group, related to the I-N phase transition. In the Eq. 1.4, the first-quadratic term describes the free energy increase due to the emergence of nematic ordering. The cubic term is allowed by symmetry and is responsible for the first-order nature of this transition, which is therefore discontinuous. The quartic term has the role of stabilising the free energy density. The coefficients A, B and C of this LdG free energy density expansion depend on the material.

Like in any LdG expansion, the parameter of the quadratic term, $A = \alpha(T - T^*_{NI})$, is temperature-dependent, and "drives" the I-N phase transition. It is responsible for the onset of the nematic order, when the temperature T decreases throughout the transition. Namely, positive value of $A > 0$ correspond to the stability of the isotropic phase with vanishing $S = 0$, since any additional value of the nematic order δS costs some free energy. When this coefficient changes sign at T^*_{NI}, and becomes negative in the nematic phase, $A < 0$, it stabilizes finite value of the order parameter, $S \neq 0$. This change of the sign of the quadratic term in the free energy expansion is therefore responsible for the spontaneous onset of the nematic ordering below T^*_{NI}.

The coefficient B of the cubic term is negative by symmetry, $B < 0$, which means that the coefficient C of the quartic term has to be positive, $C > 0$. If C were negative, $C < 0$, the free energy corresponding to the quartic term would always be negative and would thus have no lower bounds. Obviously, $C > 0$ is necessary to stabilise the free energy. Typical values of these coefficients are: $\alpha \simeq 10^5$ J/m^3K, $B \simeq -10^6$ J/m^3, and $C \simeq 10^6$ J/m^3.

If we consider an unbound (bulk) nematic liquid crystal that is also not distorted, the minimisation of the total free energy $F = \int f(\mathbf{r})dV$ is straightforward and the nematic degree of order in the equilibrium nematic phase is:

$$S_{eq} = \frac{1}{2}\left[\frac{-B}{3C} + \sqrt{\left(\frac{B}{3C}\right)^2 - \frac{8a(T - T^*_{NI})}{3C}}\right] \quad T \leq T^*_{NI} \tag{1.5}$$

The equilibrium value of the order parameter is temperature-dependent as illustrated in Fig. 1.3b and the magnitude of the order increases with decreasing temperature T. One can also see that the degree of order is the result of the balanced quartic and quadratic terms, which determine the minimum of the free energy density.

1.3 Elastic Distortions of Nematic Liquid Crystals

In the Eq. 1.4 we have not considered any variation of the order parameter tensor in space. However, in most cases of interest and in particular in nematic liquid crystals with defects, the spatial dependence of the Q-tensor is of prime importance. In reality, these spatial inhomogeneities of the Q tensor represent elastic deformations of the nematic liquid crystal. Because the nematic liquid crystal is a state of spontaneously broken rotational symmetry, it develops an "orientational rigidity", or elasticity, and is therefore able to transmit elastic torques. The situation is similar to spontaneously broken translational symmetries in solids, which are able to transmit static forces.

The elastic free energy volume density, expanded in terms of the spatial derivatives of the tensorial order parameter is [5]:

$$f_{elast} = \frac{1}{2}L_1 \cdot \frac{\delta Q_{ij}}{\delta x_k} \cdot \frac{\delta Q_{ij}}{\delta x_k} + \frac{1}{2}L_2 \cdot \frac{\delta Q_{ij}}{\delta x_j} \cdot \frac{\delta Q_{ik}}{\delta x_k} + \frac{1}{2}L_3 \cdot Q_{ij}\frac{\delta Q_{kl}}{\delta x_i} \cdot \frac{\delta Q_{kl}}{\delta x_j} \tag{1.6}$$

The expansion coefficients L_1, L_2 and L_3 are the corresponding tensorial elastic constants, accounting for the three basic deformation modes. x_i are the Cartesian coordinates and the summation over repeated indices has to be done. In the expression above, the elastic constants L_i are independent of the degree of nematic order and this will be considered further on in this section. This expansion contains a minimal number of symmetry invariants that describe the three deformation modes in a nematic liquid crystal. Other invariants are allowed by symmetry as well, and lead to more complex free energy density.

The free energy density expansion given by the Eqs. 1.4 and 1.6 has to be used when the nematic order is spatially inhomogeneous, such as in nematic colloids. In this case, the colloidal inclusions induce the appearance of the topological defects, which are the singularities of the nematic ordering field. They appear in a form

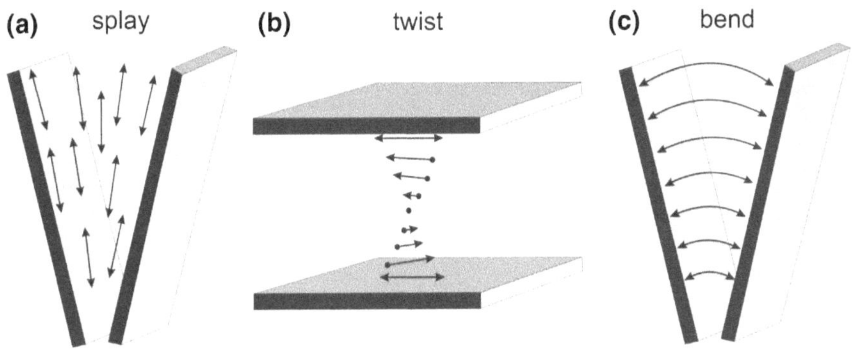

Fig. 1.4 Graphic visualisation of the basic elastic deformations in the nematic liquid crystal: **a** splay, **b** twist and **c** bend deformation. Image courtesy of M. Ravnik [5]

of points, called hedgehogs, and defect lines, called disclinations, which all have a core with a very low order parameter. The core of the liquid crystal is therefore somewhat molten, which brings about huge differences in local free energy density. However, there are also many cases when the defects in nematic liquid crystals can be completely neglected and the degree of the nematic order can be considered constant throughout the sample. In these cases, we can consider only the elastic deformation of the nematic liquid crystal without taking into account the changes in the degree of ordering, and the tensorial free energy density can be greatly simplified. Formally, it is rewritten in the director representation, also called the standard Frank-Oseen elastic free energy density (Fig. 1.4):

$$f_{elast}^{FO} = \frac{1}{2}K_1 \cdot (\nabla \cdot \mathbf{n})^2 + \frac{1}{2}K_2 \cdot [\mathbf{n} \cdot (\nabla \times \mathbf{n})]^2 + \frac{1}{2}K_3 \cdot [\mathbf{n} \times (\nabla \times \mathbf{n})]^2 \quad (1.7)$$

There are three terms in this expression, each corresponding to a specific elastic deformation with a specific elastic constant. K_1 is the splay elastic constant, K_2 is the twist elastic constant and K_3 is the bend elastic constant. The corresponding deformations are called splay, twist and bend deformation, and they are visualised in Fig. 1.2.

Having rewritten the tensorial free energy density into elastic free energy density, the elastic constants K_i must be related to the LdG expansion coefficients L_i. Indeed, the relation between the LdG coefficients and elastic constants K_i is:

$$L_1 = \frac{(K_3 + 2K_2 - K_1)}{9S^2} \quad (1.8)$$

$$L_2 = 4\frac{(K_1 - K_2)}{9S^2} \quad (1.9)$$

Table 1.1 Values of elastic constants of 5CB and 8CB nematic liquid crystals $5K$ below the clearing point

Liquid crystal	K_1	K_2	K_3	*Note*
5CB	9 pN	4.8 pN	11 pN	$T = T_{NI} - 5K$
8CB	13 pN	6 pN	13 pN	$T = T_{NI} - 5K$

Note the elastic constants are strongly temperature dependent. For 8CB K_2 and K_3 diverge when approaching the smectic [6] phase. The values of the elastic constants are taken from Ref. [6]

$$L_3 = 2\frac{K_3 - K_1}{9S^3} \tag{1.10}$$

The elastic constants K_i are to the leading order proportional to the square of the magnitude of the order parameter S with corrections. These relations can be very useful when considering the analysis of different experiments. Typical values of the elastic constants of some selected materials are presented in the Table 1.1:

1.4 Interactions of Liquid Crystals with Surfaces

Interactions of liquid-crystal molecules with surfaces are the key to the technology of liquid-crystal displays [1, 7]. In these devices, liquid crystal has to be homogeneously ordered over large surface areas, which can be achieved with different alignment techniques. When liquid-crystal molecules are in contact with a solid, a liquid or a vapour, the intermolecular interactions between the liquid-crystal molecules in the interior and on the interface to the outside of the liquid crystal are different. As in any liquid, this gives rise to the well-known surface tension phenomenon, which accounts for the free energy that is necessary for the creation of a unit surface of a given interface.

Unlike ordinary, disordered fluids, liquid-crystalline phases exhibit spontaneous molecular ordering, which gives rise to anisotropic material properties, including the anisotropy of surface tension. This means that the free energy of the liquid-crystalline phase will depend on the orientation of liquid-crystal molecules in relation to the surface of confining material and will thus be anisotropic. Formally, this is described with the introduction of an additional term in the LdG free energy expansion, which is also called Rapini-Papoular surface free energy functional:

$$f_S = \frac{1}{2} W_{surface} \cdot (Q_{ij} - Q^o_{ij})^2 \tag{1.11}$$

The surface free energy density f_S shows quadratic increase, when the order parameter tensor Q_{ij} deviates from the preferred order parameter Q^o_{ij}. The expansion coefficient $W_{surface}$ is called the uniform surface anchoring strength. For homeotropic (perpendicular) surface anchoring of liquid-crystal molecules on the interface, the easy axis is normal to the surface. In the case of planar (parallel) surface anchoring of liquid-crystal molecules, the easy axis is lying within the confining surface and points into a given (predetermined) direction. On some materials, there is no preferred direction for planar anchoring, which is then called a degenerate planar anchoring, and the molecules are free to choose any direction within the surface. Degenerate planar anchoring is difficult to achieve, a good example of such a surface is the liquid crystal-liquid interface.

In the director notation, the Rapini-Papoular free energy density is rewritten into:

$$f_S = \frac{1}{2} W_a \cdot \sin^2 \Phi \tag{1.12}$$

where W_a is the anchoring energy coefficient and Φ is the angle between the easy axis and the director. Typical values of the anchoring coefficient W_a range from 10^{-3} J/m^2, which corresponds to very strong surface anchoring, down to 10^{-7} J/m^2, which corresponds to weak surface anchoring. Typical values of a nematic liquid crystal surface anchoring on polyiimide materials, nylon or solid crystals are of the order of 10^{-4} J/m^2. Weak anchoring of nematic liquid crystals is usually obtained on surfaces covered with surfactant molecules, with typical anchoring coefficient being 10^{-5} J/m^2.

When considering the orientation of the nematic director at the interface and the nematic director field is elastically deformed, it is useful to consider the concept of the surface extrapolation length (or Kleman-de Gennes length). For example, if we are considering the homeotropic alignment of a nematic and we turn on an external magnetic or electric field, which is parallel to the surface, the molecules will tend to align along the field direction for positive magnetic (or electric) anisotropy. Far away from the interface, this alignment will be complete; however, when approaching the interface, there will be an elastically distorted region that will terminate at the interface with a finite deviation of the nematic direction from the normal. If the director profile is extrapolated beyond the interface, it will reach the undistorted value at some separation from the surface. This separation is called the surface extrapolation length and is given as a ratio of the corresponding elastic constant and the anchoring strength:

$$\xi_s = \frac{K}{W_a} \tag{1.13}$$

For a weak surface anchoring ($W_a \sim 10^{-6}$ J/m^2) and the typical elastic constant $K \sim 10^{-11}$ N, the extrapolation length is around $\xi \sim 10\,\mu$m. For very strong surface anchoring ($W_a \sim 10^{-3}$ J/m^2), the extrapolation length decreases to around $\sim$ 100 nm, but is still much larger than the typical liquid crystal molecules (several nm).

In reality, any confining interface, whether it is solid, liquid, gas or vacuum, will induce a given preferential direction of the liquid crystal director at that interface. Perhaps the best known insight into the physics of liquid-crystal interactions with surfaces is given by the Scanning Tunneling Microscopy (STM) experiments with cyanobiphenyls on crystalline graphite. It was discovered that the liquid crystal

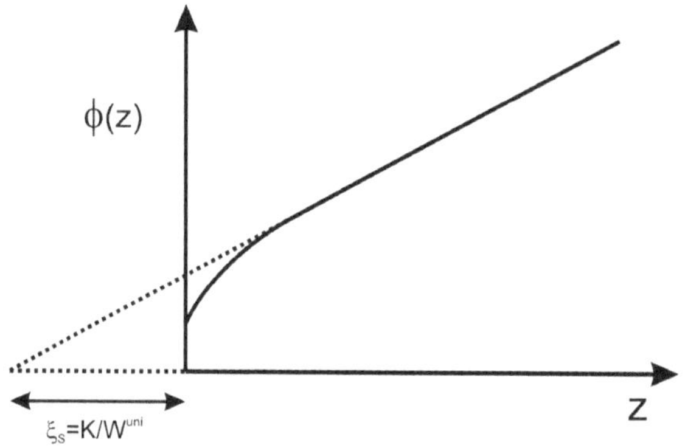

Fig. 1.5 Meaning of the surface extrapolation length ξ_S. For infinitely strong anchoring, the angle ϕ, which is measuring the distortion of the liquid crystal, would start with the value 0 exactly at the surface at $z = 0$. Because of finite anchoring strength, the distortion starts with a finite value of the angle ϕ at the surface. However, one can imagine that the distortion starts "inside" the surface at some imaginary position $z = -\xi_S = -K/W$. Image courtesy of M. Ravnik [5]

molecules are actually crystallised on the graphite because of the strong electron interaction of biphenyl rings with graphite surface, see for example Ref. [7]. This surface-induced crystallisation of the molecular order is actually observed in many experiments as a "surface memory effect" (Fig. 1.5).

In contact with any solid, liquid crystal develops the first interfacial layer, which is smectic-like or even crystalline-like [7]. When such an interface is heated into the isotropic phase and cooled back down, one can observe the same surface pattern as in the beginning. The interface therefore has a "memory", which actually originates from the first, strongly adsorbed layer of liquid-crystal molecules, which melts into the isotropic phase only at very high temperatures.

When such a solid (glass, polymer, crystal) surface is rubbed mechanically along a given direction, one observes that the liquid-crystal molecules tend to align themselves along this "rubbing direction" [7]. There were several explanations of this effect, ranging from the idea of creating micro grooves by rubbing to the deposition of extra material on the surface. Some experiments reported melting of polymer backbones due to high local temperatures, which were caused by the friction between the rubbing cloth and the substrate [7]. In all cases, external mechanical action creates a preferred direction of a different origin, which efficiently aligns liquid-crystal molecules even over extremely large distances.

In contrast to rubbing a substrate, one can also deposit organic surfactant molecules on the substrate, which can be a very efficient way of generating desired alignment. One example is the silanisation of glass surfaces with octadecyldimethyl (3-trimethoxysilylpropyl) ammonium chloride (DMOAP silane), which creates excellent homeotropic alignment of most cyanobiphenyl liquid crystals. It was shown that the DMOAP creates a dense monolayer of silane molecules covalently bound to the oxygen in the glass [7]. The first layer of cyanobiphenyl liquid

Table 1.2 Alignment of some selected nematic liquid crystals on various surfaces

Liquid crystal	Surface	Alignment	Anchoring energy coefficient (J/m^2)	Note	Ref.
5CB	Polyimide PI-2555	Parallel	–	Nissan Chemicals	[8]
5CB	Rubbed Nylon	Parallel	$10^{-5} - 10^{-4}$	W_θ	[7]
5CB	Rubbed Nylon	Parallel	$5 \times 10^{-6} - 5 \times 10^{-5}$	W_φ	[7]
5CB	Graphite	Parallel	–	STM	[7]
5CB	PVCN	Parallel	10^{-5}	W_θ	[7]
n-CB	Polyimide (BPDA-PDA)	Parallel	–	–	[7]
n-CB	Polystyrene	Parallel	–	Normal to rubbing	[7]
5CB	Clean glass	Parallel	–	–	
8CB	Rubbed Nylon	Parallel	3×10^{-6}	W_θ	[7]
8OCB	DMOAP on glass	Normal	1×10^{-4}	ABCR GmbH	[7]

Note the temperature dependence of the anchoring energy coefficients [7]

crystals was shown to be smectic-like but in the layers towards the bulk, the usual nematic ordering takes place. Table 1.2 gives some typical examples of alignment on different substrates together with surface anchoring strengths.

The surface anchoring of liquid-crystal molecules on fluid interfaces is quite different from anchoring on solid interfaces. On many fluids, including water and glycerol, the alignment is planar. However, the anchoring between two fluid phases can be modified by adding surfactant molecules. These have two different parts, each of which prefers one liquid to the other. The free energy of these molecules will be minimised after they are included into the interface, which gives us the possibility of tuning the anchoring of fluid-liquid-crystal interfaces. If sodium dodecyl sulfate (SDS) is added to the water, the planar alignment will switch to homeotropic for SDS concentration above 5 mM. Strong homeotropic anchoring of a nematic liquid crystal can be obtained on PDMS, both polymerised and non-polymerised. However, this interface is not very stable and is subject to ageing and smearing. Similar problems were observed for polystyrene-nCB interface. Another special family of materials are fluorinated oils, which form sharp interfaces both to hydrophilic and hydrophobic fluids. Liquid crystal anchoring on different fluorinated oils can be either planar or homeotropic. Table 1.3 presents some typical examples of alignment of nematic liquid crystals on water interface with added surfactants (Fig. 1.6).

Table 1.3 Type of alignment for some liquid crystal-fluid interfaces, where the amphiphilic molecules were added to achieve either normal or planar surface alignment

Sample	Liquid crystal	Amphiphilic	Anchoring	Ref.
1	5CB, K15	Pure water	Planar	[9]
2	5CB, K15	Water + Tween 60	Homeotropic	[10]
3	5CB, K15	Water + 5 mM SDS	Homeotropic	[9, 10]
4	5CB, K15	Water + 5%PVA	Planar	[10]
5	5CB, K15	Pure glycerol	Planar	[11]
6	5CB, K15	Glycerol + 2–4% lecithin	Homeotropic	[11]
7	5CB, K15	PDMS	Homeotropic	[9]
8	CCN47/55	Glycerol + 2–4% lecithin	Homeotropic	[11]
9	ZLI 2620	Water + PVA	Planar	[10]

Note that anchoring on fluid interfaces is degenerate

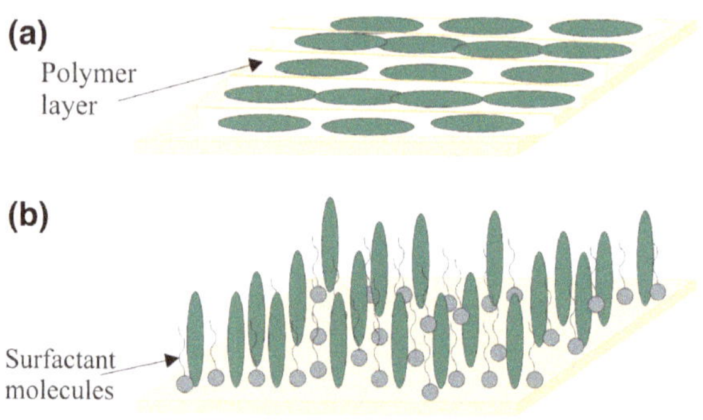

Fig. 1.6 **a** An uniformly rubbed polymer film (such as polyimide) will usually result in a strong planar anchoring. **b** When surfactant molecules with appropriate structure (i.e. an alkyl tail) are deposited on a solid substrate, strong homepotropic alignment will likely be induced. Image courtesy of M. Nikkhou [8]

1.5 Optical Properties of Nematic Liquid Crystals

Because nematic liquid crystals are uniaxialy ordered fluids with long-range orientational order, they display anisotropic optical properties [1]. This means that the speed of light propagation in a nematic liquid crystal depends not only on the direction of propagation, but also on the direction of polarisation.

Modes of light propagation in matter are governed by the Maxwell's equations of electromagnetic field [12–14], which can be combined into the wave equation for the electric and the magnetic fields:

$$\nabla \times \nabla \times \mathbf{E}(\mathbf{r}) - \frac{\omega^2}{c_o^2} \cdot \varepsilon(\mathbf{r}) \cdot \mathbf{E}(\mathbf{r}) = 0 \tag{1.14}$$

Here, $\mathbf{E}(\mathbf{r})$ is the electric field o the electromagnetic wave, $\varepsilon(\mathbf{r})$ is the dielectric constant of the material, ω is the frequency of the electromagnetic field, and c_o is the speed of light in vacuum. In an optically inhomogeneous material, the dielectric tensor $\varepsilon(\mathbf{r})$ might be a complicated function of position, while for isotropic matter, this is simply a scalar quantity, independent on position and direction of the electric field of light.

In an infinite and optically isotropic matter the eigensolutions of the wave equation (1.14) are linearly polarized and propagating plane waves:

$$\mathbf{E}(\mathbf{r}, t) = \mathbf{E}_o \cdot e^{i(\mathbf{k}\cdot\mathbf{r} - \omega\cdot t)} \tag{1.15}$$

Here, $\mathbf{k} = (k_x, k_y, k_z)$ is the wave vector of the electromagnetic wave with the frequency ω, the magnitude of which is related to the phase velocity c and frequency by $k = \omega/c$. The phase velocity of the wave is related to the dielectric constant of the material by the equation $c = 1/\sqrt{\varepsilon\varepsilon_\circ\mu_\circ}$, which follows from the eigenvalue of the wave equation.

In an anisotropic matter, ε is no longer a scalar quantity, but a tensor ε with three distinct eigenvalues ε_{xx}, ε_{yy} and ε_{zz}, the form of which is simplified in optically uniaxial materials to:

$$\varepsilon = \begin{bmatrix} \varepsilon_{xx} & 0 & 0 \\ 0 & \varepsilon_{xx} & 0 \\ 0 & 0 & \varepsilon_{zz} \end{bmatrix} \tag{1.16}$$

The solutions of the wave equation (Eq. 1.15) are now quite different compared to isotropic matter. In general, there are two distinct eigenvalues and two distinct eigenvectors for phase velocity of light, which correspond to the velocity of the ordinary (c_o) and extraordinary (c_e) electromagnetic wave, respectively.

$$c_o = \frac{c_\circ}{n_\circ} \tag{1.17}$$

$$c_e = \frac{c_\circ}{n_e} \tag{1.18}$$

The phase velocity of the ordinary wave (c_o) does not depend on the direction of propagation, and is therefore a remnant of the isotropic phase. On the contrary, the phase velocity of the extraordinary wave (c_e) depends on the direction of propagation. The corresponding refractive indices are given by (Fig. 1.7):

$$n_o^2 = \varepsilon_{xx} \tag{1.19}$$

$$\frac{1}{n_e^2} = \frac{\sin^2\alpha}{\varepsilon_{zz}} + \frac{\cos^2\alpha}{\varepsilon_{xx}} \tag{1.20}$$

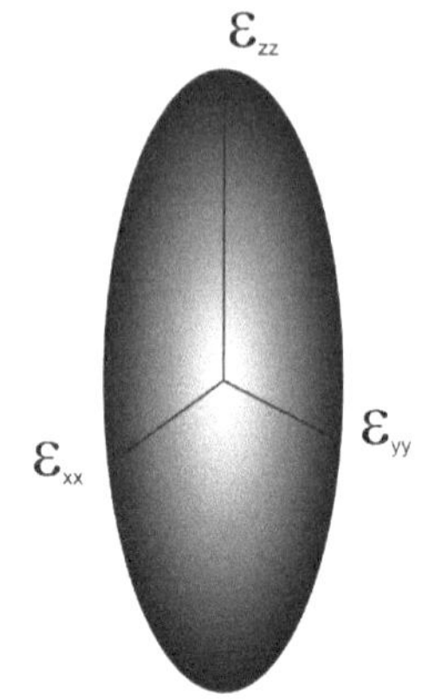

Fig. 1.7 Dielectric tensor in a liquid crystal can be visualized as a closed ellipsoidal surface with three different principal values, ε_{xx}, ε_{yy},and ε_{zz}. In a nematic liquid crystal two eigenvalues are equal and the material is optically uniaxial, with the optical axis along the z-axis in Fig. 1.2

Here, α is the angle between the direction of wave propagation and the optical axis (i.e. z-axis). Both velocities are equal only when the waves are propagating along the optical axis of the material, in our case this means along the director. The eigenvectors of both waves (i.e. their polarizations) are always orthogonal to each other. The maximum difference in the phase velocity of the extraordinary and the ordinary wave is related to the birefringence of the material:

$$\Delta n = n_e - n_\circ \tag{1.21}$$

Whereas the solutions of the wave equation in a homogeneous and infinite sample of a nematic liquid crystal are simply plane waves, the optics becomes complicated when the liquid crystal is distorted or even contains topological defects. There are no simple approaches to the optics of nematic liquid crystals in general; instead, one has to use numerical approaches to the solution of wave equations, which are known as the Jones-matrix approach [12] or the Berreman approach [15]. These approaches rely on dividing a liquid crystal into thin slices, which are birefringent and have a different orientation of the optical axis. The state of polarisation of light, propagating perpendicularly to this set of slices, is then calculated as a sequence of multiplications of the wave amplitude with a transfer matrix that reflects the local optical properties of the material. Both Jones' and Berreman's approach disregards spatial lensing effects that result in spatially inhomogeneous amplitude of propagating light. The liquid-crystal material therefore acts as a complicated phase grating with no amplitude variations. This is problematic when the birefringence of the material is high.

When the optical properties of a liquid crystal are changing gradually along the direction of light propagation with a spatial period p, the Mauguin limit of wave propagation could be used when the wavelength of the propagating light is much smaller than the product $p \cdot \Delta n$:

$$\lambda << p \cdot \Delta n \tag{1.22}$$

Within this limit, the product of the birefringence and the rate of spatial change of the optical axis is much smaller than the wavelength of the light considered. In this case, the eigen-waves are linearly polarised ordinary and extraordinary waves, which follow the orientation of the local optical axis. This is therefore a type of a wave-guiding, or adiabatic, regime, where the polarisation of light follows the structure of the medium.

1.6 Topological Defects in Nematic Liquid Crystals

Topological defects in matter and fields [16–19] are remnants of the spontaneously broken symmetry of the higher-temperature disordered phase [4]. Topological defects are ubiquitous in nature and are observed as Abricosov vortices in type-II superconductors [20], superfluid vortices in ^{3}He [21, 22] and Bose-Einstein condensates [23], as quasiparticles in the fractional quantum Hall effect, fermionic atoms in optical lattices and defects in magnetic materials [24–26] and field theories [27]. Topological defects emerge in optical vortices [28, 29] and describe the orbital angular momentum of light [30]. In all cases, topological defects are regions in 3D space, where the order parameter of the corresponding phase is singular or ill-defined. The emergence of topological defects is described within the Kibble-Zurek mechanism of monopole production across a given phase transition [31, 32]. When the system is rapidly cooled from its isotropic phase, it develops "seeds" of the low-temperature phase after the phase transition temperature is crossed. With time, these seeds grow into domains and when they finally meet each other, they necessarily form defects because of the mismatch of the fields in different domains. In later stages, this network of defects exhibits a coarsening dynamics via merging and coalescence of domains.

Nematic liquid crystals are particularly suitable for optical observation of topological defects. These appear in the form of point defects, also called hedgehogs, and disclination lines [33–36], and are characterised by a defect core, which is disordered and of the size of the molecule. Because of the softness of the nematic liquid crystal, the topological defects are large objects in nematics and are easily observable under an optical microscope. Optical analysis of defects is further simplified because of the birefringence of liquid crystals, which makes the distorted regions around the topological defects easily observable in polarising microscopy.

1.6.1 Winding Number and Topological Charge

Topological defects in nematic liquid crystals are characterised by their winding number (or strength of a defect) and their topological charge [35, 37–40]. As an example, Fig. 1.8 shows point and line defects as observed between crossed polarisers in a typical nematic liquid crystal cell with degenerate planar surface anchoring. One

(a) (b)

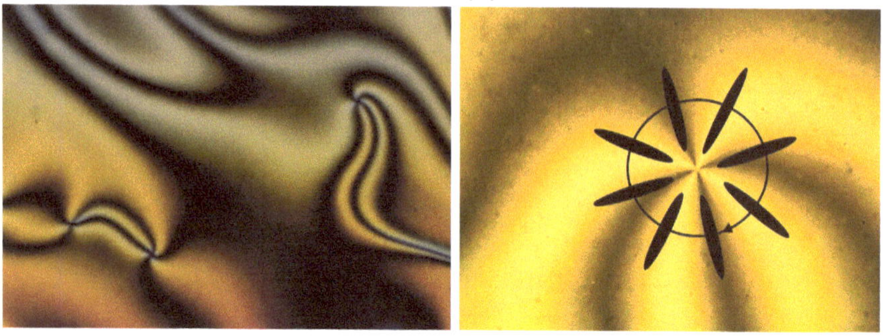

Fig. 1.8 **a** Image of a nematic planar cell taken between crossed polarisers. **b** Enlarged region around a defect with four dark brushes. The schematic drawing of possible molecular orientation is shown around this defect

can see points, which are surrounded with dark and bright brushes, as well as lines joining these points, which are also accompanied by different shades of light.

It is not quite easy to determine the in-plane orientation of liquid-crystal molecules in the dark regions. However, we know that the long axis must be either parallel or perpendicular to the polariser and the analyser in the dark region. There are several possible and different kinds of brushes around each point defect. Those with four dark brushes, such as in Fig. 1.8 might correspond to the "radial" director orientation, which is shown in Fig. 1.8b.

Let us now consider a selected point defect, as illustrated in Fig. 1.8 and we construct a circular path around this defect. In the next step, we monitor the orientation of the director on that path as we move in a clockwise manner around the defect. One can immediately see that the total angle of director rotation as we complete one circle is a multiple of π. This winding of the director as we move along the closed path encircling the defect is described by the winding number (strength) k:

$$k = \frac{\alpha}{2 \cdot \pi} \tag{1.23}$$

Here, the angle α is the total angle of the rotation of the director, as we encircle the defect. Because the director is a headless vector, the winding number can obtain fractional values $+1/2$ and $1/2$ for most of the simple cases, where the director winds for an angle π or $-\pi$, as we complete one circular path. In reality, defects with winding numbers $+1/2$ and $-1/2$ are often observed when making a cross-section of the disclination lines. The two cross-sections of a disclination line with $+1/2$ and $1/2$ winding number are shown in Fig. 1.9a and b.

Integer winding numbers $|k| = 1$ correspond to the director field, which is a cross section of a point defect in a nematic. There are three possible director configurations in this case, corresponding to radial, circular and hyperbolic hedgehog, all illustrated in Fig. 1.10 (a–c) [35]. Higher winding numbers of defects are possible; however

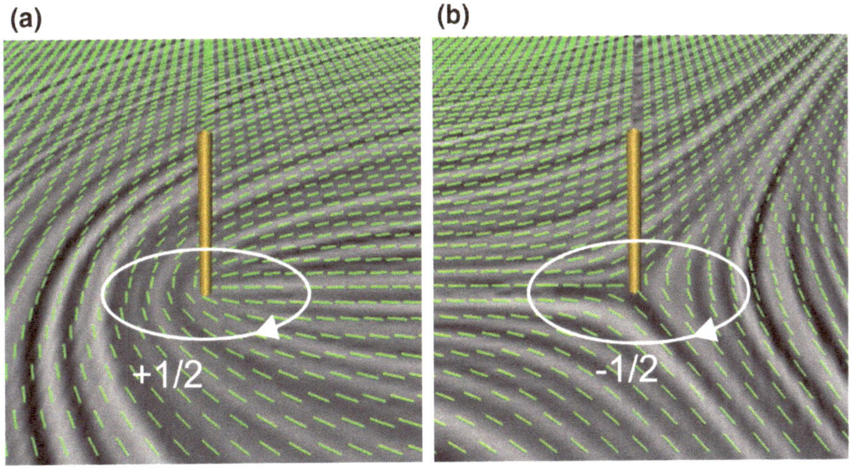

Fig. 1.9 **a** Cross-section of a disclination line with +1/2 winding number. Note the clockwise rotation of the director (*green dashed lines*), as we move in a clockwise manner along a closed path encircling the disclination core, coloured *gold*. **b** Cross-section of a disclination line with −1/2 winding number. Note the opposite, i.e. anti-clockwise, rotation of the director, as compared to (**a**). Image courtesy of M. Ravnik [5]

their total free energy increases with increasing winding number because of increased elastic deformation. This reduces the possibility of their occurence and observability in real samples. Higher winding number defects were recently observed in chiral nematic droplets by Posnjak et al. [41].

The winding number itself is a topological invariant that describes the topological properties of defects to a certain extent. However, there is another much more important quantity that is attributed to the topological defects, called the topological charge. The topological charge q of a point defect in a nematic director field is calculated as an integral over arbitrary closed surface σ embracing this defect [35]:

$$q = \frac{1}{8\pi}\oint_{\sigma} \varepsilon_{ijk} \cdot \mathbf{n}\left(\frac{\partial \mathbf{n}}{\partial x_j} \times \frac{\partial \mathbf{n}}{\partial x_k}\right) \cdot dS_i \tag{1.24}$$

The integral over the closed surface in Eq. 1.24 is well known as the Gauss law in electrostatics, where the flux of the electric field is calculated over the closed surface and gives the electric charge embraced by the surface. In Eq. 1.24 ε_{ijk} is the Levi-Cività totally asymmetric tensor and x_i are the Cartesian coordinates. Note that Eq. 1.24 is odd in the director field and as a consequence the sign of the topological charge is not well-defined since $+\mathbf{n}$ and $-\mathbf{n}$ are formally equivalent in nematics. However, it is a convention that different point hedgehogs are assigned different values of their topological charge, including the sign. The charge of a radial hedgehog is by convention $q = +1$, and the charge of a hyperbolic hedgehog is $q = -1$. The values of topological charges of defects are important because of the natural law

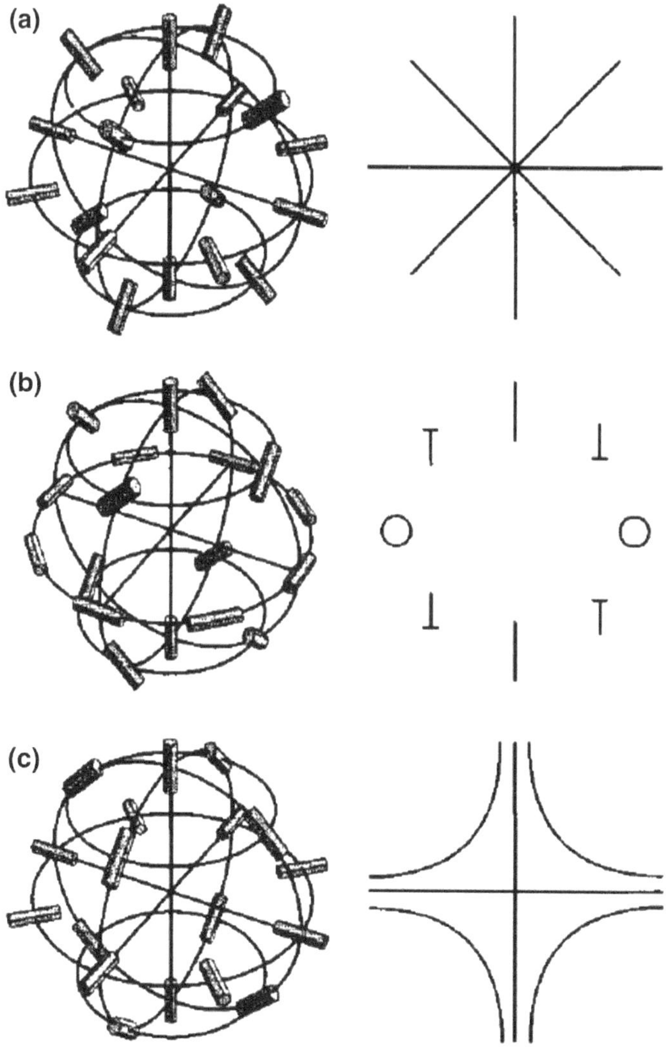

Fig. 1.10 **a** A radial hedgehog is characterised by a director field emanating from the point defect and pointing radially outward. This defect and its field are similar to an electric charge, which can be considered as a defect or singularity of the electric field. Note that the electric field is a vector field, whereas the director field is in fact originating from the tensorial field. When we move along the closed path encircling this defect, the director rotates for 2π, corresponding to the winding number $k = +1$. **b** A circular hedgehog, where the director field is encircling the point hedgehog. The corresponding winding number is also $+1$. **c** A hyperbolic hedgehog, where the director "avoids" the singular point in a hyperbolic manner. The winding number of this defect is -1. Reprinted figure with permission from T. C. Lubensky, D. Pettey, N. Currier, H. Stark, Phys. Rev. E**57**, 610 (1998). Copyright (1998) by the American Physical Society

of conservation of the total topological charge, when there are several topological charges in a given sample. For vectorial fields, the combined topological charge of two hedgehogs is simply algebraic sum of their respective charges $q_1 + q_2$. In nematics, the non-vectorial nature of the director field causes complications, because the sign of the charge has no meaning. Combined topological charge of two hedgehogs is either $|q_1 + q_2|$ or $|q_1 - q_2|$ [35].

Winding number k can also be associated to a point hedgehog by intersecting it with a measuring plane, encircling it with a closed path and performing the analysis of director rotation, as we move along that closed path. We obtain the integer value of the winding number $k = +1$ for a radial hedgehog for any chosen plane. This has an obvious interpretation: if we consider all possible orientations of a director field around a radial hedgehog, we observe that each orientation is met only once. The same applies for a hyperbolic hedgehog point defect which has the negative winding number $k = -1$. In other words, the director field wraps the full solid angle only once, hence its topological charge is $|q| = 1$.

Topological charges play a fundamental role in the classification of defects in matter and fields, and form the basis of a homotopy theory [40]. The law of conservation of topological charges is one of the fundamentals of physics. This law is responsible for the processes of creation, annihilation and transformation of topological charges. One of the well known examples are electric charges and law of conservation of total electric charge. To a certain extent, the methods of measuring both the winding number and the topological charge are somehow similar to the Ampere law from classical magnetism and the Gauss law from electrostatics. The total flux of the electric field through a closed surface embracing the electric charge is the measure of the electric charge inside that testing surface. Formally, the total flux of the electric field is obtained by the Gaussian integral across the closed surface.

Unlike simple objects, such as a sphere, topologically more complex objects, such as a torus, induce different values of the topological charge of the nematic director field, when inserted into a nematic liquid crystal as will be discussed in Sect. 2.1.

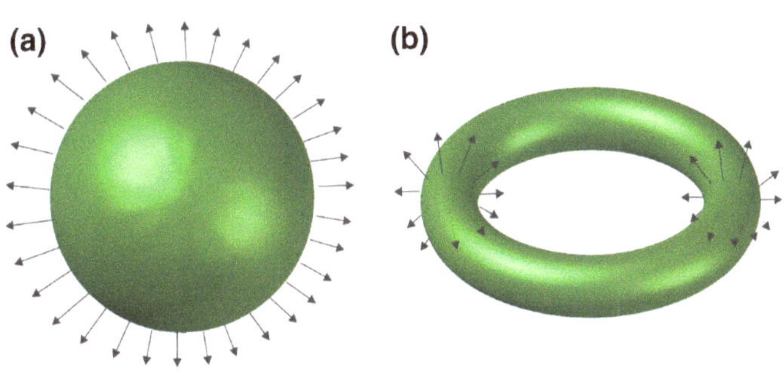

Fig. 1.11 **a** A vectorial field oriented perpendicular to the surface of the sphere wraps the full solid angle only once. **b** A vectorial field perpendicular to the surface of a torus wraps the full solid angle twice. Each direction of the field is met twice, when travelling along the closed surface. Image courtesy of U. Jagodic

This will be understood by considering how many times the director wraps the full solid angle, when we consider it on two different objects: a sphere and a torus, as shown in Fig. 1.11.

1.6.2 Euler Characteristic and Genus of Colloids

For each closed surface (such as a sphere or a torus), there is a topological invariant, called the Euler characteristic χ, which does not change under smooth transformations of that surface, which are allowed in topology [18]. Euler characteristic for a closed surface is calculated by drawing a polygonal set on the surface, followed by counting the number of vertices V_t, the number of edges E_d and the number of faces F_c (O.D. Lavrentovich [42]):

$$\chi = V_t - E_d + F_c \tag{1.25}$$

It can be shown that Euler characteristic χ does not depend on the choice of the network, but depends on the type of the closed surface chosen. Gauss-Bonnet theorem relates the Euler characteristic of the closed surface to the Gaussian curvature, integrated over a closed surface S:

$$2\pi\chi = \oint_S K dS = \oint_S d\theta d\phi \cdot \boldsymbol{v} \cdot \left[\frac{\partial \boldsymbol{v}}{\partial \theta} \times \frac{\partial \boldsymbol{v}}{\partial \phi}\right] \tag{1.26}$$

Here $K = \kappa_1\kappa_2$ is the local Gaussian curvature and κ_i are the two principal curvatures at that point on the surface. The vector $\boldsymbol{v}$ is the local normal to the surface at the point under consideration. θ and ϕ are the two angles defining the position of the chosen point on the surface. For a sphere or any closed surface, which is obtained by smoothly morphing the surface, one finds the Euler characteristic is $\chi = 2$, and

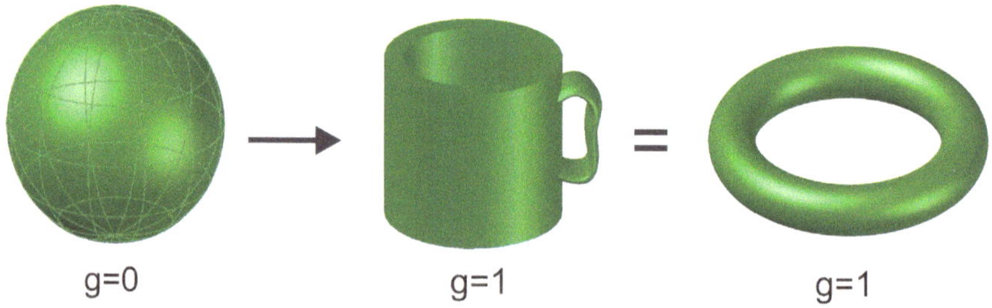

Fig. 1.12 By attaching a single handle to a sphere, one obtains a sphere with a handle, which can be smoothly transformed into a coffee mug. This is a closed surface with a genus $g = 1$, which is different from the genus of the sphere $g = 0$. In topology, it is allowed to apply smooth transformation to the coffee mug in order to obtain a torus. A torus and a coffee mug are therefore topologically the same objects, characterised by $g = 1$. Image courtesy of U. Jagodič

for a torus or torus-like closed surface, the Euler characteristics is $\chi = 0$. The Euler characteristic is also related to the "genus" g of the surface:

$$\chi = 2(1 - g) \tag{1.27}$$

The genus g is the number of "handles" attached to a sphere, resulting in the formation of "handle-bodies" (or "pretzels"). For example, a torus is obtained by attaching a single handle to the sphere and then smoothly transforming this closed surface into a torus, as shown in Fig. 1.12. By adding more handles, additional "holes"

Name	Image	Euler characteristics	Genus
Disk		1	-
Sphere		2	0
Torus		0	1
Double torus		-2	2
Moebius strip	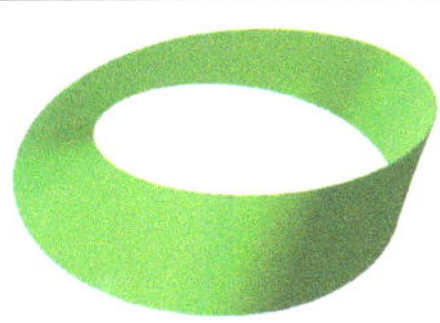	0	-
Two spheres	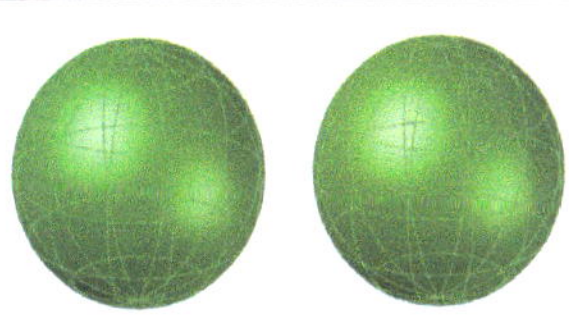	2+2=4	-

Fig. 1.13 List of some representative objects and their Euler characteristic χ

are obtained, and at a larger genus, the objects correspond to pretzels with g holes. List of some representative objects, their Euler characteristic χ and genus g is shown in Fig. 1.13.

Consider now any closed surface S, bounding a region of space V, which is inserted into a vectorial field $\mathbf{n}$ and this field is forced to align perpendicularly at all points of this inserted surface. The topological charge m_c of this region of space V bounded by the surface S is the degree of $\mathbf{n}$ along S, or, in another words, the number of times the orientation of director visits each possible direction in space. If the vector sign is attributed to the director the sign of the topological charge is assigned to that particular object. The degree $\mathbf{n}$ along S is calculated by integrating the Jacobian of $\mathbf{n}(\mathbf{r})$ over that surface (Senyuk et al. [43]):

$$m_c = \frac{1}{4\pi} \oint_S d\theta d\phi \cdot \mathbf{n} \cdot \left[\frac{\partial \mathbf{n}}{\partial \theta} \times \frac{\partial \mathbf{n}}{\partial \phi} \right] \tag{1.28}$$

Because the vector field $\mathbf{n}(\mathbf{r})$ is locally perpendicular to the surface and therefore along the local unit normal field of the surface S, this integral reduces to the total Gaussian curvature from Eq. 1.26 integrated over the closed surface, divided by 4π. The Gauss-Bonnet theorem states that the total Gaussian curvature of a closed surface is quantized in units of 4π and remains constant under all continuous deformations of the surface. This means that the hedgehog charge m_c of the vectorial field $\mathbf{n}(\mathbf{r})$ aligned locally perpendicular to the surface S is

$$m_c = \pm \frac{2\pi\chi}{4\pi} = \pm(1 - g) \tag{1.29}$$

The choice of the sign ($\pm$) for m_c depends on the assignment of the direction to the director field at the surface. Let us now come back to this picture of a closed surface S, immersed into the vectorial field, which is forced to align along the normal to the surface. We assume that far away from this surface, the field $\mathbf{n}$ is homogeneous, and therefore the net topological charge of this region will be zero. Because of that, the field $\mathbf{n}$ is forced to create topological defects close to the surface S, because it is in general not possible to transit from the vicinity of the surface to a far-away position without creating topological defects. This means that the topological charge of the surface m_c should be compensated by the topological charge of all defects m_d accompanying that surface [43],

$$m_c + m_d = 0 \tag{1.30}$$

For a *vector* field $\mathbf{n}$, The Euler characteristic of the surface χ therefore defines total topological charge m_d of all defects ($m_i, i = 1, ..., N$) of that vector field [43], which are accompanying the closed surface:

$$m_d = \sum_i m_i = \mp \frac{\chi}{2} = \mp(1 - g) \tag{1.31}$$

For a sphere, the Euler characteristic equals $\chi = 2$ and according to the Gauss theorem, there should be a single topological charge accompanying the sphere. This is indeed observed in the experiments, and the defect appears in a form of a point hedgehog or a Saturn ring, carrying the unit hedgehog charge. Gauss theorem for more complex handlebodies with higher genus was experimentally proved by Senyuk et al. [43] and will be discussed in Sect. 6.2.

1.6.3 Elastic Energy of Topological Defects

There are two kinds of topological defects in nematic liquid crystals: line defects also called disclinations, and point defects [1, 4, 35, 36, 42, 44]. The winding number of line defects is 1/2, which implies rotation of the director by π as we move around a closed path encircling the defect line. The winding number of point hedgehogs, which are usually observed in nematics is $k = 1$, although higher winding numbers and higher charges were reported recently in chiral nematic droplets [11]. Because each topological defect implies strong elastic deformations, these regions are accumulating a lot of elastic energy. Within the director free energy approach one can calculate the energy of elastic distortion for lines and points in nematics.

For the 1/2 disclinations presented in Fig. 1.9, the free energy per unit length of such a disclination is (Lubensky et al. [35, 37]):

$$\varepsilon = \frac{1}{4}\pi \ln\left(\frac{R}{r_c}\right) + \varepsilon_{core} \tag{1.32}$$

Here R is the radius of the cylindrical surface embracing the disclination and r_c is the core of the defect. Within the core of the defect, the order parameter is depressed and the free energy of the core is ε_{core}. A typical dimension of the core of a point defect is $r_c \sim 10\,\text{nm}$.

The energies of the three hedgehog configurations shown in Fig. 1.10 are calculated by embracing each hedgehog with a sphere of a radius R and then calculating the elastic energy using the director free energy density in Eq. 1.7. In a spherical region with a radius R the energies of three simple hedgehogs shown in Fig. 1.10 are [35]:

$$E_{radial} = 8\pi(K_1 - K_{24})R \tag{1.33}$$

$$E_{circ} = \frac{8\pi}{15}(3K_3 + 5K_2 + 2K_3 - 5K_{24})R \tag{1.34}$$

$$E_{hyper} = \frac{8\pi}{15}(3K_1 + 2K_3 + 5K_{24})R \tag{1.35}$$

In these expressions, the saddle-splay elastic energy term was added and K_{24} is the corresponding elastic constant. For zero saddle-splay term the hyperbolic hedgehog has the lower energy and the circular hedgehog has the most bend deformation.

Chapter 2
Dipolar and Quadrupolar Nematic Colloids

Abstract This chapter starts with discussion of long range forces between particles in nematic liquid crystals. The relation between topological defects and elastic distortion around spherical micro-particles in a nematic liquid crystal with perpendicular surface anchoring is discussed. The origin of the structural force in nematic colloids is explained, including particles of different symmetry and shape, but simple topology, such as microspheres, microcylinders and Janus particles. The interaction of nano-particles in nematic liquid crystals and the onset of ferromagnetism in nematic dispersion of magnetic nano-platelets is described in detail. The chapter concludes with the effects of chirality on the nematic colloidal interactions.

2.1 Dipolar Nematic Colloids: Elastic Dipoles with Hedgehogs

When a small microsphere is inserted into a nematic liquid crystal, the molecules are forced to align perpendicularly at all points of the closed surface, as shown in the schematic Fig. 2.1. One can immediately see that this requirement causes frustration for the liquid crystal molecules: they are naturally aligned homogeneously at distances far away from the microsphere, and they are forced to align perpendicular all along the spherical surface of the particle. As a result, a topological defect has to be formed in the vicinity of the inserted microsphere and it appears in a form of a hyperbolic hedgehog defect, carrying a topological charge of $q = -1$, as discussed previously. This topological charge has to be compensated by another topological charge in order to preserve the total charge neutrality of the whole system. This charge-compensating defect is in fact a virtual radial hedgehog, which is residing in the center of the microsphere and carries the charge of $q = +1$.

This pair of topological charges forms a topological dipole, which is attributed to the colloidal particle. It is analogous to an electric dipole, which is formed of two opposite electric charges $+e_0$, separated by a distance d. The director field around the inserted colloidal particle is strongly distorted and thus carries elastic energy, which is why this topological dipole is often called the elastic dipole.

I. Muševič, *Liquid Crystal Colloids*, Soft and Biological Matter,
DOI 10.1007/978-3-319-54916-3_2

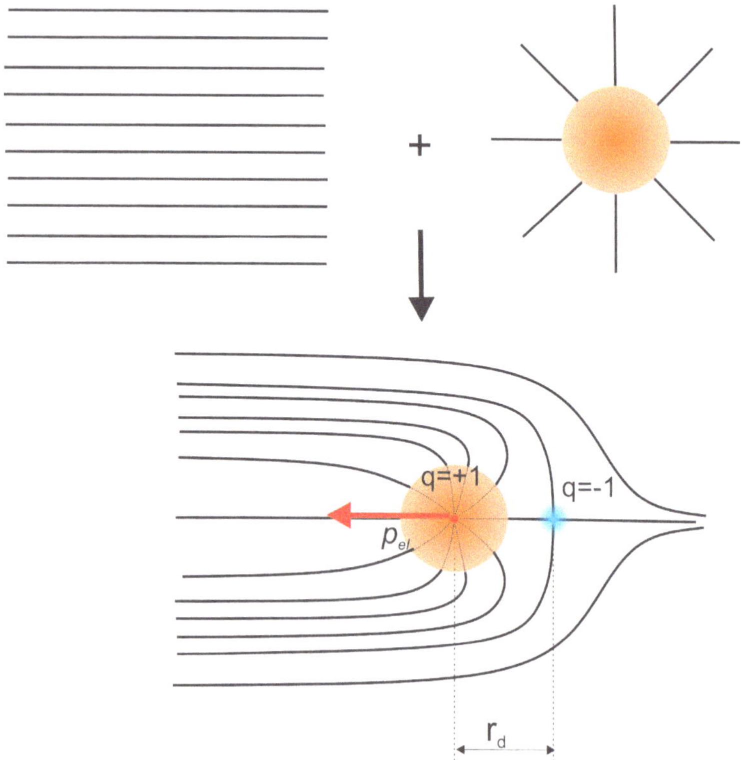

Fig. 2.1 Formation of a hyperbolic hedgehog defect (*blue*-haze object), when a microsphere with perpendicular surface anchoring is inserted into a nematic liquid crystal. There are actually two topological defects, which form spontaneously during the insertion of the microsphere: a virtual radial hedgehog carrying the topological charge $q = +1$ is formed in the center of the sphere (*red dot*) and a hyperbolic hedgehog defect, indicated by the *blue*-haze dot, is formed at a distance r_d from the center of the sphere. The hyperbolic hedgehog carries a topological charge of $q = -1$, which compensates for the topological charge of the virtual radial hedgehog in the center of the microsphere. Together they form a topological dipole, also named elastic dipole which is compensated in topological charge

The dipolar colloidal particles were first reported by Poulin et al. [45, 46] in experiments, where small water droplets were dispersed into a nematic liquid crystal, as shown in Fig. 2.2a–c. They observed rather strong forces between water droplets, which were of long range and strongly anisotropic. If left free, several isolated water droplets spontaneously formed chains, which were directed along the far-field director, as shown in Fig. 2.2a.

A closer inspection of a chain of water droplets (Fig. 2.2b) reveals dark water droplets with remnant of the characteristic Maltese cross due to the crossed polarisers, which occurs due to the birefringent nematic liquid crystal surrounding the optically isotropic water droplets. However, in between each pair of water droplets, one can clearly see in Fig. 2.2b a dark and distorted cross, marked with red arrow. This is a

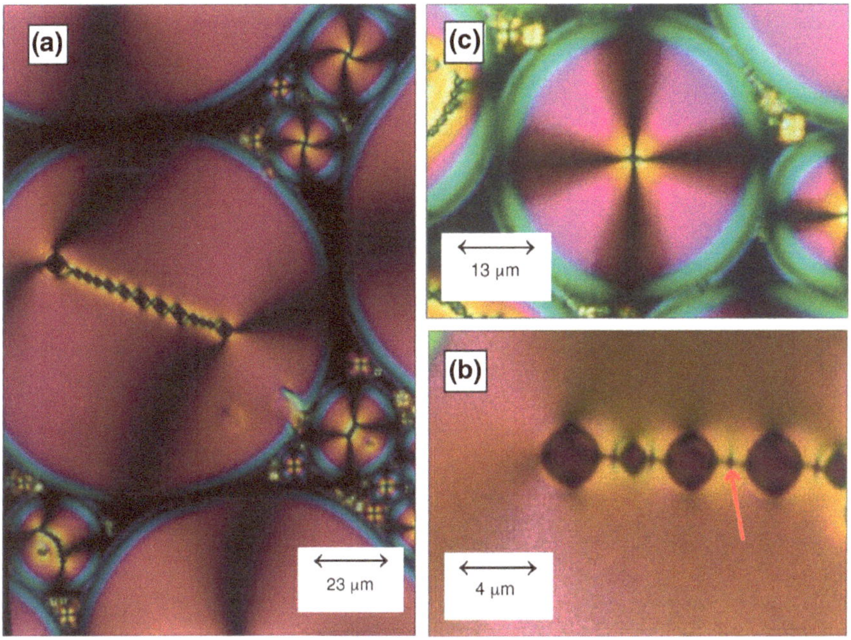

Fig. 2.2 **a** Microscope image of a nematic-water emulsion taken between crossed polarisers. A chain of water droplets residing in a larger nematic droplet is distinguished. **b** A magnified view of the water chains. Note the dark crosses residing in between each water droplet, marked by *red arrow*. **c** A magnified view of a nematic droplet containing a single water droplet. From P. Poulin, H. Stark, T.C. Lubensky, and D.A. Weitz, Science **275**, 1770 (1997). Reprinted with permission from AAAS

hedgehog point defect, created in the nematic host after water droplets were inserted. The separation between two droplets increases with droplet radius, and there is always one hedgehog defect fewer than there are water droplets. These water chains are stable and robust structures and do not coalesce spontaneously. This indicates that the hedgehog defects, residing in between each pair of water droplets, have the role of a topological and energy barrier that prevents spontaneous coalescence of water droplets.

Observation of a series of hedgehog defects separating the water droplets can be qualitatively understood by considering the total topological charge q in a nematic liquid crystal. Because the boundary conditions on the director field are uniform, the total topological charge of a uniform nematic liquid crystal is $q = 0$. If the nematic liquid crystal is confined to a droplet surrounded by water (as it is in our case, shown in Fig. 2.2a and c), its topological charge is equal to $q = +1$. Normal boundary conditions at the liquid crystal-water interface therefore force the creation of a radial hedgehog at each droplet. However, if a water droplet is added into the interior of the nematic droplet, it creates its own topological defect, a hedgehog with a charge $q = -1$, which compensates the radial hedgehog to the total charge of 0. We then

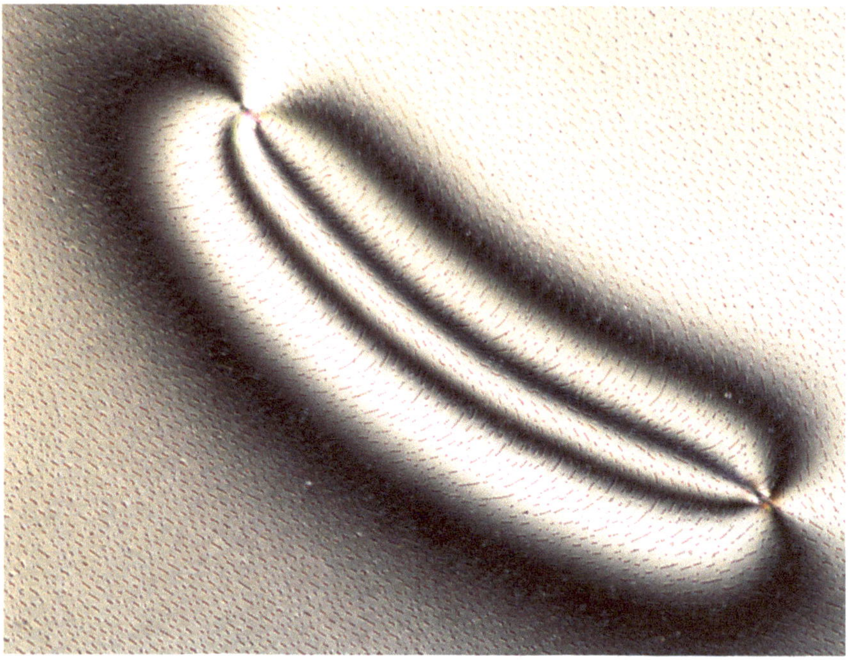

Fig. 2.3 Microscope image of a nematic liquid crystal with small droplets of glycerol, as observed between crossed polarisers. The microdroplets spontaneously form chains of particles, which follow the local orientation of the nematic director on a free interface towards the air. This makes a pair of topological defects clearly visible. Image courtesy of V.S.R. Jampani

have a situation, when there is a single water droplet in the center of the nematic droplet and no defects are needed (Fig. 2.2c).

However, when the second water droplet is added into a nematic droplet, it brings an additional hyperbolic hedgehog defect, which has to be positioned in between the two water droplets. The total topological charge is therefore conserved, because the second droplet brings already compensated topological charge. It is clear that a system of N water droplets in a nematic droplet induces $N - 1$ hyperbolic hedgehog defects in the nematic host.

Agregation of particles in the nematic liquid crystal, which is similar to this spontaneous formation of water chains in the host liquid crystal was observed already in the 1970s, when the topological defects in liquid crystals were studied by Cladis et al. [47] and Rault [48]. They dispersed microscopic gas bubbles in a liquid crystal and analyzed the free interface of the nematic liquid crystal with air. Surprisingly, they observed spontaneous organisation of micro-bubbles into well-ordered chains, which were following the local orientation of the liquid crystal molecules. This decoration of the director lines helped to visualise the surface ordering of a nematic liquid crystal and the details of topological defects, as shown in Fig. 2.3. After those experiments and prior to the publication of Poulin et al. in 1997 [45], there were

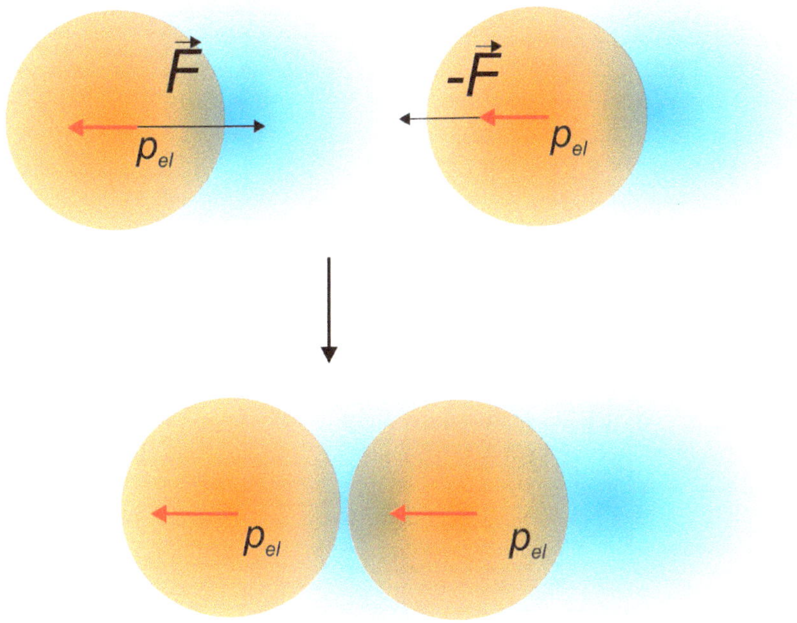

Fig. 2.4 The elastic interaction between two dipolar colloidal particles. If their elastic dipoles $\mathbf{p}_{el}$ are collinear and pointing in the same direction, they will be attracted by a pair of elastic forces F to minimize the distorted volume

several studies of particle aggregation, as discussed by Poulin et al. [49] and later by Raghunathan et al. [50] on the dispersion of 60 and 120 nm latex particles in lyotropic nematic solution. The conclusion of these studies was that there is clearly a power-law attractive force between colloidal inclusions in the lyotropic nematic liquid crystals, but the smallness of the particles prevented the observation of topological defects. The first line of studies was initiated by the theoretical paper by Brochard and de Gennes [51] on possible collective ferromagnetic order of magnetic nanoparticles in the nematics, whereas the second experiment on nematic colloids at that time was the study of Kreuzer et al. [52] on the erasable optical storage effect in the nematic liquid crystal filled with 7–40 nm silica nanoparticles.

The spontaneous formation of water or colloidal chains in the nematic liquid crystal is a clear indication of forces between colloidal particles due to the liquid crystal. The force between a pair of colloidal particles (solid, liquid or gas-like) in a liquid crystal can be explained in terms of the elastic distortion of the liquid crystal surrounding each colloidal particle, as shown in Fig. 2.4. This distortion is strongly anisotropic due to the hyperbolic hedgehog and is concentrated in its vicinity, as illustrated by the blue region in Fig. 2.4. If two such particles are brought closer together, their regions of elastic distortion will start to overlap. This overlapping of elastically distorted regions can be either in favour of both particles, or not (in terms of total free energy). When the elastic distortion is energetically favourable for both particles (meaning that it lowers the total free energy of the system), they will tend to share that region as much as possible. It is then clear that there will be a pair of attractive forces between the two particles, and by moving the particles closer to each other, they will lower their total free energy. This will result in the attraction of the two dipolar colloidal particles with their elastic dipoles parallel and collinear, as in Fig. 2.4.

2.2 Quadrupolar Nematic Colloids and Saturn Rings

In the experiments with nematic emulsions, Poulin et al. [45] observed water droplets, which were accompanied by hyperbolic point hedgehog defects and there was no evidence of ring-like topological defects, encircling the droplets, which were predicted a couple of years before the experiment. These ring-like defects, also called Saturn ring defects, were predicted in 1995, by E.M. Terentyev [53] and his prediction was based on numerical simulations. Director structure around a colloidal particle suspended in a nematic liquid crystal was soon analyzed also by Kuksenok et al. [54] and H. Stark [55]. The reason why a ring topological defect should exist instead of a point hyperbolic hedgehog is the topologically allowed transformation of a hyperbolic hedgehog into a hyperbolic ring, as shown schematically in Fig. 2.5a.

The transformation of a hyperbolic hedgehog point defect into a –1/2 disclination ring is performed by opening the point into a ring and filling in the vertical lines of the director field. When considering the far field of the nematic director, the disclination ring is still "seen" as a hyperbolic hedgehog. This means that the ring is assigned the same topological point charge $q = -1$. Similarly, a radial point defect could be smoothly transformed into a $+1/2$ disclination ring, as illustrated in Fig. 2.5b.

While from the point of view of topology this smooth transformation of points into loops is allowed, the energies of points and loops might be quite different, not only because of the size but also because of the type and degree of deformation. Detailed analysis of the energies of points and loops is thoroughly discussed by H. Stark [37]. It is evident that opening point defects into rings increases the elastic energy of distortion, which means these are rare objects in nematic colloids. Quite recently, Wang et al. [56] were able to determine the structure of topological defects at a nanoscale by templating defect cores with amphiphilic molecules and imaging them with TEM microscopy.

Indeed, the first experiments by Poulin et al. [45] found no evidence of spherical colloidal particles accompanied by defect rings, although they were predicted already in 1995 by Terentyev [53]. In this work, it is predicted that a suspended spherical particle with perpendicular surface anchoring of liquid crystal molecules introduces on its surface a closed disclination loop, which was called the Saturn ring, shown in Fig. 2.7.

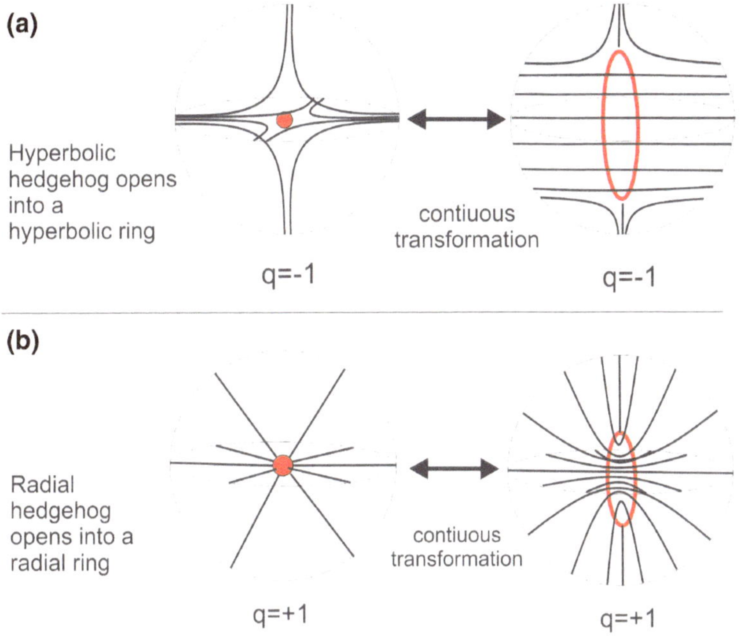

Fig. 2.5 **a** A schematic drawing of the director field around an isolated hyperbolic hedgehog defect. This defect can be opened into a hyperbolic defect ring, which preserves the total topological charge of $q = -1$. Note that the winding number changes from $k = -1$ for a point to $k = -1/2$ for a hyperbolic ring. **b** A radial hedgehog defect with charge $q = +1$ and winding number $k = +1$ can be opened into a $k = +1/2$ defect ring, which preserves the total topological charge $q = +1$

As it happens many times in science, it turns out that the Saturn rings were observed much earlier than 1995, when they were predicted. In the year 1990, Pratibha and Madhusudana [57] were studying mixtures of rod-like (NR) and disc-like (ND) nematic liquid crystals in the range of coexistence of two nematic phases, NR and ND. They observed radial point defects of charge $q = +1$ in NR droplets in an otherwise continuous ND medium, as shown in Fig. 2.6a. However, they noticed an additional ring surrounding the NR droplets in the disc nematic liquid crystal, surrounding the NR droplet. This ring has the opposite charge of $q = -1$ and therefore compensates the charge of the radial point defect, as shown in Fig. 2.6b

Detailed experimental analysis of the conditions for emergence of Saturn ring defects showed that they can be observed in four different cases: (i) For weak surface anchoring, a point hedgehog is transformed into a Saturn ring encircling the colloidal particle. This was demonstrated by Mondain-Monval in 1999 [58] for thermotropic nematic liquid crystal E7 in water with surfactants (SDS, Pluronic F68). By adjusting the amount of SDS, dipolar configuration with point hedgehogs could be transformed into the Saturn ring for weak surface anchoring. (ii) Saturn ring defects around colloidal particles in the nematic are observable when particles with homeotropic anchoring are confined to thin planar cells. This was demonstrated by

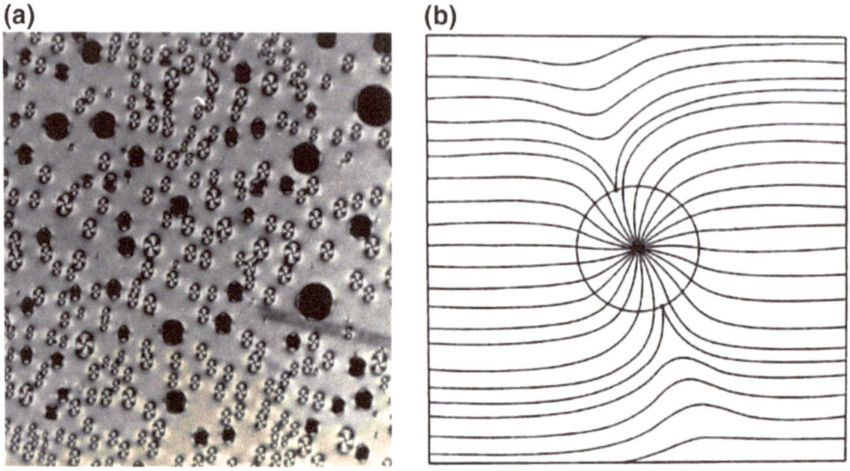

Fig. 2.6 **a** Coexistence of NR and ND phases in a mixture with 30 mol % of C7OHBT in 3CHP3B. Note the bright cross in the centre of each droplet indicating radial point defect and the ring-like halo surrounding the droplet. **b** Schematic diagram of the director configuration with a radial point defect in the NR droplet and the −1/2 ring defect surrounding the NR droplet. Image courtesy of M. Pratibha

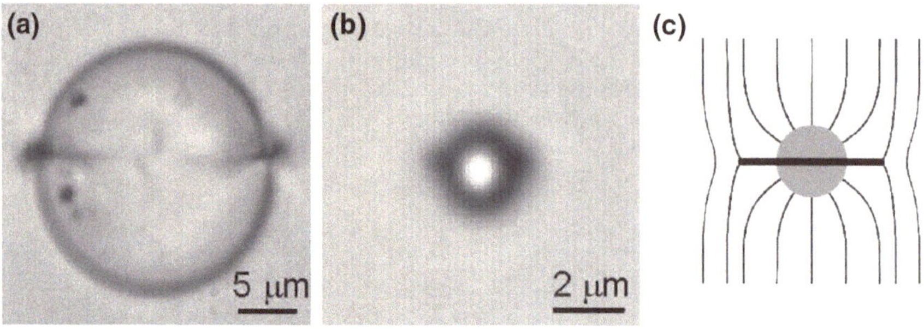

Fig. 2.7 Saturn rings around colloidal particles in thin planar cells. **a** Saturn ring is clearly visible, when larger colloidal particles with homeotropic surface anchoring are observed in a planar nematic cell under an optical microscope. **b** With smaller microspheres, the ring is barely observed and appears in a form of two dark spots. **c** Schematic drawing of the director field around Saturn-like colloidal particle

Gu and Abbott [59] for larger, 100 μm diameter glass spheres in a 120 μm thick planar liquid-crystal cell, as shown in Fig. 2.7. (iii) An external electric field applied along the dipolar axis of an elastic dipole induces a transition to an elastic quadrupole with the Saturn ring [60, 61]. (iv) If the diameter of the colloidal microsphere is decreased, a transition from a dipolar configuration at large diameters to quadrupolar configuration is induced at some critical diameter. This was demonstrated by Völtz et al. [62] in experiments with gas bubbles of variable diameter in nematics.

Using an external electric field, a dipolar colloidal particle could be transformed into a quadrupolar one, as demonstrated by Loudet et al. [61]. Once the electric filed has been turned of, the Saturn ring relaxes back into the hedgehog point defect. Typical ring velocities for this topological transformation are of the order of several μm per second.

2.3 Forces Between Spherical Colloidal Particles in Nematic Liquid Crystals

When nematic colloidal dispersion is put in a thin layer between two glass plates, one can immediately observe that the colloidal particles have spontaneously assembled into irregular structures, which, however, show some local ordering such as chains, shown in Fig. 2.8.

This observation clearly demonstrates the existence of a force between a pair of (or several) colloidal particles, which are inserted into the nematic liquid crystal. If the sample with a nematic dispersion is heated into the isotropic phase of the

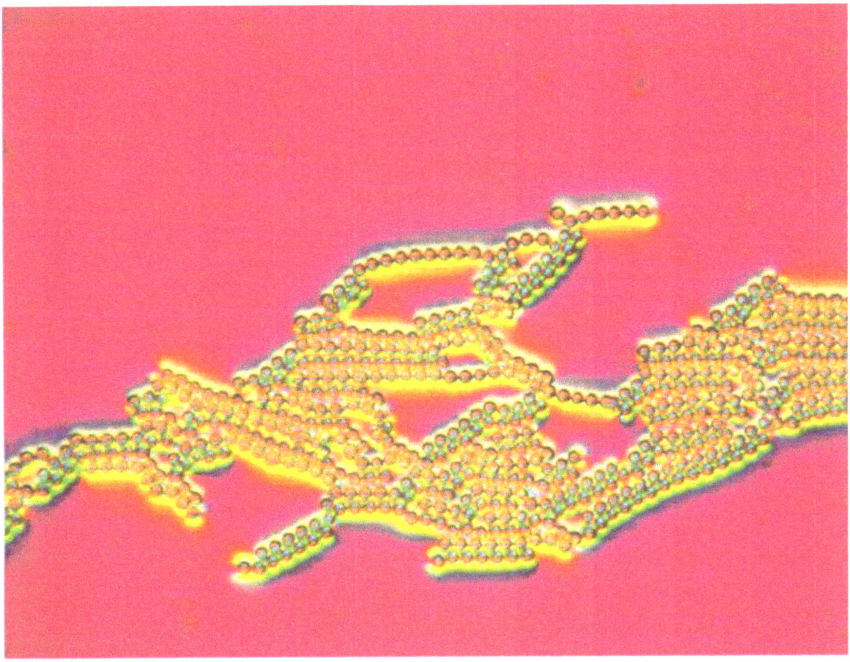

Fig. 2.8 Microscope image of spontaneously assembled structures of dipolar colloidal particles in a thin nematic liquid-crystal cell. The diameter of silica microspheres is 2.3 μm and their surface is treated with DMOAP for good homeotropic alignment

liquid crystal, the colloidal clusters spontaneously disassemble due to the Brownian motion of individual particles. This is an evidence of the colloidal force, which has its origin in the nematic liquid crystal surrounding the particles. As we have seen in the beginning of Sect. 2.1, colloidal particles in nematic liquid crystals are not only accompanied by a topological defect (in a form of a point or a loop) but also surrounded by a large area where the liquid crystal is strongly distorted. The reason for this distortion is liquid-crystal alignment along the closed surface of the colloidal particles, which has to transform in the far-field into a homogeneous nematic liquid crystal. Because of the orientational long-range order, this elastic distortion is also of long range and spreads out from the surface as a power-law.

Consider now two colloidal particles together with their elastically distorted regions, each of which surrounds the colloidal particle up to the distance of several particle diameters. When two particles are far away from each other, these elastically distorted regions, originating from each particle, do not overlap significantly. However, when the particles are brought together, their distorted regions start to overlap. This overlapping could be either in favor or not in favor of both particles. When in favor, the particles will tend to share this common region as much as possible, which will drive the particles closer together. It is clear that there will be an attractive force due to the elastic deformation between the two particles. We call it structural force for evident reasons. On the contrary, the particles will be repelled from each other by the elastic distortion when the overlapping is not in favor and the energy increases by approaching the particles.

It is then clear that the elastic distortion of a nematic liquid crystal, created by the insertion of the particle into the liquid crystal, generates forces between particles. In terms of physics, this is a generalized force and has its origin in elastic deformation. It can be expressed as the total derivative of the free energy F with respect to the separation between the particles d:

$$\boldsymbol{F} = -\frac{\partial F}{\partial d} \tag{2.1}$$

Alternatively, this structural force has a topological origin. By inserting particles into a continuous nematic ordering field, topological defects are created, which mediate the force between the inserted particles. Topological defects are therefore the generators of forces between colloidal inclusions.

2.3.1 Forces Between Spherical Dipolar Colloidal Particles in Nematics

Sharing of fields is a common concept in physics and could be used to explain, for example, the electrostatic attraction or repulsion between the electric charges. This simple analysis of nematic colloidal forces using hand-waving arguments is nicely illustrated in a simple experiments with two dipolar nematic colloidal particles in a

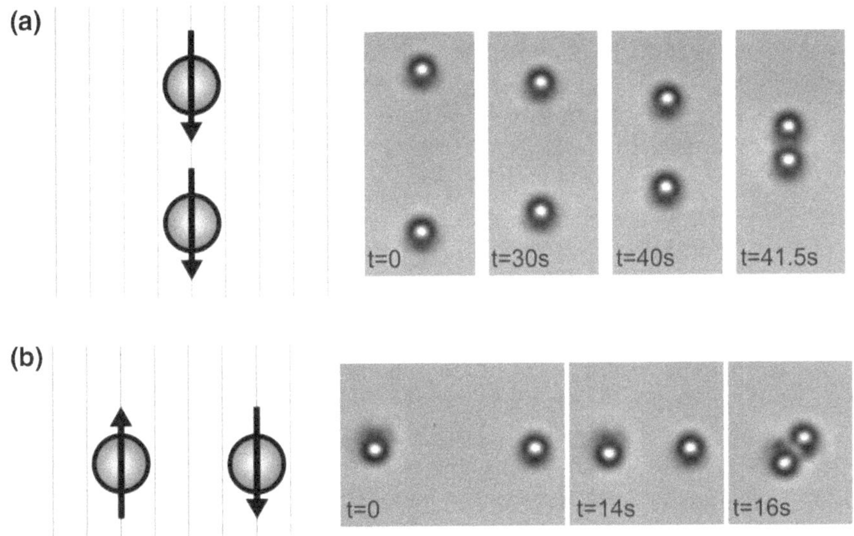

Fig. 2.9 Attraction of particles in a nematic liquid crystal. **a** Selected snapshots from a video showing the attraction of two dipolar colloidal particles placed collinearly with respect to the far-field director. Both dipoles are pointing into the same direction. **b** The dipoles are now anti-parallel, but they are also placed side-by-side, not collinearly as in **a**. Snapshots of colloidal motion show their attraction

planar nematic cell, as shown in Fig. 2.9. We select two colloidal particles by the laser tweezers and we bring them close together. There are several options of positioning the particles, and we should consider only two: the particles will either be placed along the far-field director of the planar cell (collinear position) or they will be placed along the line perpendicular to the far-field director (side-by-side position). In addition, we can choose the direction of each dipole, as they can be set either parallel or perpendicular to each other. One can see from Fig. 2.9 that the interaction between two elastic dipoles is similar to the interaction of electric dipoles. This analogy was indeed developed in the theory of elastic forces in nematic colloids.

Figure 2.9a shows selected snapshots from a video showing the interaction of two parallel and collinear dipoles. One can see the rather slow motion of colloidal particles when they are well separated. However, with gradual approaching, the speed of approaching increases and reaches maximum when the two particles bind together into a chain. Figure 2.9b shows a different situation for the two dipolar colloidal particles, which were brought together with their dipoles anti-parallel, but they are placed side-by-side. This manipulation was done with laser tweezers and will be explained in Chap. 3.

The mathematical description of the structural force between two colloidal particles in a nematic liquid crystal is quite difficult problem because of nonlinearity of the system. The resulting equations are highly nonlinear and the interaction of several particles is a many-body problem. Another complication is the anisotropy of elastic deformation of a nematic liquid crystal and large difference between the elastic constants. If one considers the interaction of two colloidal particles, one can immediately see that different types of interactions will be more pronounced for different interaction scenarios. Finally, the ordering field is a tensorial field, where the local degree of the order parameter in principle depends on the particle separation. Because the elastic constants depend on the degree of order, this complicates the problem even further. In more simple cases, we use the director approach, where the degree of order remains unchanged throughout the system at all times. In the other extreme, one can consider fully tensorial Landau-de Gennes description, where the degree of order is variable. However, one usually uses a one elastic constant approximation to keep the numerical difficulties solvable.

For the purpose of this book, we shall briefly review the approach based on the electrostatic analogy and multipole expansion, as introduced by Lubensky et al. [35]. Within this approach the interactions between the particles are linearized because the nonlinear regions with strong elastic deformation close to the particles are excluded from the analysis. These regions are properly consider to construct the "far field" behaviour, where spatial gradients of the director field are small and the corresponding equations governing the elasticity are linear.

When a colloidal particle with well-defined surface-anchoring conditions is inserted into the infinite and homogeneous nematic liquid crystal, it will locally distort the nematic director field and induce the formation of topological defects, which are the singularities of the ordering field. Because of the constraint of topological charge conservation, which should be kept zero at all times, the distorted director field should smoothly approach the homogeneous nematic director field at far separation from the particle. In addition, we have to consider the invariance of the director field direction, as $+\mathbf{n}$ and $-\mathbf{n}$ are indistinguishable. We should also consider the far-field form of the distortion induced by the colloidal particle. These deviations of the local director from the far-field director are considered to be small. At large separations $|\mathbf{r}|$ from the colloidal particle, the full nonlinear Frank elastic free energy Eq. 1.7 is replaced by the harmonic expression:

$$F_{har} = \frac{1}{2}K \sum_{i=x,y} \int d^3r(\nabla n_i)^2 \tag{2.2}$$

Here, the director components are described by the notation n_i, $i = x, y$ for the components of the director $\mathbf{n}$ perpendicular to the far-field director $\mathbf{n}_\circ$. The Euler-Lagrange minimisation leads to the Laplace equation for the director field components n_i:

$$\nabla^2 \mathbf{n}_i = 0 \tag{2.3}$$

At large separations $\mathbf{r}$ from the particles, the solutions to the Laplace equation can be expanded in multipoles.

$$n_i = \frac{A^i}{r} + \frac{\mathbf{p^i} \cdot \mathbf{r}}{r^3} + \frac{c^i_{kl} r_k r_l}{r^5} + \cdots \tag{2.4}$$

Here A^i, p^i and c^i_{kl} are expansion coefficients and r_i are Cartesian coordinates of the position vector $\mathbf{r}$. Because of the invariance with respect to the rotation above the z- axis, and the absence of the azimuthal component to $\mathbf{n}$ (i.e. without the twist of $\mathbf{n}$ around the z-axis), the monopole term A_i equals zero. Furthermore, the coefficient of the second dipolar term $(\mathbf{p^i} \cdot \mathbf{r})$ equals to $(\mathbf{p} \cdot \mathbf{n}) \cdot \mathbf{e}^i$. Here, $\mathbf{p}$ is identified as the vector of the dipole moment of the particle-defect configuration, whereas $\mathbf{p} \cdot \mathbf{n}$ is the z- component of this dipole. The parameter c^i_{kl} of the third term in Eq. 2.4 is identified as the amplitude of the quadrupole moment tensor of the droplet-defect combination.

Based on the multipole expansion and electrostatic analogy, each colloidal particle creates far-field distortions of the director $\mathbf{n}$, determined at large distances by the equations:

$$n_x = p_z \frac{x}{r^3} + 2c \frac{zx}{r^5} \tag{2.5}$$

$$n_y = p_z \frac{y}{r^3} + 2c \frac{zy}{r^5} \tag{2.6}$$

This far-field director interacts with the director fields of other particles, leading to an effective particle-particle interaction that can be expressed as pair-wise interaction between dipole and quadrupole densities. The resulting pair interaction energy $U(\mathbf{R})$ between two colloidal particles at positions $\mathbf{r}$ and $\mathbf{r}'$ $(\mathbf{R} = \mathbf{r} - \mathbf{r}')$, with their respective dipole and quadrupole moments (p_z, p'_z, c, c') is thus:

$$U(\mathbf{R}) = 4\pi K \left[p_z p'_z V_{PP}(\mathbf{R}) + \frac{4}{9} cc' V_{CC}(\mathbf{R}) + \frac{2}{3}(cp'_z - c' p_z) V_{PC}(\mathbf{R}) \right] \tag{2.7}$$

where:

$$V_{PP}(\mathbf{R}) = \frac{1}{R^3}(1 - 3\cos^3\theta) \tag{2.8}$$

$$V_{CC}(\mathbf{R}) = \frac{1}{R^5}(9 - 90\cos^2\theta + 105\cos^4\theta) \tag{2.9}$$

$$V_{PC}(\mathbf{R}) = \frac{\cos\theta}{R^4}(15\cos^2\theta - 9) \tag{2.10}$$

The three terms clearly correspond to dipole–dipole (V_{PP}), quadrupole–quadrupole (V_{CC}) and mixed dipole-quadrupole (V_{PC}) pairwise interaction energy, and they all display the well-known power-law dependencies, similar to the interaction of electric multipoles.

Using this *pair-interaction potential* between dipolar, quadrupolar and mixed densities, one can calculate *the force* between the two particles as a function of their separation R. By remembering that the force is spatial derivative of the pair-interaction energy, the force between two dipolar colloidal particles should exhibit the $1/R^4$ dependence on their separation R. Similarly, two quadrupolar colloidal particles should interact with a force showing the $1/R^6$ power law dependence on

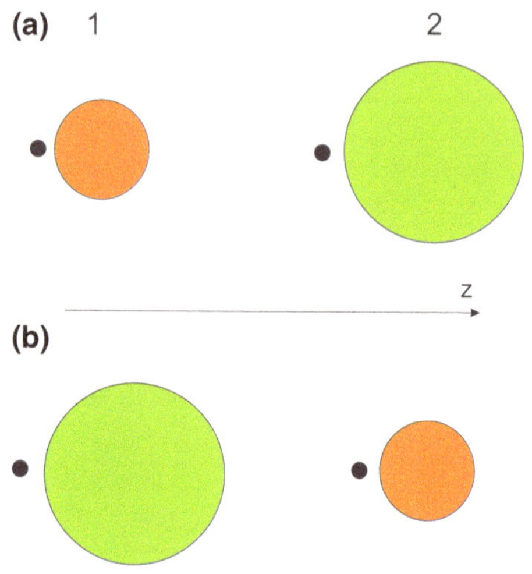

Fig. 2.10 **a** The large particle is to the *right* of the smaller particle. **b** The inverse configuration. The attractive force in case (**a**) is larger than in case (**b**). Adapted from Lubensky et al., PRE 1997

their separation R, whereas a dipole-quadrupole force should fall as the $1/R^5$ power law.

Let us now consider this pair-interaction potential to determine the force between two microspheres as a function of their mutual separation, as shown in Fig. 2.10. The particles are labeled 1 and 2, respectively and they have different radii, a_1 and a_2. The particles are placed collinearly at a separation R and along the far-field director, as illustrated in Fig. 2.10. The pair interaction force $\boldsymbol{F}$ between two colloidal particles, normalized to the elastic constant K, is:

$$\frac{\boldsymbol{F}}{4\pi K} = -\alpha^2 a_1^2 a_2^2 \frac{6}{R^4} + \beta^2 a_1^3 a_2^3 \frac{120}{R^6} - \alpha\beta a_1^2 a_2^2 (a_1 - a_2) \frac{24}{R^5} \tag{2.11}$$

Here, α and β are the dipolar and quadrupolar coefficients within the dipole ansatz. The centres of particles are separated by R, and we have used one elastic constant approximation with elastic constant K. The leading term in the Eq. 2.11 is the attractive dipole–dipole force, which is proportional to R^{-4}. It is therefore analogous to the electrostatic force between two electric dipoles. The quadrupolar part of the force depends as R^{-6}, whereas the mixed dipole-quadrupole force scales intermediately as R^{-5}.

This comprehensive analysis of pair interaction of elastic multipoles and in particular topological dipoles by Lubensky et al. [35] stresses the importance of topology and discusses the role and structure of topological hedgehog defects in dipolar colloidal interaction, which has proven much later to be fundamental in the mechanism of topological entanglement of colloids. Further mean field approaches to nematic colloids include investigations of power-law forces between particles in nematics by Ramaswamy et al. [63], the stability of colloidal clusters by Lev and Tomchuk [64] and the effects of confining walls on colloidal pair interaction by Fukuda et al. [65, 66]. While these studies concentrated on director approach, thus discarding the effects of spatially varying the degree of order, Fukuda et al. [65, 66] have used fully tensorial Landau-de Gennes approach to the analysis of colloidal pair interaction in the nematic liquid crystal. Pergamentschik and Uzunova have used a refined approach to the colloidal nematostatics [67, 68]. They observed that, in spite of the analogy to the electrostatics, the three-dimensional colloidal nematostatics is substantially different in both its mathematical structure and its physical implications. They formulated a general tensorial structure of the elastic multipoles that allows for a classification of different types of nematic colloids. In this approach, the elastic multipoles have one extra tensorial index, so an elastic dipole is characterized by three coefficients: (i) isotropic strength, (ii) anisotropy and chirality, and (iii) a two-component vector along the unperturbed director [69, 70]. Instead of a single electric dipole, they found several different pure and mixed types of elastic dipoles. A review of this approach can be found in Ref. [68]. Chernyshuk et al. [71] have considered the theory of colloidal elastic interaction in the presence of an external electric or magnetic field using the Green's function method.

The first experimental confirmation of the power law dependence of the force between two dipolar colloidal particles in the nematic were given by Poulin et al. [72]. Instead of solid colloidal particles, they were using droplets of a ferrofluid, dispersed in the nematic liquid crystal. A ferrofluid is water-based dispersion of tiny super-paramagnetic particles. These particles have zero permanent magnetic moment, but are easily magnetised. In a small external magnetic field, each droplet will therefore develop a rather strong magnetic moment, directed along the external magnetic field.

When such a paramagnetic droplet is inserted into a nematic liquid crystal, their interfaces can induce dipolar topological configuration. If one observes a planar nematic cell with a multitude of super-paramagnetic droplets, one can easily find pairs of such droplets, which form chains because of their dipolar attraction. If an external field is now applied in a direction perpendicular to the sample, it will induce a magnetic moment in each super-paramagnetic droplet. Because their induced magnetic moments are parallel and they are placed side-by-side, they will start to repel because of the repulsion of their parallel magnetic moments. This repulsive force will be stronger if the magnetic field is increased. Taking all forces into account, one observes that two dipolar colloids will be more and more separated when the external magnetic field is increased up to $\sim 100G$, as shown in Fig. 2.11.

Once the particle separation is increased beyond several particle diameters (to around 20 μm), the magnetic field is turned off and the structural force due to elastic attraction draws the droplets back together as illustrated in Fig. 2.11d and e. The initial velocity just after switching off the field is rather small, but the speeds of the two droplets increase as they approach each other. The trajectory of the approach is a straight line joining the centres of the two droplets and is parallel to the far-field director.

In the experiment, one determines the elastic attractive force as a function of particle separation using a standard method of video microscopy. The process of attraction is filmed at a sufficiently high frame rate and is captured from the start at large separations to the end when the droplets come into close contact and have zero velocity. The method is fully described in Chap. 3. Here, let us just mention the

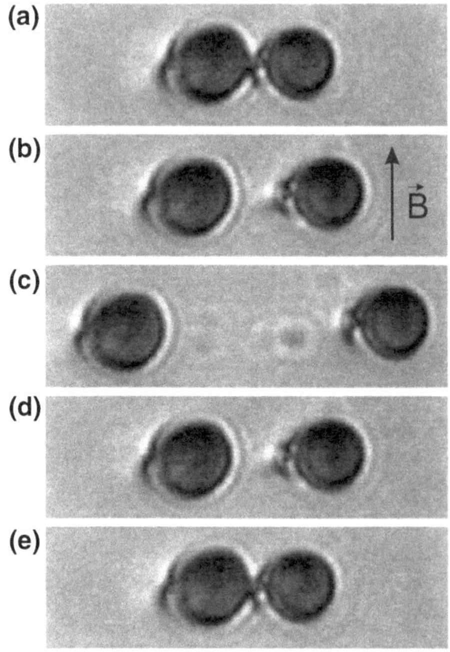

Fig. 2.11 **a** In zero magnetic field, the two colloids are attracted by elastic force. When an external magnetic field is applied, the colloids are repelled from each other by magnetic repulsion of their magnetic moments, as shown in **b** and **c**. After switching off the magnetic field in **c**, the elastic attraction pulls the superparamagnetic droplets together, shown in **d** and **e**. Reprinted figure with permission from P. Poulin, V. Cabuil, D.A. Weitz, Phys. Rev. Lett. 79, 4862 (1997). Copyright (1997) by the American Physical Society

most important steps in determining the force. It turns out that the motion of each sphere is strongly overdamped because of the high viscosity and high Stokes drag force on a sphere. This means that the acceleration in the second Newton law is zero at all times at the force acting on a given sphere is balanced by the viscous force on that sphere due to the Stokes drag. Because the viscous force is proportional to the velocity of the microspheres, one needs to know the velocity during the approach of the particles at all times. This is determined by numerically deriving the trajectory of the particles as a function of time. The trajectory is simply reconstructed from the taken video frames by using standard video particle tracking software. One also needs to know the Stokes drag coefficient for that particular microsphere, which is determined with an independent experiment in which the Brownian motion of the sphere is monitored and analysed.

Using this simple method, Poulin et al. [72] were able to measure the attractive structural force on a selected microsphere as a function of particle separation, which is shown in Fig. 2.12 for three different diameters of the microspheres. The log-log plot in Fig. 2.12 clearly shows the power-law dependence of the attractive force between two dipolar and collinear microspheres in a nematic liquid crystal. The solid lines are the best fits to the data and have a slope of −4. This clearly confirms the predicted power-law dependence for the collinear attraction of two elastic dipoles in a nematic liquid crystal, $F(r) = R^{-4}$. Because the force is expected to scale as a fourth power of the particle size, $F \sim a^{+4}$, the inset to Fig. 2.12 shows the separation-dependence of the measured attractive force divided by the fourth power

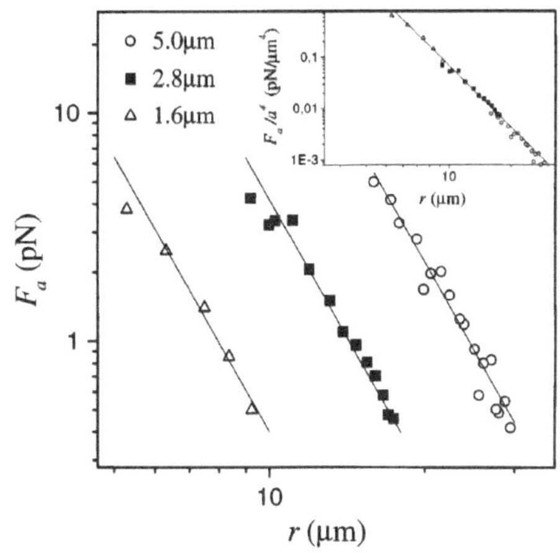

Fig. 2.12 Force F, acting on a dipolar ferro-fluid microdroplet as a function of separation r for three different diameters of the pairs of droplets. *Solid lines* are best fit to the power law, with the slope equal to -4. The *inset* shows the scaling of the force with the particle size. Reprinted figure with permission from P. Poulin, V. Cabuil, D.A. Weitz, Phys. Rev. Lett. **79**, 4862 (1997). Copyright (1997) by the American Physical Society

of the particle diameter. One can clearly see that all three independent measurements fall on the same line, which confirms the predicted size-dependence of the elastic force between dipolar nematic colloids. It should be noted that this observation of the power-law dependence of the attractive elastic force between nematic colloids is quite limited in the interval of measured separations. This interval is rather narrow, and is typically less than one decade in experimentally accessible separations. At small separations, the two particles are practically in contact and the attractive elastic force is balanced by a strong repulsion due to near contact of the two surfaces. On the other hand, if the two particles are separated for more than several particle diameters, the elastic attraction becomes very small and is comparable to the Brownian noise due to thermal fluctuations of the positions and orientational fluctuations of the nematic. There is therefore a limiting maximum separation, determined by the noise level in the system, which prevents measuring colloidal interaction forces in a separation interval larger than one decade. This experimental limitation is present for practically all measurements of the structural forces in liquid-crystal colloids.

The elastic force for simple collinear attraction along the far-field director of a planar cell is of course the simplest possible geometry, easily realised by using simple experimental set-up. However, it is expected that the two dipolar colloidal particles will show complex pattern of attraction or repulsion if they are positioned at an arbitrary angle with respect to the far-field director. However, in order to be able to measure the anisotropy of the dipolar colloidal force, one has to be able to position the particles at an arbitrary and predetermined mutual angle of attraction. This can be realised by using the laser tweezers as a practical tool to grab, move, and precisely position an individual particle at a predetermined position. This technique is described in detail in Chap. 3.

Figure 2.13a–c shows an example of the measurement of the *pair-interaction potential* (not the force itself, but its integral along the trajectory of the particle) for two dipolar silica microspheres in a nematic liquid crystal 5CB [73]. The sequence of microphotographs in Fig. 2.13 shows the attraction between the two dipolar

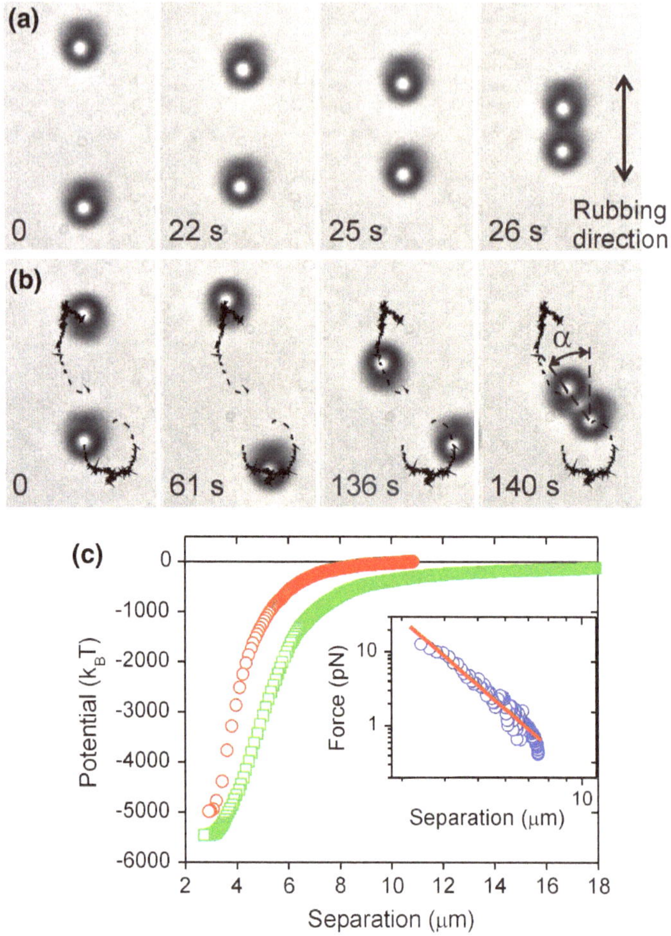

Fig. 2.13 The pair-interaction between two dipolar colloidal particles. **a** Collinear attraction between two dipoles that were aligned "head-to-tail". **b** The direction of the *upper* dipole was reversed, resulting in a strong repulsion at the beginning of the experiment. This was followed by a curved motion of both particles, indicating strongly anisotropic repulsion and attraction. In the final state, the two particles have anti-parallel dipoles and their centres are tilted with respect to the far-field director. **c** The pair-potential, as measured along the trajectories of both particles. The potential is calculated by integrating the work of the attractive force along the trajectory of a selected particle. The *inset* shows the dipolar force for parallel and "head-to-tail" attraction. The *solid line* is the $1/R^4$ power-law for the elastic force

colloids for two possible orientations of their dipoles. In both cases, the particles were positioned collinearly along the far-field nematic director, however, their two dipoles were oriented either in the same (a) or in the opposite direction (b). In the first case, the two particles are always attracted and start approaching each other from the beginning of the experiment. In the second scenario in Fig. 2.13b, they are first strongly repelled from each other (see the dark trace, which is the trajectory of particles). After some time they find themselves side-by-side and start attracting because of their anti-parallel dipoles. The inset to Fig. 2.13c shows the force of attraction for parallel and "head-to-tail" attraction of two dipolar colloidal particles in Fig. 2.13a, which clearly shows the expected $1/r^4$ power law attraction.

Focusing on the small particle separations, Noel et al. [74] analysed the short-range repulsive force between two dipolar iron colloidal particles in a nematic liquid crystal, resulting from the presence of a hedgehog defect between the two particles. If one considers two collinear nematic dipolar colloids and the magnetic field is applied along their axes parallel to the line joining their centers, the induced magnetisation will induce stronger magnetic attraction between the two colloidal particles. The particles will be therefore forced to approach closer together and the magnetic force will be balanced by an elastic short-range repulsion between two colloidal particles. This repulsion is due to the hyperbolic hedgehog defect, located in between two dipolar colloids. By monitoring and measuring the positions of the two particles as a function of the magnetic field, one is able to calculate the short-range elastic repulsion. It was found that at a given threshold force, the point hedgehog defect, which is residing in between the two colloidal particles, was expelled and transformed into a Saturn ring located between the particles. In this way it provided enough space for the particles to come closer together. Before the threshold for this sudden transformation, the short-range repulsion showed a quadratic dependence on the particle separation. The elastically distorted point hedgehog therefore acts as an elastic spring separating the two particles [74]. The experimental measurements of the separation dependence of the structural force between dipolar colloids in nematics have been reported by several groups [75–77]. The inelastic collisions of nematic dipolar colloidal particles using a bi-directional hydrodynamic flow of the nematic was studied by Pishnyak et al. [78].

2.3.2 Forces Between Spherical Quadrupolar Colloidal Particles in Nematics

The force between two nematic colloidal particles with quadrupolar symmetry of the elastic distortion was first measured by Smalyukh et al. [79]. They were using laser tweezers to manipulate small fluorescently labeled microspheres made of melamine resin with the diameter of 3 μm. The microspheres were treated to produce tangential degenerate alignment with a very small azimuthal anchoring coefficient. The positions of microparticles were monitored by using a fast Fluorescent Confocal Polarising Microscope (FCPM). This made it possible to not only follow the positions of the particles (and determine their velocity and the force acting on them), but also allowed for the visualisation of the director distortion around the particles, which is shown in Fig. 2.14.

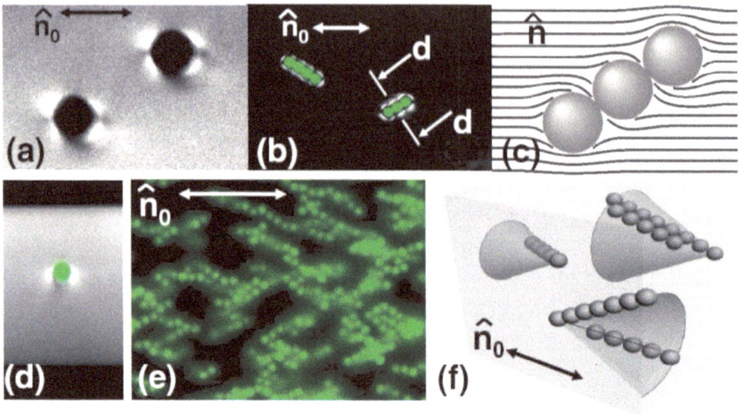

Fig. 2.14 Colloidal aggregation of quadrupolar microspheres with degenerate tangential surface anchoring. **a**, **b** Polarized images of textures around a pair and several colloidal particles. One can clearly see two surface defects on each particle, called "boojums". **c** Scheme of the director distortions around the microspheres in tilted chains (**e**, **f**). Reprinted figure with permission from I.I. Smalyukh, O.D. Lavrentovich, A.N. Kuzmin, A.V. Kachynski, and P.N. Prasad, Phys. Rev. Lett **95**, 157801 (2005). Copyright (2005) by the American Physical Society

Isolated microspheres with tangential anchoring create distortions of the quadrupolar type (symmetry), with two surface boojums (see Fig. 2.14c) located at the interface between the microsphere and the nematic liquid crystal. The elastic distortion rapidly decays with distance from the surface and vanishes at the distance of approximately one particle diameter. If the spheres are free to move around due to the Brownian motion, they spontaneously attract and form chains of particles, which are oriented at 25–35° with respect to the far-field director, as shown in Fig. 2.14b, f. In thick cells, this chains are free to glide along the conical surface, embedded with its axis along the far-field director. Two individual optical traps were used to grab and position two spheres in a predetermined geometry and at different angles with respect to the far field director. By holding one sphere fixed with a strong beam, the second sphere was moved by the second optical trap on a circular trajectory encircling the fixed sphere. The trap stiffness (see the explanation in Chap. 3) of this slowly circulating beam was kept low-enough to allow for radial deviation of the second particle, when being moved around the fixed one. The interaction between the two particles resulted in deviation of the second particle from the centre of the trap. By knowing the trap elastic constant, one could therefore determine the force acting on the circulating sphere at a given separation. This measurement was repeated for several radii of circulation. The experiments indeed revealed a strong angular anisotropy of the quadrupolar interaction force and also confirmed the predicted $1/R^6$ dependence of the quadrupolar elastic force for the direction of maximum attraction.

The application of laser traps to colloidal trapping and manipulation might have some side effects, which could influence local structure of the nematic liquid crystal within the focal region of strongly focused laser light. These side effects in measuring colloidal forces in nematic liquid crystals could be bypassed by using magneto-optic traps to control the force between colloidal particles. Whereas completely transparent (non-absorbing) colloidal particles are used in laser tweezer experiments, superparamagnetic colloidal particles are used in magneto-optical trapping. Because they are based on iron and similar magnetic materials, these superparamagnetic particles are strongly absorbing the laser light and are not used in laser trapping experiments. Besides superparamagnetic solid microspheres, liquid droplets filled with superparamagnetic dispersion of small particles can be used (see the experiment of Poulin et al. [72]).

Kotar et al. [80] used superparamagnetic spheres with 4.5 μm diameter with tangential anchoring of the 5CB nematic liquid crystal in a thick homeotropic cell. They were using the magneto-optic tweezers, which provided a rather homogeneous magnetic field, which could be oriented along arbitrary direction in space. The magnitude of the field could be varied up to tens of mT, and the direction of the field could be varied with a frequency of several hundreds Hz. If the magnetic field is aligned perpendicularly to the cell surface and therefore parallel to the nematic liquid-crystal molecules inside the cell, the induced magnetic moments of the spheres are parallel. This results in a strong magnetic repulsion between two superparamagnetic microspheres. In the experiments, Kotar et al. [80] used rotating magnetic field, which was in plane with the surfaces of the cell. This kind of rotating magnetic field induces an effective attractive interaction between the two spheres, which forces the microspheres very close together. After the magnetic field was switched off, the two microspheres repelled from each other and the trajectories of the two particles were monitored as a function of time up to several tens of seconds, as shown in Fig. 2.15a.

Regardless of the initial separation, the trajectories of the particles were practically identical for different experiments. Knowing the Stokes coefficients for the spheres and the magnetic force between the spheres allows us to calculate the colloidal force, mediated by the nematic liquid crystal. This structural force was determined over two decades of force strength, i.e. in the interval between 0.1 and 10 pN. The interval of separation was between one and two diameters of the spheres, which is a rather narrow interval of separation. Nevertheless, the authors could determine that the repulsive force between two quadrupolar and collinear nematic colloids, separated by x, decays as a power law:

$$F = \frac{4\pi W^2 x_\circ^8}{K R^6}\left(1 - \frac{W r_\circ}{56K}\right) \approx \frac{C}{x^6} \tag{2.12}$$

One can see from the solid fit in Fig. 2.15b that the force between two quadrupolar colloidal particles decays as $F \propto 1/x^6$, which confirms theoretical predictions by Ruhwandl and Terentjev [81] and Lubensky et al. [35].

Whereas these experiments were all done in a rather thick nematic liquid-crystal cells, the effects of confinement on the inter-particle pair-potential were investigated by Vilfan et al. [82]. They were using magneto-optic tweezers to induce a strong attractive force between two super-paramagnetic micro-spheres with tangential boundary conditions, with diameter of 4.4 μm, in homeotropic cells with different

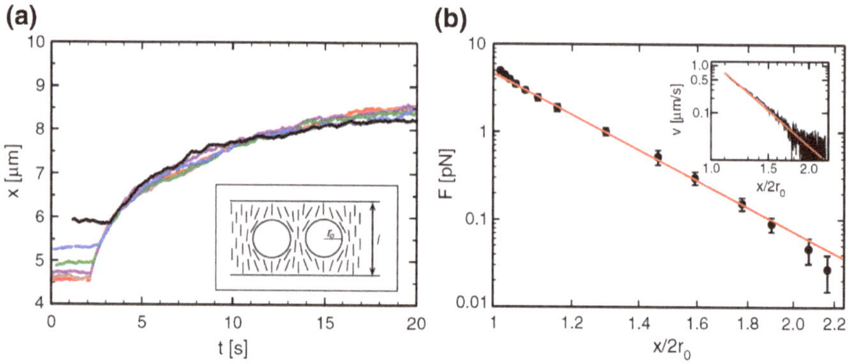

Fig. 2.15 **a** Time dependence of the separation x between the two quadrupolar nematic colloids. They were brought together by the assistance of external magnetic field and then released. There is a repulsive elastic force between the particles due to quadrupolar symmetry of distortion. The inset shows schematic director field. **b** The elastic repulsive force between two colloidal particles with tangential anchoring of liquid crystal molecules. The force shows a power law dependence $F \propto x^{-6}$ (*solid line*). The *solid line* is the best fit to Eq. 2.12. The *inset* shows the velocity of the particles as a function of their separation with the best fit of $v \propto 1/x^6$. Reprinted figures with permission from J. Kotar, M. Vilfan, N. Osterman, D. Babič, M. Čopič, I. Poberaj, Phys. Rev. Lett. **96**, 207801 (2006). Copyright (2006) by the American Physical Society

thicknesses (8 and 6.5 μm). The inter-particle force F was measured by balancing the liquid-crystal-mediated repulsive force and the calibrated and attractive magnetic force. The magnetic force was set by an electric current through the coils of the magneto-optic trap and the equilibrium separation between the two microspheres was measured. The magnetic field was reduced step by step and the corresponding micro-sphere separation was measured. In this way, they could measure the magnetic force in the interval of centre-to-centre separation between 1.05 and 2.7 diameters of colloidal particles. The resulting force was then integrated along the particle separation in order to get the inter-particle pair potential. This pair potential between two quadrupolar colloidal particles in a thin homeotropic layer of 5CB nematic liquid crystal is shown in Fig. 2.16 for two different thicknesses of the cell.

For small separations, the quadrupolar force decays as a power law with an exponent $\beta \approx 5.1 - 5.4$, which is close to, but different from the predicted exponent $\beta = -6$. However, this power law is valid for only for a limited range of interparticle separations. The decay of the force is definitely faster for particle separation approaching the value equal to the cell thickness. This is interpreted in terms of the screening of the inter-particle force by the confining walls. Because the molecules of liquid crystal are strongly anchored at the confining walls, their influence competes with the elastic distortion generated by each of the micro-spheres. At smaller microsphere separation, the influence of the two walls cannot penetrate the narrow space between the two beads. However, when this space increases, the influence of the two walls gets stronger. Based on the length-scale arguments, we expect that the effect of the surfaces, which tend to make the liquid crystal between the

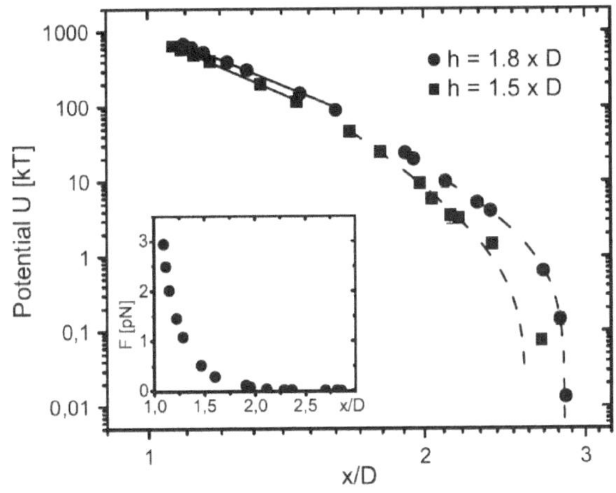

Fig. 2.16 The interparticle potential shown in the log-log plot as a function of normalized colloidal separation x/D for different thicknesses h. The diameter of beads is $D = 4.4\,\mu$m. For small separation x/D, the potential shows power law dependence (*solid lines*), whereas for larger separation, it show exponential decay (*dashed lines*). Reprinted figure with permission from M. Vilfan, N. Osterman, M. Čopič, M. Ravnik, S. Žumer, J. Kotar, D. Babič, I. Poberaj, Phys. Rev. Lett. **101**, 237801 (2008). Copyright (2008) by the American Physical Society

micro-spheres uniform, will become important when the micro-sphere separation equals to the wall separation. At much larger separations, most of the liquid crystal between the two spheres will be forced by the walls to be uniform. The free energy will not depend on separation of the particles, which means there will be no force of nematic origin.

Such an exponential screening of the inter-particle interaction is also expected from the analogy with classical electrostatics. Here, the colloidal particles are analogous to electric quadrupoles placed between two parallel conductive plates. The homeotropic anchoring of liquid-crystal molecules on the confining walls is analogous to the electric field at the conductive wall surface, which is perpendicular to the conductive surface. The analogy with electrostatics gives an effective decay length λ of the screened structural force of $\lambda \sim 0.16h$, which is in excellent agreement with experimentally observed screening length of $\lambda \sim 0.14h$. Here h is the thickness of the sample. Screening of the interaction of colloidal particles with weak homeotropic surface anchoring was analyzed experimentally by Sung-Jo Kim and Jong-Hyun Kim [83].

The interaction of quadrupolar nematic colloids with homeotropic surface anchoring of liquid-crystal molecules was studied by Škarabot et al. [84]. Silica spheres of 2.32 and 4.7 μm diameters were used in the experiments. Their surfaces were covered with a monolayer of silane (DMOAP), which ensures a very strong homeotropic surface anchoring of a nematic liquid crystal. In thick cells of 5CB nematic liquid crystals, these colloids usually obtain dipolar configuration because of the very strong surface anchoring of the 5CB. However, if these colloids are confined in a very thin

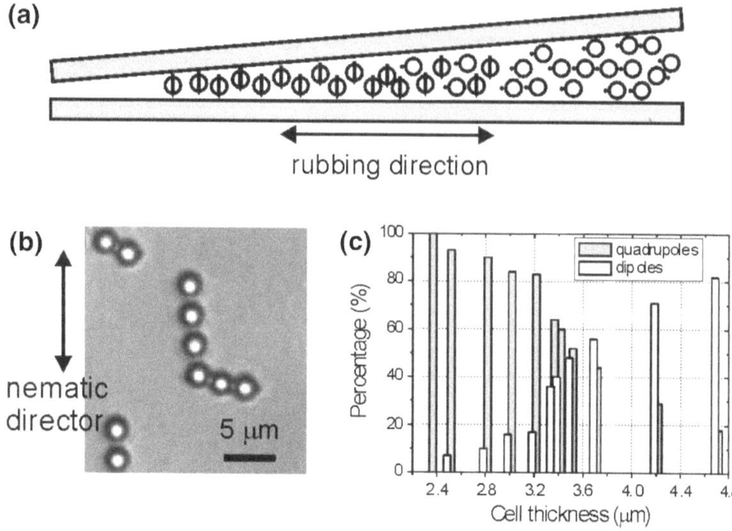

Fig. 2.17 Colloids with homeotropic surface anchoring in wedge-type 5CB cell. **a** Schematics: in thinner parts, the particles are of quadrupolar symmetry, whereas in thicker parts of the cell, the colloids are dipolar. **b** At a critical cell thickness, both types of particles are observed. Dipolar colloids form linear chains along the nematic director (defined by the rubbing direction), whereas quadrupolar colloids form kinked chains perpendicular to the rubbing direction. **c** Histogram presenting the number of dipolar and quadrupolar colloidal particles at different cell thicknesses. The diameter of particles is 2.32 μm

planar cell, one observes two types of symmetry of the colloidal particles: dipolar and quadrupolar nematic colloids. This is most easily obtained in wedge-type planar cells, where the cell thickness varies from hundreds of nanometers to several micrometers. An example of colloidal particles in a wedge-type cell is shown in Fig. 2.17.

At the critical thickness of the cell, which is 3.5 μm for 2.32 μm homeotropic silica microspheres, both types of particles are observed, as shown in Fig. 2.17b. At larger thickness, the number of dipolar colloids increases, whereas at smaller thickness the percentage of quadrupolar particles increases. The interactions of quadrupolar particles were studied by simply selecting two or several microspheres in the region abundant with quadrupoles. Using the laser tweezers, the particles were positioned at a predetermined separation and released, as illustrated in Fig. 2.18.

These experiments showed that quadrupole–quadrupole interaction is much weaker than the dipole–dipole interaction. The pair-binding energy (which equals to the work of the force separating the particles from the bound state to infinity) are of the order of several hundreds of $k_B T$. This is an order of magnitude weaker than the dipole–dipole interaction, which reaches several thousands of $k_B T$ for the same size of particles.

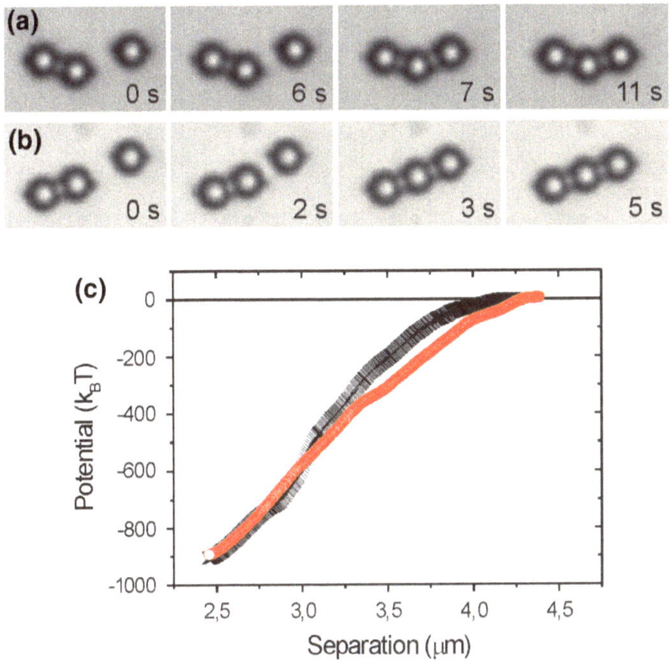

Fig. 2.18 Interaction of an isolated quadrupolar colloid with a cluster of quadrupolar colloids. **a, b** In this position, the particle is attracted to the pair of quadrupolar particles from the side, which results in a growth of chains. **c** The quadrupole–quadrupole interaction as a function of separation for a quadrupole approaching a pair of already assembled quadrupoles

There are numerous very detailed studies of force between colloidal particles in nematic liquid crystals, mainly performed by optical tweezers. Takahashi et al. [75, 85] measured the inter-particle force for two elastic dipoles and found the r^{-4} dependence, in agreement with theoretical predictions. Screening of the dipole–dipole force in cells with different thickness was measured by Kondo et al. [76]. These experiments confirmed exponential screening of the pair-interaction colloidal force in very thin nematic liquid crystal cells.

2.3.3 Mixed Interaction: Dipolar Spherical Particles Interact with Quadrupolar Particles in Nematic

The experiments that we have analysed in the previous two sections clearly confirmed the power-law nature of the nematic-mediated force between colloidal particles in nematic liquid crystals. For dipolar colloids, this force scales as $F \sim R^{-4}$, whereas the force between two quadrupoles scales as $F \sim R^{-6}$. Although all of this was proved for a limited interval of particle separation, it basically confirms theoretical

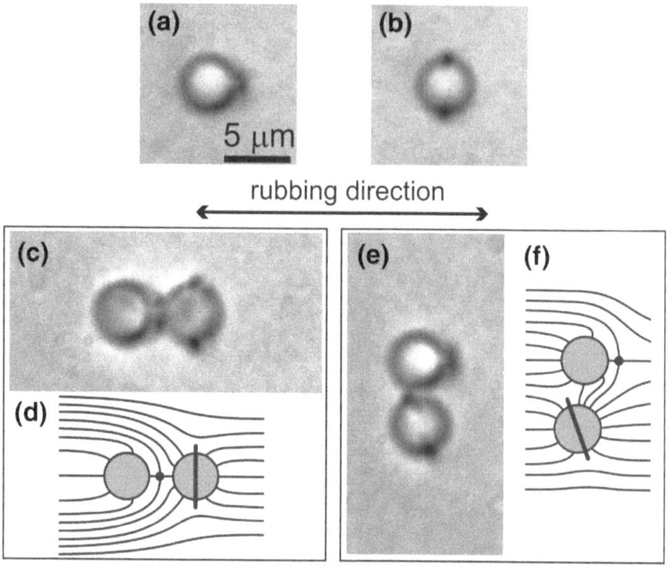

Fig. 2.19 Microscopic images of (**a**) dipolar and (**b**) quadrupolar 4 μm colloidal particles. Dipolar colloids are characterised by their dark hyperbolic hedgehog defect; quadrupolar colloids have the Saturn ring around them. **c** A pair of dipole-quadrupole particles is attracted along the nematic director. Schematics of the director field is shown in **d**. **e** Dipole-quadrupole pair is attracted in a direction perpendicular to the far-field director. **f** The schematics of the corresponding director

predictions (Lubensky et al. [35], Ruhmwandl and Terentjev [86]). Theory also predicted that the interaction between elastic dipoles and elastic quadrupoles should scale as a natural intermediate between the two power laws, resulting in $F \sim R^{-5}$ separation-dependence of the nematic-mediated force between an elastic dipole and an elastic quadrupole.

This theoretical prediction of the mixed type of interaction was experimentally confirmed by Ognysta et al. [87–89]. They were using wedge-type planar nematic cells with their thickness tuned in such a way that the region around the critical thickness was experimentally accessible. For 4 μm particles of acrylic resin, surface-coated with DMOAP silane, the critical thickness was in the range $6 - 7$ μm. In that region, they could easily find isolated dipolar and quadrupolar colloidal particles, as shown in Fig. 2.19a.

It was found that a single pair of dipolar and quadrupolar colloidal particles has three stable equilibrium arrangements, presented in Fig. 2.19c and e. Figure 2.19c shows the first stable configuration, where the dipolar colloidal particle is attracted directly along the director to the quadrupolar colloidal particle. There is a hyperbolic hedgehog point defect sitting in between the particles and stabilising them. In the second case, the dipole-quadrupole pair is stabilised in a side-wise fashion, as is shown in Fig. 2.19e. The third stable configuration is a mirror image of the second

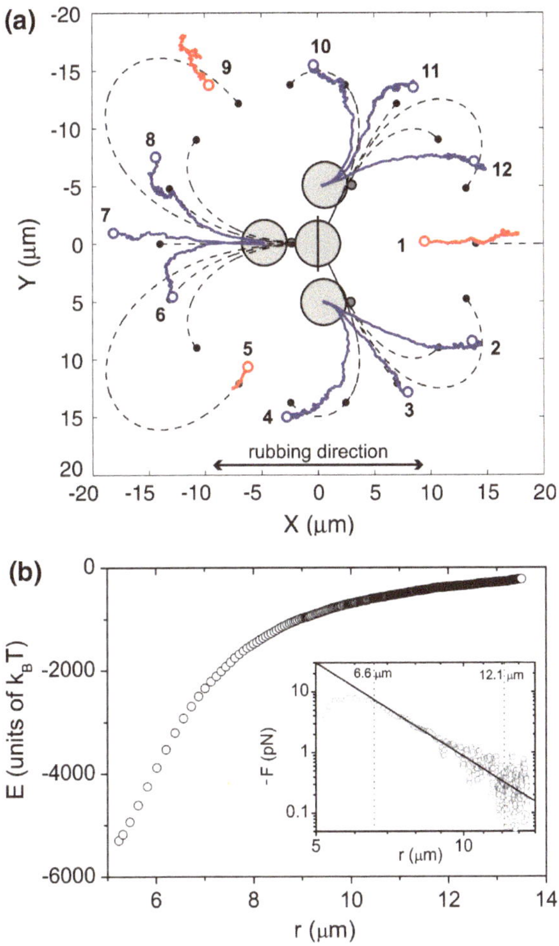

Fig. 2.20 **a** Trajectories of the dipolar colloidal particle after it was released from the optical trap near the quadrupolar colloidal particle, which is placed at the origin. *Blue lines* (2–4, 6–8, 10–12) show the attractive trajectories, red lines (1, 5, 9) show repulsive trajectories, and *black dashed lines* are calculated by using an electrostatic ansatz. Open circles and *black dots* indicate the starting positions. The Saturn ring orientation corresponds to trajectories 5–9. It is tilted like in Fig. 2.19f for trajectories 10–12 and in the opposite direction for trajectories 2–4. The thickness of the cell is $h = 7\,\mu$m. **b** The interaction potential as a function of separation between a pair of a dipolar and a quadrupolar colloidal particle for the starting position 7. *Inset* log-log plot of the interaction force with the best fit to a power-law exponent of $\beta = -5.1 \pm 0.25$. Vertical *dashed lines* indicate boundaries of the fitting region; their values are indicated at the *top*

one. In these two cases, the Saturn ring of the quadrupolar particle is tilted away from its configuration in the isolated state at an angle of ~25°.

The force between an elastic dipole and elastic quadrupole was measured by positioning a quadrupolar particle to a fixed point, whereas the dipolar particle was positioned with a second optical trap at different separation from the quadrupolar particle and also at different angles with respect to the far-field nematic director, as labeled 1–12 in Fig. 2.20a.

After the dipolar particles was released from the particle trap, it either approached to or was repelled from the quadrupolar particle, which is illustrated by differently labeled trajectories in Fig. 2.20a. From some positions (labeled 1, 5 and 9), the particle was repelled from the quadrupolar neighbour, whereas for other starting positions, it was attracted to it. These attractive trajectories (in blue colour in Fig. 2.20a) are curved, except for the collinear approach from the position No. 7 in Fig. 2.20a.

To test the predicted $1/R^5$ separation-dependence of the dipole-quadrupole interaction force, the force along the trajectory was determined experimentally, using the standard method described in Chap. 3. For this case, the interparticle force shows a power-law dependence over the centre-to-centre separation, which is shown in the log-log plot in the inset to Fig. 2.20b. Although the range of separation is less than one decade, one can clearly see that the observed dipolar-quadrupolar colloidal interaction follows the $F \sim R^{-5.1}$ power law, which is close to the expected exponent for the dipole-quadrupole interaction. This power-law exponent does not depend significantly on the thickness of the cell. The pair interaction potential between the elastic dipole and the elastic quadrupole is shown in Fig. 2.20b. The pair interaction energy is quite large (6000 k_BT), which is due to the relatively large diameter of the particles. Note that the force and the corresponding inter-particle potential depend on the fourth power of the particle diameter. It was also observed that this pair-interaction energy depends on the thickness of the nematic liquid-crystal layer.

2.4 Forces Between Micro-rods in a Nematic Liquid Crystal

We have already mentioned that rod-like colloidal particles are topologically equivalent to spherical colloidal particles, since they both have zero handles and the corresponding genus $g = 0$. We expect a single defect close to a micro-rod in a nematic liquid crystal, similar to the single topological defect accompanying the sphere in a form of a hyperbolic point hedgehog or a hyperbolic –1/2 Saturn ring. Figure 2.21 shows for comparison a spherical homeotropic colloidal particle with the Saturn ring encircling the sphere, and to the right is a micrometer-diameter homeotropic micro-rod with a Saturn ring encircling it. This is therefore a direct proof of topological equivalency of a sphere and a cylinder.

The simplest material in the form of elongated cylinders are glass micro-rods, which are produced by cutting alkaline-free glass fibres and are mainly used as spacers that maintain constant gap and constant thickness of a liquid crystal in liquid crystal displays. For technological reasons, the distribution of their diameter is very precise (within tens of nanometres) but their length distribution might be quite broad. An example of glass micro-rods of 1.5 μm diameter (PF-15S Nippon Electric Glass) is shown in Fig. 2.22. Distribution of the lengths of the micro-rods is quite broad and

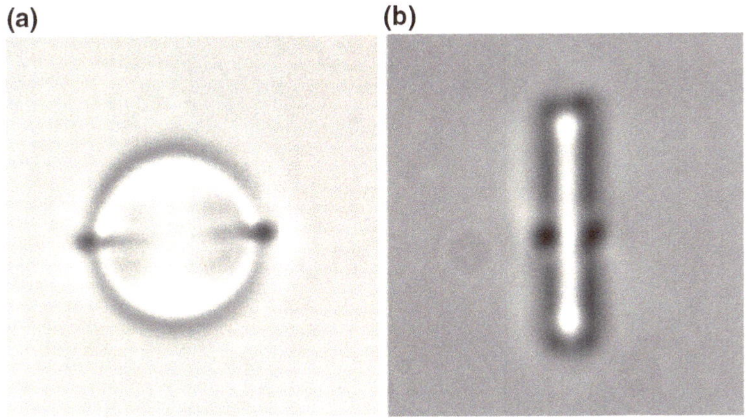

Fig. 2.21 **a** Saturn ring defect is encircling a homeotropic glass micro-sphere in a planar nematic cell. **b** Saturn ring is also encircling a homeotropic glass micro-rod in a planar nematic cell

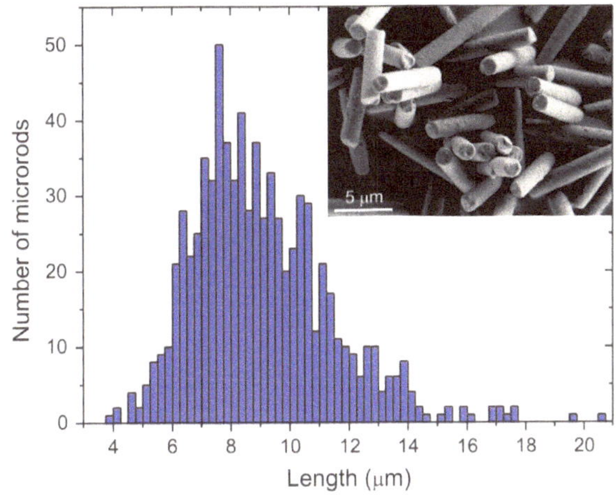

Fig. 2.22 Distribution of lengths of 1.5 μm diameter micro-rods, as determined by measuring the lengths of 660 micro-rods under an optical microscope. The insert shows a SEM image of micro-rods

is centred around approximately 8 μm, so that a typical micro-rod has an aspect ratio of 1:5.

Rod-like colloidal particles in nematic liquid crystals should therefore behave quite similarly to spherical colloidal particles. The director structure around elongated colloidal particles, topological defects and the resulting structural forces were studied theoretically by a number of authors [90–98]. On the other hand there are only few experimental studies of rods in nematic liquid crystals [99, 100].

If glass micro-rods of μm diameter are treated to induce perpendicular surface alignment (Tkalec et al. [99]) and inserted into a planar aligned nematic liquid crystal cell, one always observes only two characteristic types of director pattern, which are shown and illustrated in Fig. 2.23a–f. The first kind of director pattern is similar to the dipolar configuration around micro-spheres in a nematic liquid crystal and is called

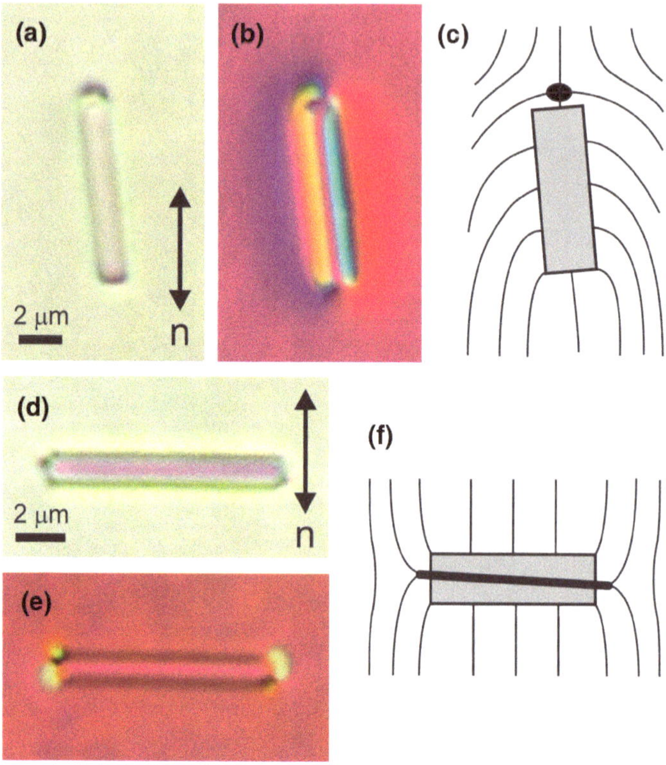

Fig. 2.23 **a**–**c** Micro-rods of dipolar symmetry. **a** A dipolar micro-rod with homeotropic surface anchoring in a planar cell of 5CB, observed without polarisers. **b** Using a red plate (i.e. λ-plate for 530 nm), the differently coloured regions indicate different orientations of the liquid-crystal molecules. **c** The director, as reconstructed from **b**. **d**–**f** Quadrupolar micro-rods. **d** Unpolarised image of a micro-rod, which was spontaneously oriented perpendicularly to the nematic cell director. **e** The same micro-rod, as observed using the *red* plate. **f** Schematic drawing of the director, as deduced from **e**

a dipolar micro-rod. Figure 2.23a is a non-polarised micrograph of such a dipolar micro-rod, where the hedgehog point defect is located on the top-end of the rod.

The director orientation can be easily reconstructed using the optical technique with a λ-plate (also called the red-plate), which is inserted in between the polarizer and the sample. The correspondence between different colours and different molecular director orientations under observation using the red plate is explained in Fig. 2.24. In all images, the red color corresponds to vertical orientation of the long axes of molecules whereas bluish and yellowish colours correspond to clockwise and anticlockwise rotation of the director, respectively. This enables one to reconstruct the director patterns around arbitrary-shaped colloidal particles.

The second type of rods are oriented perpendicular to the overall director field in the nematic cell and their appearance under optical microscope is shown in

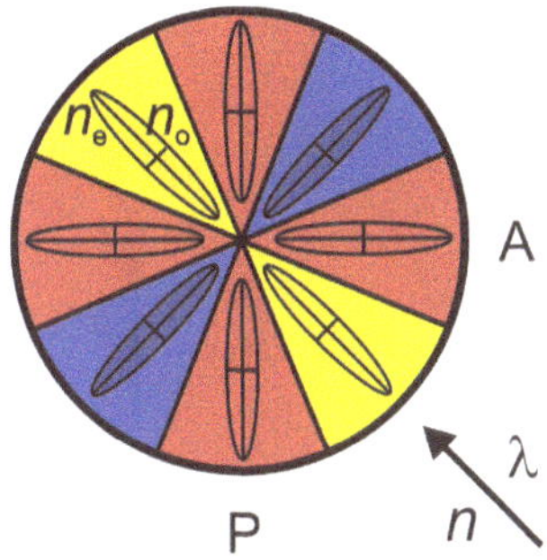

Fig. 2.24 Relation between the orientation of the liquid crystal molecules and the color observed under a polarizing microscope, with the λ-plate inserted at 45° with respect to polariser P and crossed analyser A. This correspondence between the colours and molecular orientation is observed when the birefringence of the liquid crystal is small and we are in the first order of retardation. By increasing the retardation, interchanged colour scheme is observed

Fig. 2.23d, e. It is immediately recognised that this kind of rod is in fact equivalent to a quadrupolar micro-sphere and there should be a Saturn ring encircling such a rod along its longer dimension. This could indeed be tested by the laser tweezers, where one is able to grab the Saturn ring and pull it away, as discussed in more detail in Chap. 5.3, which explains topological charge production and entanglement on a fibre in a nematic liquid crystal.

The simple optical observation and reconstruction of the director field confirms without a doubt the topological equivalence of micro-rods and micro-spheres with homeotropic surface anchoring of nematic liquid crystals. Similarly to micro-spheres, the micro-rods are accompanied either by a hyperbolic hedgehog or a Saturn ring. In the first case, the rods will be oriented along the overall director in the planar nematic cell and the point hedgehog will be located either on the top- or on the bottom-end of the rod. In the second situation, there will be a Saturn ring encircling the micro-rod all along its axis, and the micro-rod will be oriented perpendicularly to the overall director. In reality, one indeed observes both types of director distortions.

Defects on short micro-rods can be manipulated using the laser tweezers (Tkalec et al. [99]) and the ring can be shrunk into a point or vice versa. Very long fibers with length orders of magnitude larger than the diameter are a special case with a very rich topology. In such case, one is able to create, with the use of a localised temperature quench, a variety of topological defects attached to the fiber (Nikkhou et al. [101]), which will be discussed in Sect. 5.3.

Detailed analysis of a larger number of micro-rods in a nematic liquid crystal shows that majority of micro-rods are of dipolar type and are oriented along the director, as shown in Fig. 2.25. The blue colour of the histogram of their orientation corresponds to the dipolar and the red colour corresponds to quadrupolar micro-rods. Dipolar micro-rods have a rather narrow distribution of around 10° with respect to the director. On the contrary, quadrupolar micro-rods have rather broad distribution and are oriented in a relatively large interval of 60–90° with respect to the director.

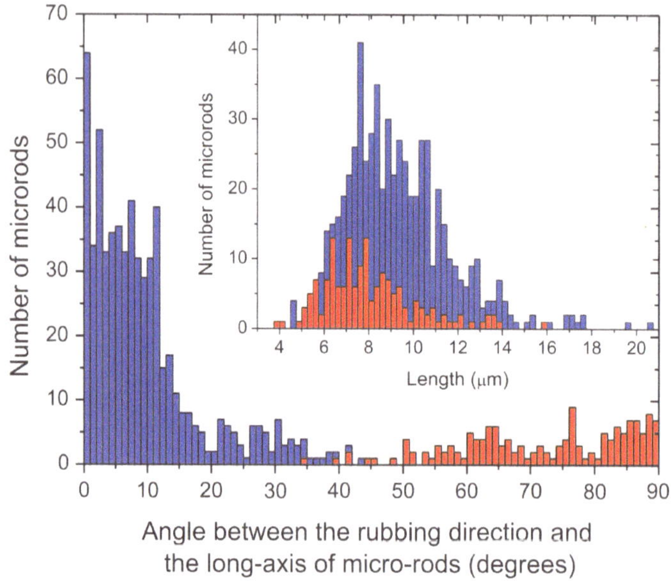

Fig. 2.25 Distribution of angles between the rubbing direction and the long-axis of 1.5 μm diameter micro-rods with homeotropic surface anchoring. The *blue colour* corresponds to dipolar micro-rods, the *red colour* represents the orientation of quadrupolar micro-rods. The cell thickness is 4 μm, which is very close to the critical thickness of 3 μm, where dipolar and quadrupolar rods coexist. The *inset* shows the same data, but here the number of dipolar (*blue*) and quadrupolar (*red*) rods is shown as a function of their length

It seems that this broad distribution is due to the imperfect ends of the micro-rods, which are produced by a random breaking of the glass. This produces geometric irregularities, which are very important for pinning of topological defects that appear exactly at both ends of the particle.

Topological defects on these fibres can be manipulated to a certain degree with the laser tweezers. Figure 2.26 shows an example where the point defect was opened into a Saturn ring, encircling the micro-rod along its shorter dimension in a planar nematic cell. This position is energetically unfavourable and causes the defect to slowly slide towards the end of the fibre where it closes back to a point defect. Hyperbolic point defect could also be grabbed and stretched, as shown in Fig. 2.27.

There are several interesting questions related to micro-rods, such as how the micro-rods interact, what is the strength of their binding, and what are the topological rules to be obeyed during the interaction. We should note that the interaction of micro-rods is important to understand because there was a great interest in the interaction of nano-fibres in liquid crystals. On micro-rods, the interaction can be easily studied by optical methods, whereas this is not possible for nano-fibres with a diameter of hundreds of nanometres.

Before the interaction forces are measured, one needs to determine the viscosity coefficients, which are quite different for such an anisotropic particle. This is done

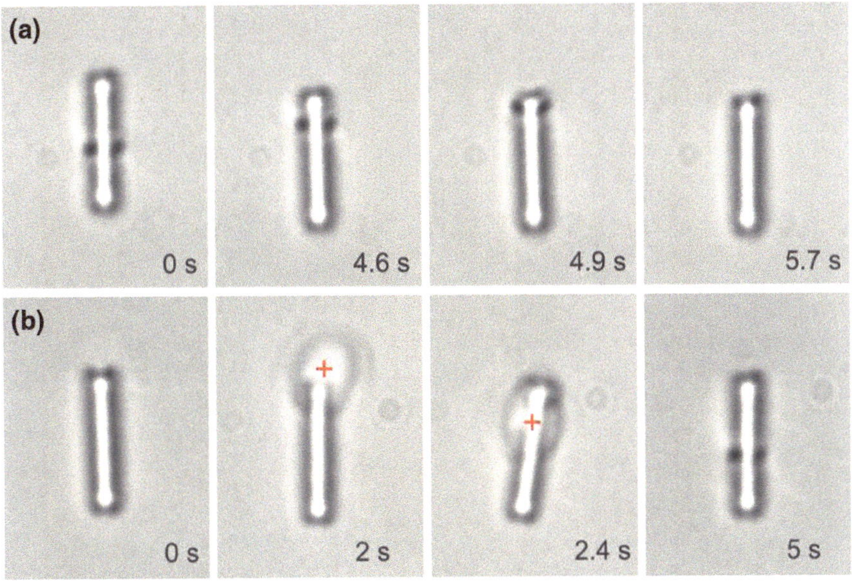

Fig. 2.26 **a** The position of the Saturn ring encircling the fibre in a planar nematic cell is unstable, the ring spontaneously slips back into the point defect, sitting at the top-end of the micro-rod. **b** Using the laser tweezers, one can grab and force the point defect of the dipolar micro-rod to open into a Saturn ring, visible in the middle of the rod in the last image

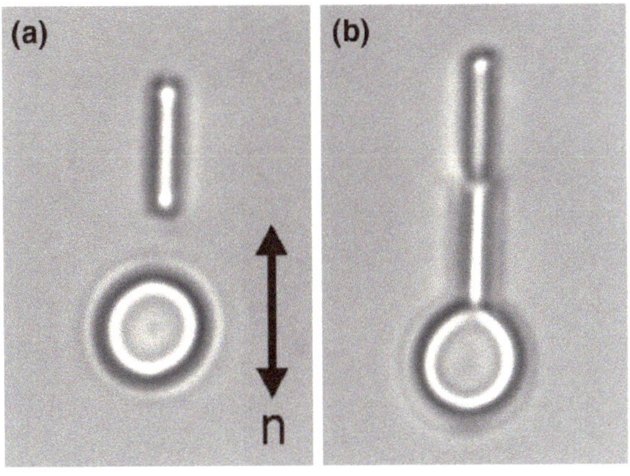

Fig. 2.27 **a**, **b** Point hedgehog of a dipolar micro-rod can be grabbed and stretched into a line defect by the laser tweezers

by performing the Brownian motion experiment where the micro-rod is let free to perform a random Brownian walk in a planar cell. It is intuitively expected that this random thermal motion is easier along the longer dimension of the rod because of the lower resistance to flow. This is indeed observed in the experiments, where thousands of video frames are taken during several minutes of the thermal motion of the micro-rod and the trajectory of the particle is determined.

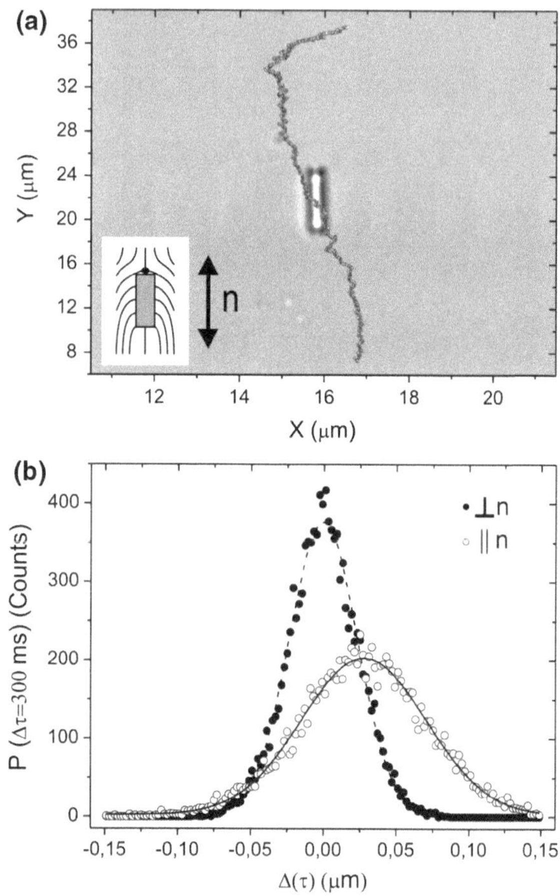

Fig. 2.28 **a** Brownian motion of a dipolar micro-rod in the nematic liquid crystal. Note the extreme anisotropy of the movement, which is partially caused by macroscopic flow of the liquid crystal. **b** For each time interval, the length of the movement of the micro-rod is measured and plotted in a histogram, where the number of steps of chosen length are presented. Note that due to macroscopic flow, probably caused by external stimuli, the rod is carried in one direction and the histogram presenting movements along **n** is displaced along $+y$

Figure 2.28 shows an experiment on a micro-rod, which shows that their motion is highly anisotropic. The micro-rod experiences much more vivid motion along the director than in the perpendicular direction. The anisotropy of motion and the possible flow of the liquid crystal (this is always a problem in tracking experiments) could easily be analysed by showing the distribution of the length of displacement, which the micro-rod performs in a precisely determined time interval, which in this case is 300 ms. The y-axis in Fig. 2.28b presents the number of such steps out of the total 11 000 steps recorded in the experiments. There are two such distributions in Fig. 2.28b, one for the direction along the director and the second in the direction perpendicular to it. The distribution of steps along the director is obviously displaced for some length, which indicates macroscopic flow of the liquid crystal and is not related to the self-diffusion of this particle. The widths of these two distributions tell us the self-diffusion coefficient D for each direction. By using the Stokes-Einstein relation, we can determine the product of the characteristic dimension of the particle and the viscosity, which in this case are $0.8 \cdot 10^{-6}$ and $2.2 \cdot 10^{-6}$ kg/s respectively. The anisotropy of the self-diffusion coefficients is approximately 1:3, which is close to the aspect ratio of the particles, 1:5.

Once the self-diffusion coefficients D are determined for each direction, one is able to calculate the force on the particle, which is acting during the interaction of two micro-rods in the nematic liquid crystal. The standard procedure of video tracking of particles' positions is used, which is described in detail in Sect. 3.3. There are two characteristic geometries of the attraction of the two micro-rods: (i) the two rods are positioned side by side and parallel to each other, (ii) the two rods are positioned along the director and they interact collinearly.

A series of snapshots of two parallel dipolar micro-rods interacting side by side is shown in Fig. 2.29a. Indeed, they behave similarly to spherical colloidal dipolar particles with anti-parallel orientations of their dipoles. They slowly attract and bind into a pair of rods, which are inclined by an angle of approximately 10° with respect

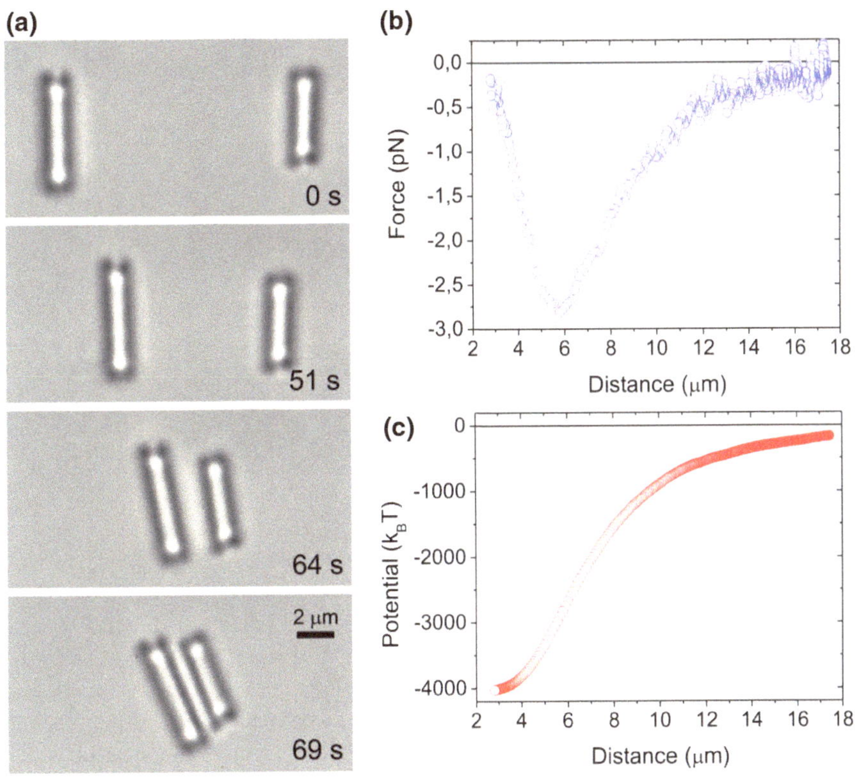

Fig. 2.29 **a** Side-by-side attraction of a pair of micro-rods with opposite direction of their topological dipoles. **b** The dipolar force versus micro-rod separation, extracted from a series of video images taken from **a**. **c** Lateral dipolar potential of two micro-rods with antiparallel direction of their topological dipoles

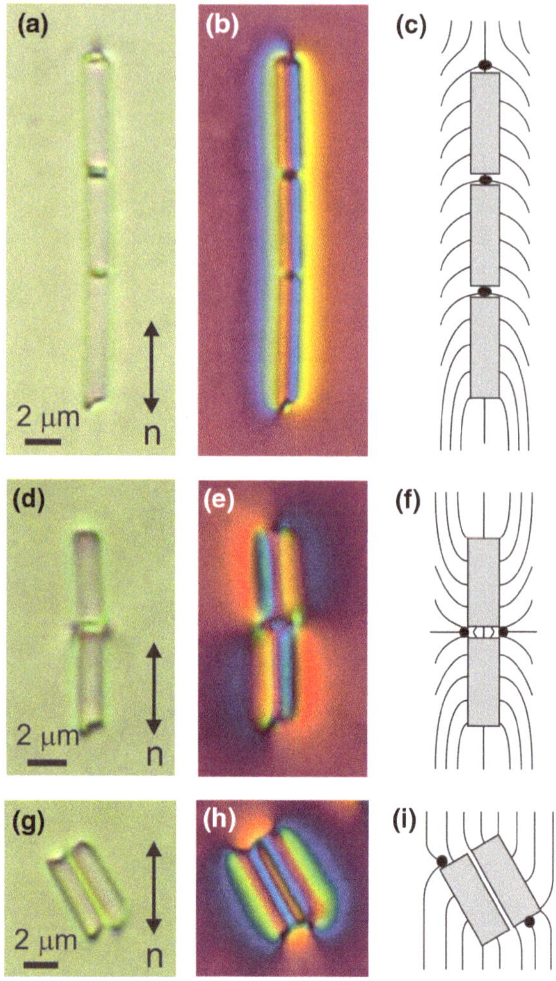

Fig. 2.30 Overview of stable configurations of dipolar micro-rods in planar cells. **a** Chains of micro-rods of dipolar symmetry. **b** The same structure under the *red-plate*. **c** Schematics of the director field. **d**–**f** "Bubble-gum" configuration of the two originally antiparallel and collinear dipolar micro-rods. **g**–**i** Side-by-side stable configuration of two antiparallel dipolar micro-rods

to the overall director. This inclination is due to the even-odd effect, characteristic for interactions of dipolar colloids, described in Sect. 4.1. The instantaneous force of interaction as a function of micro-rods separation is shown in Fig. 2.29b. The magnitude of the force is typical for the interaction of micrometre-sized colloidal particles. It reaches the highest magnitude of −2.5pN at approx. 6 μm separation. For closer separation the force diminishes and is balanced by repulsion in the equilibrium position. By integrating this interaction force along the trajectory of the particle, one is able to calculate the interaction energy, which is shown in Fig. 2.29c. In this case as well, the obtained binding energies are typical for micrometre-diameter nematic colloids and are of the order of several 1000 $k_B T$.

Similar to the interaction of two collinear spherical colloidal particles of dipolar symmetry, the dipolar rods can also interact collinearly, provided that their dipoles

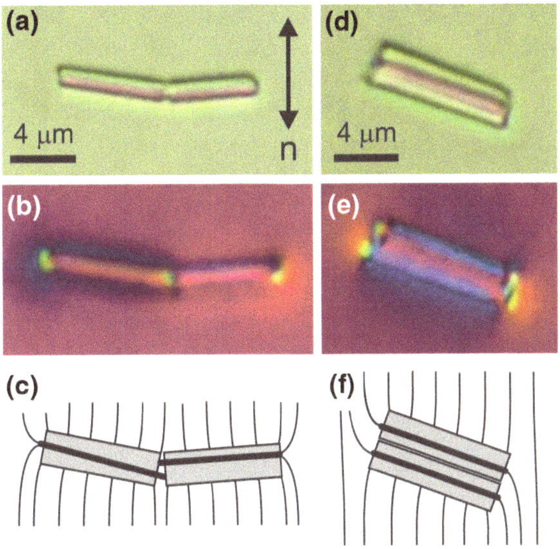

Fig. 2.31 **a–c** Side-by-side stable configuration of two quadrupolar micro-rods in a planar cell. **d–f** Top-down configuration of two quadrupolar micro-rods

are pointing in the same direction. The overview of possible structures, assembled from dipolar rods by using the laser tweezers, are shown in Fig. 2.30.

The overall similarities between dipolar rods and dipolar spherical colloids are evident. The panels (c), (f) and (i) on the far right of Fig. 2.30 show the schematics of the director field around the assemblies of rods on the nematic liquid crystal. In contrast to dipolar micro-rods, quadrupolar micro-rods exhibit only two stable configurations, which are shown in Fig. 2.31. The first one is a side-by-side attraction of two quadrupolar micro-rods and the other is a top-down configuration.

Understanding the mechanism of interaction of cylinder-like objects in the nematic liquid crystal is important for possible applications in dispersions of nanofibers. It is known that uniformly aligned assemblies of nanofibres exhibit very distinguished material properties, such as tensile strength. There is a relatively large number of theoretical studies of nematic dispersions of cylinders. Andrienko et al. [92–94] used molecular dynamics and Monte Carlo simulations, whereas Hung et al. [102] used fully tensorial LdG theory. In both cases, the actual sizes of the particles are relatively small for the reason of computational complexity. As a consequence, the symmetry of the equilibrium order parameter field around each cylinder is quadrupolar. For this reason, theoretical predictions could be compared only to experimentally studied interactions of quadrupolar rods. Hung et al. [102] succeeded in reproducing the –1/2 defect line encircling the micro-rod and being in the plane parallel to the long axis of the rod. Interestingly, there is another solution for the topological defect line which is distributed around the elongated particle, thus forming an "axially symmetric coat". From the point of symmetry, these solutions are equal to the experimentally observed ones. The difference between the theoretical predictions and experimental observations is that in our case, the defect around the micrometre-sized colloidal rod is unstable and may transform into a point defect, located at the end of the rod. Whereas the interactions of micro-rods are understood in 2D, it is not clear what are the possible 3D structures, assembled from dipolar micro-rods. Most likely this would be an equivalent to the tetragonal 3D lattice observed for spherical particles. However, the problem with micro-rods is their non-uniform length, which might add a considerable amount of disorder to any 3D micro-rod assembly.

2.5 Janus Colloids and Platelets in Nematic Liquid Crystals

The geometrical shape of colloidal micro-particle with genus $g = 0$ determines to a large extent the symmetry of the distorted nematic director around the particle. However, there is a special class of particles, called Janus particles, which have different parts of the surface treated with different surfactants. An example of a Janus particle is a glass micro-sphere, which is half-coated with a thin layer of metal, such as gold and the other half is silanized with DMOAP (Conradi et al. [103]). If the silica part of the surface is coated with DMOAP silane, inducing perpendicular liquid crystal alignment, we obtain a Janus colloidal particle with two hemispheres: one induces homeotropic alignment and the other tangential alignment. It actually turns out that the strength of the anchoring depends on the sequence of deposition steps. If the glass sphere was silanized first and then Au was deposited afterwards (DMOAP/Au), planar anchoring on Au was stronger than the homeotropic anchoring on DMOAP. If Au was deposited first and then the spheres were silanized (Au/DMOAP), the homeotropic anchoring of DMOAP was stronger than planar anchoring on Au. SEM images of Au-capped silica micro-spheres are shown in Fig. 2.32.

Fig. 2.32 SEM image of 4.3 μm silanated silica particles coated with 3 nm of Cr and 50 nm of Au

For an isolated Janus microsphere in a nematic liquid crystal, there are four different director configurations, as found in the experiments. In the first configuration, shown in Fig. 2.33, DMOAP was deposited first and this was followed by Au vacuum deposition. The micro-sphere is oriented with its Au-silanated interface parallel to the planar nematic cell, and there are two point hedgehogs located on the two opposite sides of the hemisphere. These point hedgehogs sitting on the surface are called boojums. This indicates that the whole sphere has adopted planar surface alignment, as the Au anchoring is stronger and prevails the weaker homeotropic anchoring on DMOAP hemisphere.

The second configuration is obtained with capped colloids, where the Au layer was deposited first and DMOAP layer was applied afterwards (i.e. Au/DMOAP colloids). This configuration is shown in Fig. 2.34 and one can see that the Au cap of the Janus (Au/DMOAP) microsphere is rotated by 90° with respect to the configuration in Fig. 2.33a and a hyperbolic point defect is residing at the pole of the Au cap. This leads to the conclusion that the homeotropic anchoring on DMOAP is very strong and prevails over the weaker planar anchoring on Ah hemisphere, which was deposited before the DMOAP. The observed elastic dipole of a capped particle is almost indistinguishable from an elastic dipole formed around a non-capped, only silanated particle.

When the dipolar Au/DMOAP capped colloidal particles in the nematic liquid crystal were manipulated with a laser tweezers, irreversible reorientations into two additional stable states were observed. The first configuration is shown in Fig. 2.35a and b, which is a capped colloid with a Saturn ring, positioned roughly at the border line between the homeotropic and planar hemisphere. The second stable state is shown in Fig. 2.36, and represents a mixed configuration with a surface boojum at the planar site (golden hemisphere) and a Saturn ring analog, positioned at the border line between the homeotropic and planar hemisphere. In both stable states, the border line between the two hemispheres remains positioned perpendicular to the rubbing direction and the original dipolar structure could not be reestablished by any possible means. This implies, that the original dipolar colloidal state was a metastable state, most probably induced by a liquid crystal flow during filling the cell with dispersion.

Topologically, capped colloidal particles and surrounding nematic defects are a combination of surface and bulk point defects. Defect loops are topologically equivalent (homotopic) to points. Point defects are topologically characterized by their topological charge q. Topological charge of a bulk hedgehog defect is calculated by performing the integral in Eq. 1.24 over a closed surface that surrounds the defect. For surface defects, the nematic director profile exists only in half-space. The integration in Eq. 1.24 is therefore performed only over the liquid crystal-filled hemisphere, that surrounds the point defect and q. For a capped particle being dispersed in a uniform planar cell, the total topological charge of a single particle and surrounding defects should equal 0. Figure 2.37 shows stable and metastable states of Janus microspheres, as calculated using numerical LdG approach. Homeotropic hemisphere imposes topological charge of magnitude 1/2, which is fully compensated by either a surface 'boojum-ring' (0° profile in Fig. 2.37a) or a bulk −1/2 defect arc (90° profile in Fig. 2.37a), i.e. half of the Saturn ring, depending on the orientation of the particle. Planar hemisphere can generate only surface boojum defects

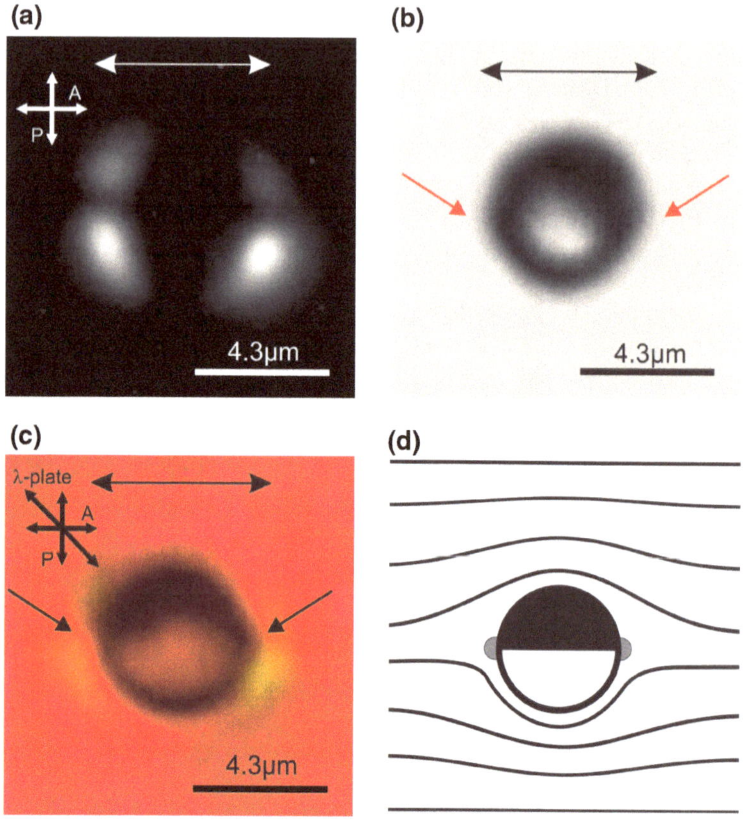

Fig. 2.33 Microscope images of a 4.3 μm DMOAP/Au capped microsphere in a 10 μm thick planar cell of 5CB. The particle was first completely covered with DMOAP and then one hemisphere was sputtered with Au. The arrows indicate rubbing direction of the planar cell. **a** The capped colloidal particle between crossed polarizers. **b** No polarizers are used. *Red arrows* indicate two symmetrically positioned boojum defects. **c** Polarized image with an inserted λ-retardation plate. The directions of crossed polarizers and retardation plate are indicated. **d** Schematics of the director field. Gold-coated hemisphere is on *top* and DMOAP-covered hemisphere on *bottom*. Boojum defects are drawn in *gray*

and together they also have a net zero topological charge. When calculating the net charge from different contributions (characteristics), the rules of summation have to be applied in accordance with the nematic director being an improper vector with the $\mathbf{n} \rightarrow -\mathbf{n}$ symmetry, to give proper total topological charge.

When one attempts to qualitatively generalize the behavior of colloidal particles with patterned surfaces, one should consider them as superpositions of only-homeotropic and only-planar "single-anchoring particles". Volume pieces of the single-anchoring particles' director fields corresponding to proper surface regions are to be stuck together to tile in the surface of the patterned particle in order to qualitatively obtain the director profile around the chosen patterned particle. At the intersections between different volume pieces typically defects form. Equilibrium

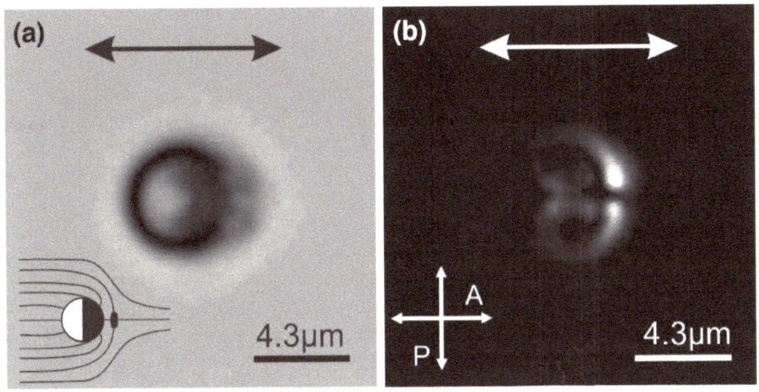

Fig. 2.34 Microscope images of a 4.3 μm Au/DMOAP capped microsphere in a 10 μm thick planar cell of 5CB. **a** Dipolar capped colloidal particle with slightly opened hedgehog defect, as observed with no polarizers. The *inset* shows a schematic drawing of the director field. **b** Dipolar colloidal capped particle between crossed polarizers

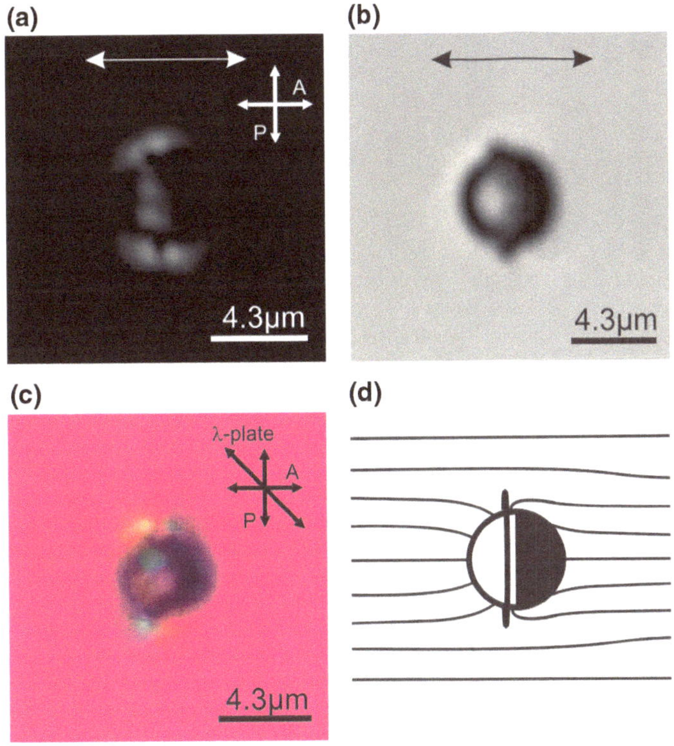

Fig. 2.35 Au/DMOAP capped microsphere with a Saturn ring defect. The *arrows* indicate rubbing direction of the planar cell. **a** The quadrupolar capped colloidal particle between crossed polarizers. **b** No polarizers are used. **c** Polarized image with inserted retardation plate. **d** Schematics of the director field with the Saturn ring encircling Janus microsphere. Particle diameter is 4.3 μm and cell thickness 10 μm

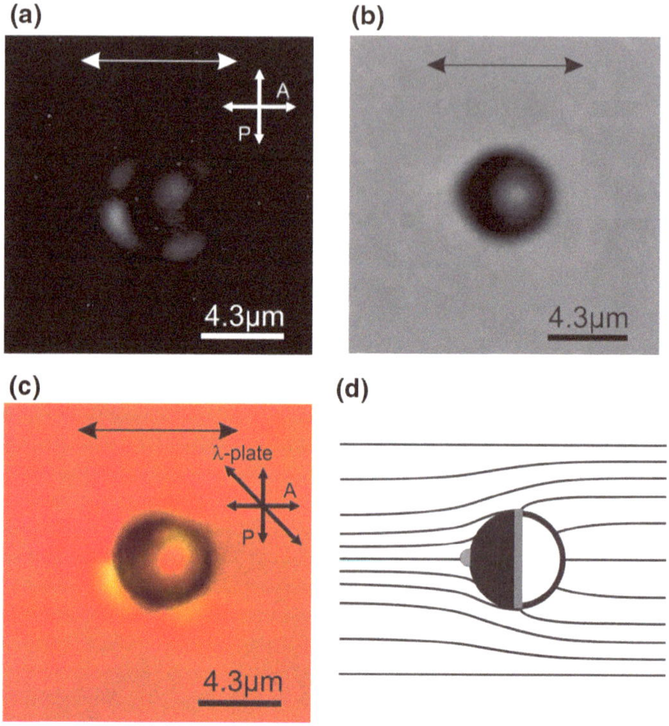

Fig. 2.36 Microscope images of a 4.3 μm Au/DMOAP capped microsphere of a boojum-ring configuration in a 10 μm thick planar cell of 5CB. The *arrows* indicate rubbing direction of the planar cell. **a** The capped colloidal particle between crossed polarizers. **b** No polarizers are used. **c** Polarized image with inserted retardation plate. **d** Schematics of the director field. Surface boojum and boojum-ring defect are visualized in *gray*

particle orientations are sensible to relative strengths of the homeotropic and planar anchoring which could be used to optimize the desired particle orientations.

In addition to micro-spheres and micro-rods, a variety of differently shaped microparticles were studied in the nematic liquid crystals, including polyhedra, nanoprisms [104], ellipsoidal particles [105] and different platelets [106]. In all cases, the number of topological defects accompanying the particle is determined by the topological charge conservation. However, complicated geometry of the particle induces a rather complicated defect structure. Such defect structures are difficult to distinguish experimentally, but have been predicted numerically using 3-dimensional numerical modeling of topological defects on triangles, squares, pentagons and hexagons in nematics (Dontabhaktuni et al. [107]).

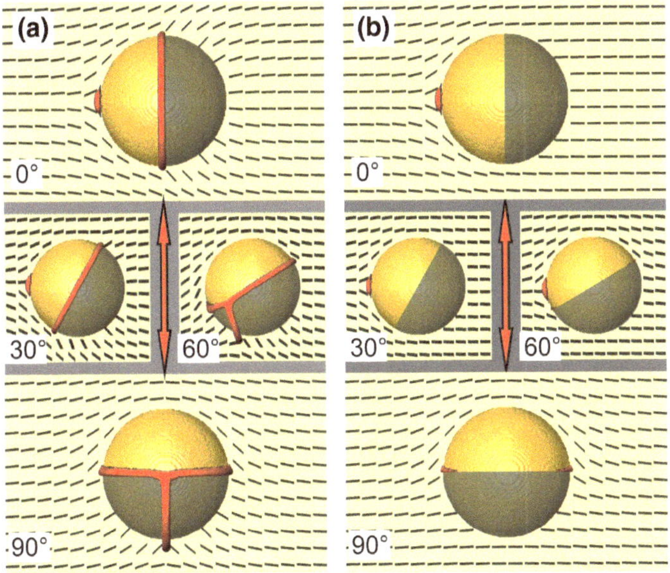

Fig. 2.37 LdG simulations of capped colloidal particle at different fixed orientations with respect to the non-perturbed far-field director (**a**) for strong planar and strong homeotropic anchoring $W_P = W_H = 10^{-2}\,\mathrm{J/m^2}$ and (**b**) for strong planar $W_P = 10^{-2}\,\mathrm{J/m^2}$ and weak homeotropic anchoring $W_H = 10^{-5}\,\mathrm{J/m^2}$. Defects are visualized in red as isosurfaces of the nematic degree of order $S = 0.51$ (bulk $S = 0.533$). Homeotropic hemisphere is darkened. Image courtesy of Miha Ravnik

2.6 Nanoparticles in a Nematic Liquid Crystal

Dispersions of nanoparticles in a nematic liquid crystal have attracted a lot of attention because of possible applications as plasmonic materials and metamaterials in photonics. I addition, dispersions of ferromagnetic solid nanoplatelets in nematic liquid crystals have shown for the first time ferromagnetic response of a liquid (Mertelj et al. [125, 126]), which is discussed in Sect. 2.8.

In plasmonic dispersions, small particles of a metal exhibiting large plasmonic response are dispersed in the nematic liquid crystal. It is characteristic of plasmonic nanoparticles (such as gold, silver, etc.) to absorb strong light within a given frequency band. This effect is due to forced oscillations and resonance of a cloud of nearly free electrons within the plasmonic nanoparticles. The external electric field of the light drives the electrons in motion inside the metal nanoresonator via the electric force, which becomes resonant at some characteristic resonant frequency, which is also called the plasmonic frequency. The nanoparticle and its cloud of electrons can be viewed as a large electric dipole, driven into mutual motion by the electric force of the light. Because of resonance, energy is dissipated in the material in the nanoparticle, meaning that the plasmonic nanoparticles strongly absorb light and induce heating of the environment, where they are embedded. Due to the particles' (and the resonator's) small size, the resonant frequency depends on two important factors: *(i)* the shape and the size of the particle, *(ii)* the electric constant of the environment. For this reason, the plasmon frequency will be shifted and the plasmonic nanoparticles aggregate because of the interaction of oscillating dipoles of the neighbouring nanoparticles. Plasmonic materials are therefore useful as absorbers and nanosensors, which are sensitive to dielectric properties of their surroundings. There are several studies of the optical properties and assembly of plasmonic nanoparticles in nematic liquid crystals, including gold nanoplatelets of high aspect ratio [108], plasmonic gold nanorods [109], gold nanodots [110] and silver nanoparticles [111]. Shape-dependent oriented optical trapping of plasmonic nanoparticles in various forms was analyzed by Senyuk et al. [112]. A comprehensive review of nanoparticle ordering in liquid crystals was written by Blanc et al. [113]. Surface plasmons on the surfaces of metals could lead to extremely miniaturized photonic circuits because of their interaction with light and surface propagation wavelengths, which are orders of magnitude smaller compared to optical wavelengths [114].

Metamaterials are synthetic, i.e. artificial materials, which have a negative reflective index. They were theoretically described by Pendry et al. [115] and are also called the "left-handed materials". This expression is related to the three fundamental vectors, which describe the propagating electromagnetic plane wave: the vector of the strength of the electric field **E**, the vector of the strength of the magnetic field **H**, and the wave vector of the electromagnetic field **k**. Normal materials have a positive dielectric constant and a positive magnetic susceptibility, and these three vectors form the right-handed triad. In left-handed materials, the dielectric constant is negative and the three vectors form the left-handed triad. It is clear that negative dielectric constant could be obtained only in some sort of resonance, where there is strong change in the phase of the oscillation. The same applies to the negative magnetic susceptibility, and both could be realised in artificially modelled and created structures, such as the *horse shoe split ring resonators*. The interest in the fabrication of metamaterials is their possible application in super-lenses and cloaking material [116], which could render a material object invisible in illuminated light.

While plasmonic and metamaterials are nanoparticle nematic dispersion that are interesting for photonics, another class of very interesting materials are the ferromagnetic nematic liquid crystals. These are nematic liquid crystals that exhibit spontaneous magnetization and were recently discovered in nematic dispersions of small magnetic nano-platelets. They will be described in detail in Sect. 2.8 of this Chapter.

The basic requirement for the size of plasmonic dispersions, metamaterials and ferromagnetic nematic liquid crystals is that their basic building blocks be much smaller than the wave lengths of the light in question. Usually this is the visible part of the spectrum, which implies that the basic building blocks are of the size of tens of nanometres. Nematic dispersions of nanoparticles are therefore a natural candidate for plasmonic materials, metamaterials and ferromagnetic nematic liquid crystals. This means that we have to understand the strength and the nature of the forces between nanoparticles in the nematic liquid crystals, if we want to understand the nature and stability of these materials.

It is clear from basic arguments that the interaction between two colloidal particles (with perpendicular surface anchoring) in the nematic liquid crystal vanishes in the two limiting cases: *(i)* for a given surface extrapolation length, the pair interaction force will approach zero for particle sizes smaller than this surface extrapolation length. In the case when the particle is smaller than the surface extrapolation length, there will be effectively no surface interaction and the particle will be invisible to the surrounding liquid crystal. *(ii)* For very strong, i.e. infinite surface strength, the colloidal pair interaction will tend to zero as the colloidal diameter goes to zero. It is then clear that the surface anchoring strength is the crucial point, which determines the minimum size of the particles that would still interact in the nematic liquid crystal. For weak surface anchoring ($10^{-6}\,\mathrm{J/m^2}$), the surface extrapolation length is of the order of a micron, which means that sub-micron particles in the nematic will not interact. One could realise very strong surface anchoring in the experiments, for example by using silica particles and silane surface agents, which give very strong homeotropic surface anchoring of the order of $10^{-4}\,\mathrm{J/m^2}$. In this case, the surface extrapolation length is of the order of several nanometers, and one could expect that nanoparticles in the nematic liquid crystal will significantly interact, in spite of their vivid Brownian motion.

In the following we will describe a series of experiments on silica particles, ranging in diameters from several micrometres down to 20 nm, performed by Škarabot and Muševič [117] and Ryzhkova and Muševič [118, 119]. Because they are perfectly spherical and they could all be chemically functionalised in the same way, this series of particles represents a good systematic study of the effect of the particle size on a colloidal pair interaction. The experiments have been performed with fluorescently labelled silica nanoparticles with diameter from 22 to 450 nm (Micromod). These particles contain a high amount of covalently bound rhodamine B, which is excited at 569 nm, and they emit fluorescent light at 585 nm. The fluorescence of these particles is extremely stable in organic solvents and buffers. The scanning electron microscope images of these particles are presented in Fig. 2.38, together with histograms of the particles' sizes, showing a standard deviation of $+/-10\%$. Strong surface anchoring of the nematic liquid crystal was insured by functionalizing the particles with DMOAP silane. After dispersing silanated particles in the nematic liquid crystal 5CB, the nano-particles tend to form clusters. It is therefore necessary to break up these clusters by thoroughly sonicating the dispersion in an ultrasound bath. The duration of sonication has to be increased for smaller particles, where several hours are needed to break up clusters of 22 nm colloids.

Due to their small size, nano-particles are difficult to observe in the nematic liquid crystal dispersions. The optical limit is set by the Abbe's law of diffraction, stating that two point sources cannot be distinguished with an optical microscope when their separation is smaller than the wave length of the light used for the observation, divided by the numerical aperture of the microscope. The numerical aperture can be as high as $N.A. = 1.6$ for index-matched objectives of optical microscopes. If blue light is used for microscope observation, the theoretical limit of the resolution of a very good optical microscope (not considering STED microscopes) is close to 300 nm. This means that sub-micron particles are difficult to observe in the nematic

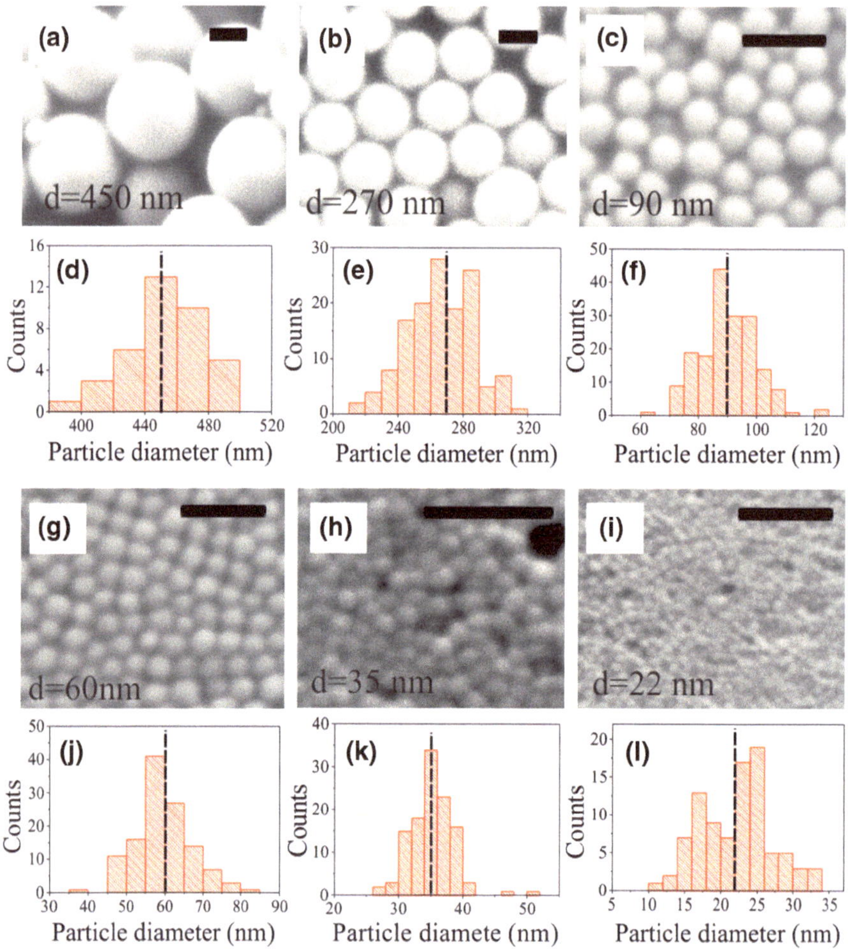

Fig. 2.38 SEM images of silica nanocolloids and histograms of colloidal size distribution. **a** 450-, **b** 270-, **c** 90-, **g** 60-, **h** 35-, and **i** 22-nm-sized particles functionalised with DMOAP. **d**–**f** and **j**–**l** show histograms of colloidal size distribution, derived from analysis of SEM images. The calculated mean values of colloidal diameters are **d** 450, **e** 270, **f** 90, **j** 60, **k** 35, and **l** 22 nm. The scale bars are 200 nm in all panels

liquid crystal, which is even more difficult because of the poor contrast of the refractive indices of the particle with respect to the liquid crystal.

There are two solutions to this problem, which could facilitate observation and tracking of the positions of nanoparticles in the nematic liquid crystal. The first solution is using fluorescent nanoparticles, which could be easily traced by following their emitted light. The problem is the rather low level of emitted fluorescent light, which implies photon counting techniques. The second possible solution is the *dark-field microscopy*, which uses a specially structured light for the illumination of small particles. Using a specially designed optical system, the light used for illumination is structured in a form of a thin and hollow cone, which enters the optical illumination system. The optical system is in fact a strongly focusing objective, which focuses this cone of light to the focal point, where the particle under observation is located. The particle is therefore illuminated by a hollow cone of light, which illuminates this particle practically from all sides at high incident angles. The light which is collected by the microscope objective, used to observe the particle, is only due to the scattering of this illuminated light cone and there is no direct, i.e. background illumination. Because only scattered light is collected, the nanoparticle appears as a bright object on a dark background, hence the expression dark-field microscopy.

Dark-film microscopy is therefore a natural candidate for observing nanoparticles in the nematic liquid crystal. The source of scattering of light in this dark-field microscopy is double: *(i)* the light is scattered at the interface between the particle and the liquid crystal due to different refractive indices, *(ii)* the light is scattered form the elastic distortion of liquid crystal around the particle (if any left because of the smallness of particles), which could be quite extended. Figure 2.39 shows two examples of 125 nm silica nanocolloids in 5CB, observed with polariser only and between crossed polarisers.

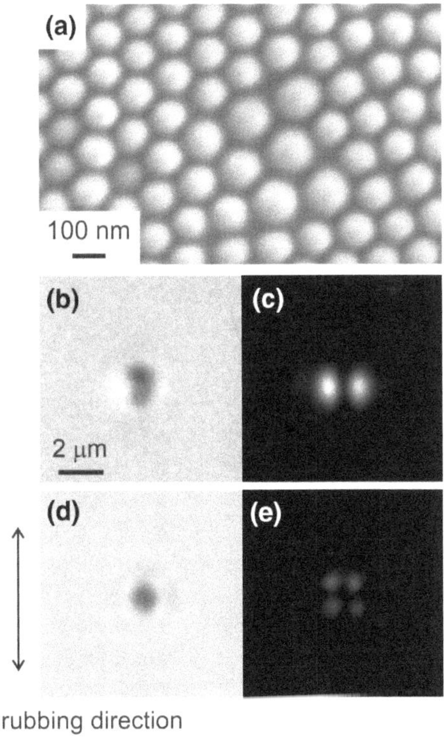

Fig. 2.39 **a** SEM image of silica nanospheres with average diameter of 125 nm. **b–e** DMOAP-silanated single nanosphere of diameter 125 nm in the 3 μm thick layer of planar nematic 5CB. **b** Micrograph of a 125 nm nanocolloid in 5CB, with polariser only inserted. **c** The same particle as in **a** but between crossed polarisers. **d** Non-polarised micrograph of another 125 nm nanocolloid in 5CB. **e** The same particle as in **c** but now between crossed polarisers

As shown in Fig. 2.39, 125 nm diameter silica nanospheres are readily observable in transmission optical microscopy either without polarisers or using crossed polarisers. This figure clearly demonstrates that 125 nm colloidal particles are stable either as dipolar colloids (panels (b) and (c)) or quadrupolar colloids. The difference can be observed even in unpolarised images, as the dipolar colloidal particle in (b) has one dark point (on top), which is the point defect. The symmetry of the colloidal particle is most easily distinguishable in crossed polarised images, shown in panels (c) and (e). The dipolar particle in panel (c) has one dark line running along the symmetry axis of the dipolar configuration of this particle. On the contrary, quadrupolar symmetry of the particle is visible from the dark cross, superposed to the bright image of the particle between crossed polarisers in Fig. 2.39e.

The reason why the same diameter of particle can appear either as a dipole or a quadrupole is in the process of preparation of particles using wet chemistry and surface cleaning. There is always some distribution of surface properties, which finally determines whether the particle appears as a dipole or a quadrupole. The diameter of approximately 100 nm is also the limiting diameter, where fine features of the particle between crossed polarisers can be distinguished. For smaller sizes, which are well below the optical defection limit, the polarised pattern becomes blurry, and one is not able to distinguish whether a single nanocolloid is a dipole or a quadrupole. In this range bellow 100 nm, one can use the aggregation properties of nanoparticles for distinguishing them, because dipolar nanoparticles form chains along the director, whereas quadrupolar nanoparticles tend to form kinked chains perpendicular to the director. Note that the aggregation energies of quadrupolar nematic colloids are an order of magnitude smaller than the dipolar. Several examples of aggregated submicrometre colloids with homeotropic surface anchoring in 5CB planar nematic cells are shown in Fig. 2.40.

2.6.1 Self-diffusion and Pair Interaction of Nanocolloids in the Nematic Liquid Crystal

Sub-micrometre-sized silica particles in the nematic liquid crystal are truly Brownian particles. They exhibit vivid and random Brownian motion due to their interaction with molecules of the liquid crystal. It is possible to follow and video-record their motion by using fluorescence signal from fluorescently labelled nanocolloids, or by observing them in dark-field microscopy, therefore collecting light that the particles scatter. These two techniques were used for tracking nanocolloidal particles in liquid crystals by Ryzhkova et al. [118, 119]. Figure 2.41 shows Brownian trajectories of various sizes of silanised silica nanocolloids in a nematic liquid crystal 5CB. Particles with diameter larger than 100 nm were video-monitored by using bright field method. Typically, particle trajectories of 18 000 snapshots are recorded for many experiments, and their trajectory is extracted from the video frames by using

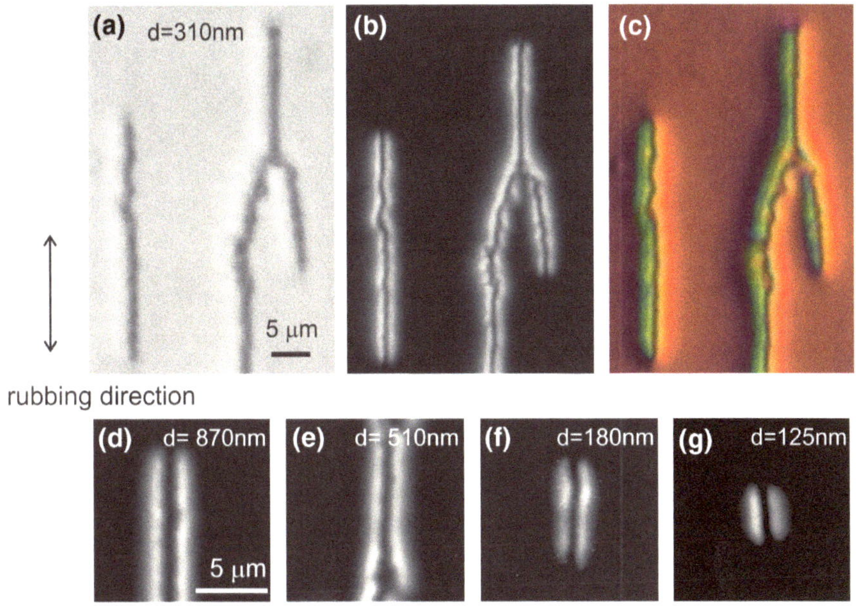

Fig. 2.40 Sub-micrometre colloids with homeotropic surface liquid crystal orientation in the planar nematic cell. **a–c** Colloids with diameter $d = 310\,\text{nm}$ spontaneously self-assemble in dipolar nanocolloidal chains parallel to the rubbing direction. The orientation of the liquid crystal around colloidal chains can be deduced from the polarised image (**b**) and by inserting the red-wave plate, (**c**). *Blue* and *yellow* stripes indicate opposite tilt of liquid crystal around the chain in **c**. **d–g** Nanoparticles with diameters ranging from 870 nm down to 125 nm are assembled into dipolar colloidal chains

the particle tracking method described in Sect. 3.3. Particles smaller than 100 nm were successfully recorded using dark-field microscopy.

The analysis of a random walk of the particle enables one to calculate the diffusivity along ($D_{||}$) and perpendicular ($D_{\perp}$) to the liquid crystal director. The analysis of the distribution of random walks shows that the displacement probability distribution versus the length of a Brownian step can be ideally approximated by the Gaussian function. The width of this distribution is then measured and is directly related to the diffusivity of the particle.

The dependence of the self-diffusion coefficients of silica nanospheres on the inverse colloidal diameter is presented in Fig. 2.41g. As expected, diffusivity is larger for smaller colloidal particles, and there is a difference between dipolar and quadrupolar particles. The particles' diffusion is also very anisotropic and is larger along the nematic director.

For a colloidal particle in the nematic liquid crystal, both diffusion coefficients $D_{||}$ and $D_{\perp}$ correspond to a colloidal diameter that is larger than the real diameter of the particle. This is due to the elastic distorted region of the nematic liquid crystal around the particle, moving together with the particle, thus effectively increasing its size. The effective size of the particle is therefore a sum of the real colloidal diameter and

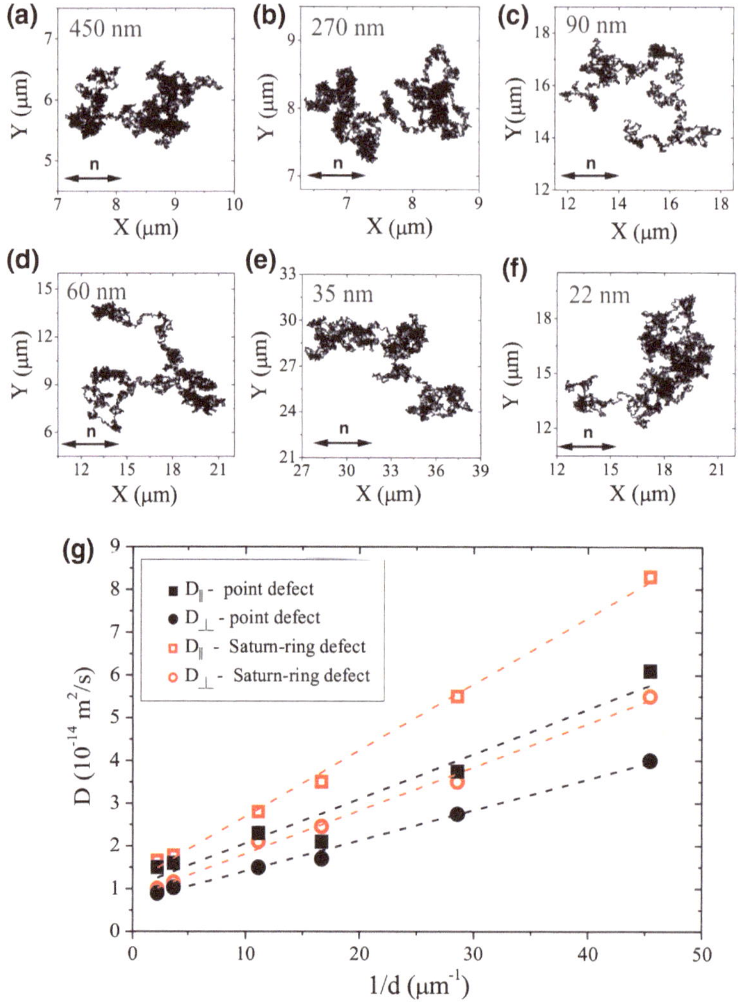

Fig. 2.41 Brownian trajectories of silanised silica nanocolloids in a nematic liquid crystal 5CB. The particle mean diameter is **a** 450, **b** 270, **c** 90, **d** 60, **e** 35, and **f** 22 nm. The time between steps is 5, 3.52, 20, 20, 20, and 20 ms, respectively. Videos **a** and **b** are acquired with bright-field microscopy; **c**–**f** are taken with dark-field microscopy. 18 000 trajectory steps are recorded in each experiment. **g** Dependence of diffusion coefficients of quadrupolar and dipolar colloids along $D_{||}$ (*black* and *red squares*) and perpendicular $D_{\perp}$ (*black* and *red circles*) to the liquid crystal director are shown

the thickness of the distorted nematic region around the particle. Using the Stokes-Einstein relation between the diffusivity and the diameter of the particle for the motion in a plane, one can estimate the effective particle diameter, which is plotted in Fig. 2.42. Dipolar colloids always appear bigger than the quadrupole colloids of the same diameter. This is a consistent result in view of the fact that dipolar configuration

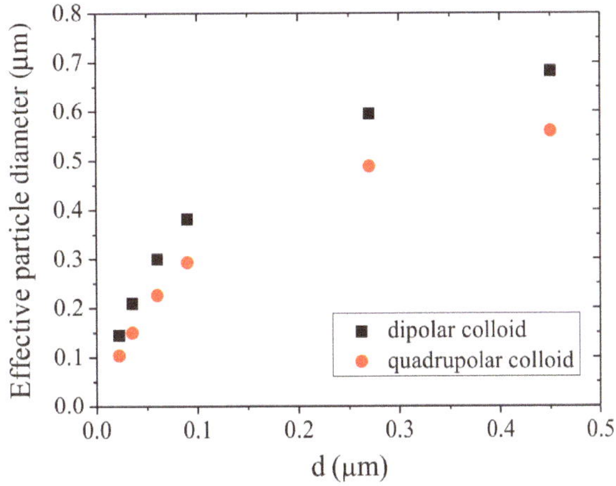

Fig. 2.42 Nanoparticles appear bigger than they actually are. Dependence of the effective particle size on the real colloid diameter. *Black squares* are data for dipolar colloids with point defect; *Red dots are* data for quadrupolar colloids with Saturn-ring defect. The effective colloid diameter is calculated from corresponding diffusion coefficients

is stable for strong surface anchoring, which has smaller extrapolation length and therefore larger region of distorted nematic around the colloidal particle.

The separation dependence and strength of the nanocolloidal pair interaction is determined using the following steps. First, two nanocolloids are brought together using the laser tweezers. They are then released and their motion is video-monitored, making thousands of snapshots of their mutual attraction and eventually binding. Several examples of snapshots from such movies are shown in Fig. 2.43. Second, the diffusion coefficients for the same time of particles are measured by following and recording the Brownian motion of the isolated colloidal particle. Third, the force of the pair interaction is obtained by calculating the Stokes drag force on one of the two particles during the interaction. Forth, this force is then integrated over the path of the attraction and one obtains the pair binding energy for different colloidal sizes.

Figure 2.44 shows the calculated pair binding energies $W_\circ$ in units of $k_B T$ for silanised silica particles with diameter ranging from 22 to 450 nm. The binding energies are bigger for the pair of dipoles and nearly permanent binding could be observed even for 35 nm diameter silica nano-colloids. The corresponding binding energies of particles bigger than 100 nm are exceeding 400 $k_B T$, whereas below this diameter, the pair binding energy falls down to zero. The smallest size of nanocolloids for which an association was observed was of the order of 22 nm, an example is shown in Fig. 2.43f. The observed pair binding energy of nano-colloids is in fact surprising, as it was a general opinion that colloids smaller than 100 nm would never interact significantly. However, these surprising results explain many of the experimental difficulties observed with time-stability of as prepared nematic nanocolloidal dispersions. Namely, one usually observes that after several hours there is practically no freely moving nanocolloidal particle in the nematic dispersion, and they all either aggregate in clusters or attach to the surfaces of the container cell. The observed strong interaction of even sub-100 nm nano-colloids explains this phenomenon as it amplifies itself: once two nano-colloids bind into a cluster, this cluster is bigger and interacts strongly with surrounding isolated nano-colloids. The cluster therefore

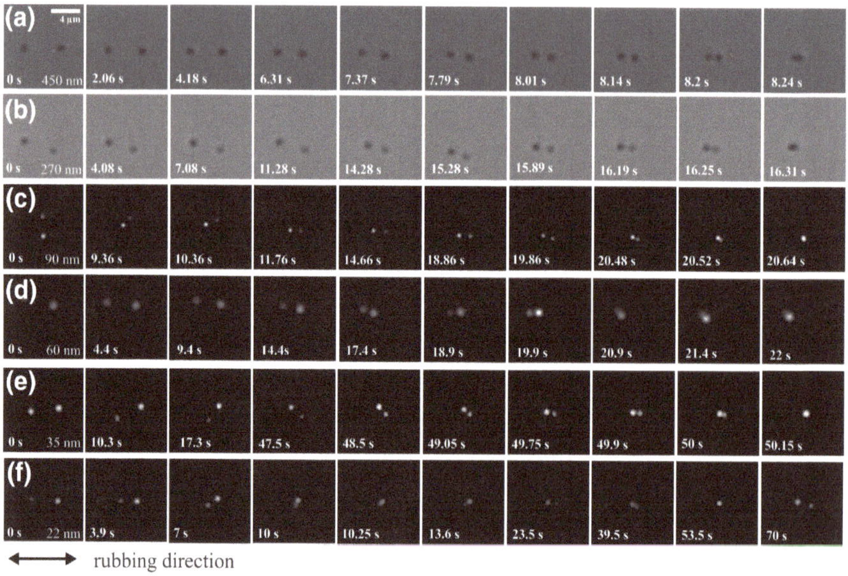

Fig. 2.43 Series of snapshots of two nanocolloids attracted into a dipolar pair. Mean diameter of colloids: **a** 450, **b** 270, **c** 90, **d** 60, **e** 35, and **f** 22 nm. Images **a** and **b** are acquired with bright-light microscopy; images **c–f** are taken with dark-field microscopy. Image size is $11 \times 11\ \mu m^2$

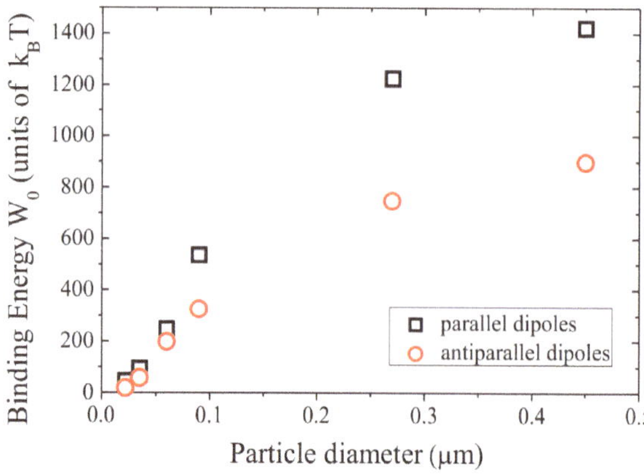

Fig. 2.44 Dependence of dipolar colloidal pair-binding energy on particle diameter: *Squares* indicate pair-binding energy of parallel dipolar particles; *circles* indicate pair-binding energy of antiparallel dipolar particles. The measured binding energy of the dipolar pair is decreasing by reducing the colloid size

collects nano-colloids and grows in size until all the available and freely floating nano-colloids are collected.

It was recently shown by Ryzhkova et al. [119] that silica nanocolloids are electrically charged in the nematic liquid crystal 5CB. This was observed by applying an external electric field and monitoring the motion of an individual 22 nm diameter silica microsphere. Upon the reversal of the electric field, the velocity of the particle was reversed as well, as shown in Fig. 2.45. By measuring the velocity of the particle,

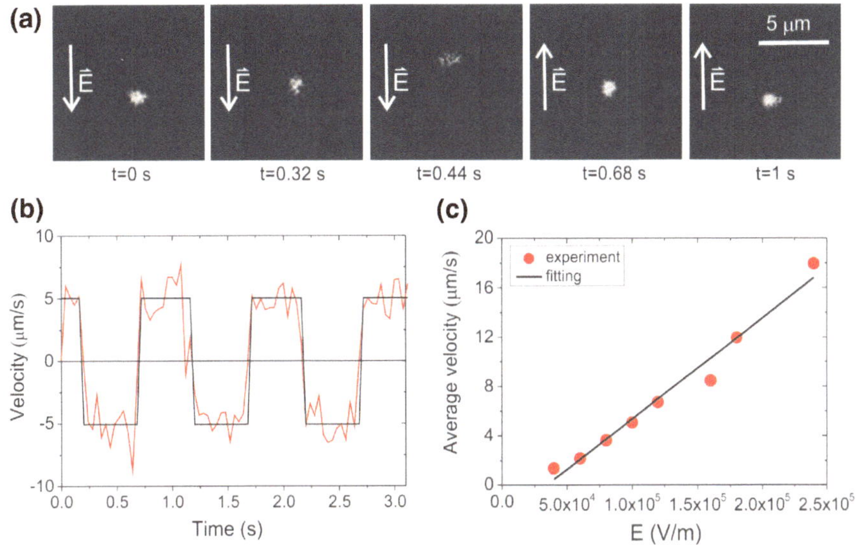

Fig. 2.45 **a** Dark-field images of 22 nm DMOAP-coated silica particles moving along the applied electric field. Note that the particles are moving opposite to the field. **b** Time dependence of the particle velocity. The parameters of the applied electric field are the following: the electric-field strength is 105 V/m, the wave form is rectangular, and the frequency is 1 Hz. **c** The dependence of the average particle's velocity on the applied electric-field strength

it was possible to determine the average electric charge of the particle, which turned out to be quite significant, as shown in Fig. 2.46.

It turns out that a typical 22 nm silica nanoparticle carries approximately 35 electron charges, giving a surface charged density of $\sim 10^{-3}\ \mathrm{As/m^2}$. Electric charging of silanated glass surfaces was also independently observed in a previous study of the electric force studies between a flat silanated glass surface and a silanated silica microsphere attached to the measuring cantilever of an Atomic Force Microscope in the force spectroscopy mode [7]. This surface charge density corresponds to rather large electric fields at the surfaces of nanocolloidal particles of the order of 25 V/μm. This field might be strong-enough to influence the surface orientation of liquid crystal molecules. For positive dielectric anisotropy of the nematic liquid crystal, such as 5CB, the total surface anchoring might be amplified compared to electrically neutral surface. Interestingly, surface charge density is much lower for larger diameter particles, and the surface charge density falls with the inverse particle size, as shown in Fig. 2.46b.

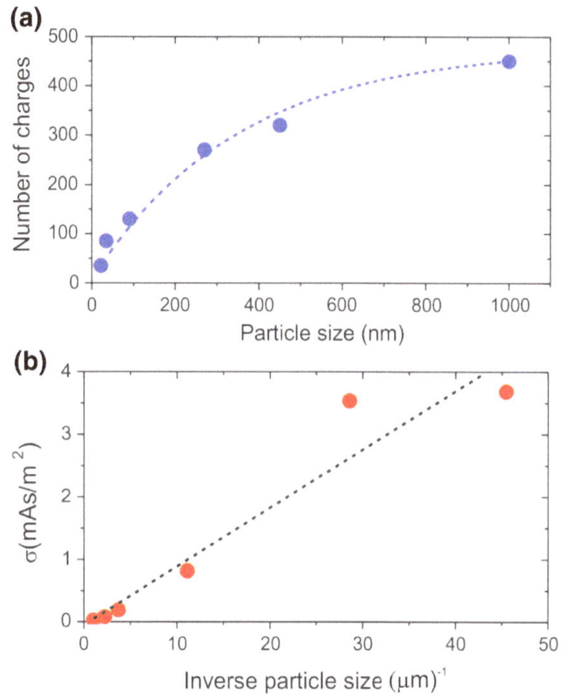

Fig. 2.46 **a** A total electric charge of DMOAP silanated silica nanoparticle as a function of its diameter. Dotted line is a guide to the eyes only. **b** Surface charge density of a particle, determined from data in panel **a**, as a function of inverse diameter of the particle. The *dotted line* is the best linear fit

2.6.2 Interaction of Nanocolloids with Topological Defects

The segregation of nano-colloids in topological singularities is interesting because of possible application in artificially engineered types of matter with new physical properties, such as metamaterials for superlenses and invisibility cloaking. Topological defect are easily created and controlled in the nematic liquid crystals using various approaches, such as particle inclusions, surface patterning, and light manipulation. In principle, complexly structured 3D arrays of topological lines could be created; the question therefore is whether these topological defect lines could act as trapping sites for small particles like nano-colloids, quantum dots, polymers, fluorescent molecules etc. The physical mechanism behind this nano-colloidal trapping by topological singularities is simple. The singularity itself presents a high energy region, where the order parameter is depressed and the liquid crystal is elastically deformed, and this all requires a lot of free energy. If some part of this energetically unfavourable region is replaced by foreign particles, free energy is decreased. This decrease of energy therefore generates a structural force, which should drive small particles into topological singularities.

There are several experimental studies and numerical simulations predicting and showing the effects of this mechanism. In one of the studies, polymers were used to fill the singular disclination lines of the blue phase of liquid crystals in order to stabilise

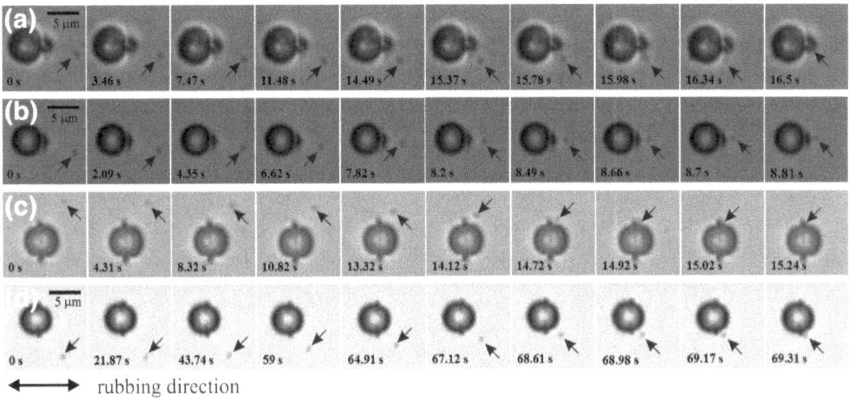

Fig. 2.47 Series of snapshots of nanocolloids attracted into microparticle-induced topological defects. The distorted region around nanocolloids exhibits **a**, **c** dipolar and **b**, **d** Saturn ring (quadrupolar) configuration. Nanocolloidal size is 90 nm. Micrograph size is $13 \times 13\,\mu m^2$

this phase, therefore increasing the phase temperature into the isotropic phase. If the included monomers are polymerised, this results in a cross-linked polymer structure following the blue phase disclination lines [5, 120]. Even more, the liquid crystal can be removed and infiltrated with another liquid crystal, causing the temperature range of the blue phase to increase significantly. In a different study, 2.3 μm colloids were infiltrated into the Saturn ring of a bigger colloidal particle, resulting in a necklace-like decorated Saturn ring, electrically equivalent to a split ring resonator.

Details of the mechanism of inclusion of nanoparticles into defect lines were studied by Ryzhkova et al. [118] with 90 nm diameter silica colloids, interacting with Saturn rings and hedgehog defects around 5 μm silica micro-spheres. Laser tweezers were used to position 90 nm diameter nanoparticles in the vicinity of the bigger silica micro-sphere where they were released. Because these bigger particles generate either a Saturn ring or a point hedgehog defect, nano-colloids are attracted towards these topological singularities, as presented in Fig. 2.47. The upper set of panels shows snapshots of a small nanocolloid (indicated by the arrow) which is attracted towards the hedgehog point defect on the right side of the bigger colloid. This attraction is anisotropic, and the small particle, which is of dipolar symmetry, is driven from the side into the point defect core. For quadrupolar particles, the symmetry of the interaction is quite different, as the particle is attracted directly to the core of the point defect, as shown in Fig. 2.47b.

The lower two panels (c, d) in Fig. 2.47 show the interaction of a dipolar (Fig. 2.47c) and a quadrupolar nanoparticle (d) with the Saturn ring of the bigger colloidal particle. In both cases the nanocolloids are attracted to the Saturn ring at an angle of approximately 45°, and there is no detectable difference in the dynamics of either the dipolar or the quadrupolar nanocolloids.

The binding energies of nanocolloids into the singularities were determined by the particle tracking technique and are plotted in Fig. 2.48 for 450, 270 and 90 nm col-

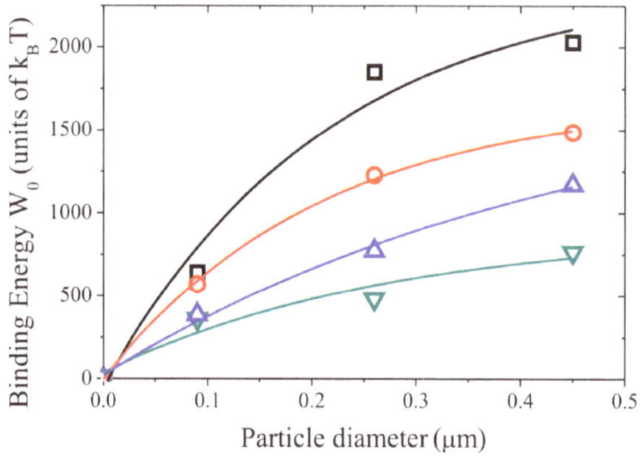

Fig. 2.48 Dependence of nanocolloid binding energy with microparticle-induced point and Saturn-ring defects in 5CB

loids. For 90 nm nanocolloids, which are the smallest colloids that could be measured in this geometry, the binding energy is still significant and ranges from 300–600 k_BT. Smaller particles are practically not traceable, because of the strong scattering of illuminating light from the disclination lines. Nevertheless, even these results clearly show that sub-100 nm particles could easily be trapped into the Saturn rings. An interesting study of the interaction of colloidal nanoparticles of different shapes was performed by Senyuk et al. [112]. Here, the higher reflectivity of metal particles allowed for a reliable and easier observation of aggregation of nanocolloids in liquid crystals.

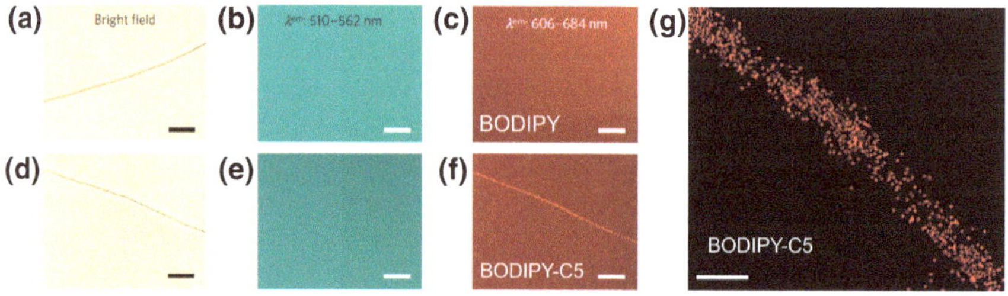

Fig. 2.49 BODIPY amphiphilic molecules self assemble in liquid crystal defects. **a**, **d** The –1/2 disclination shown in bright-field micrograph. **b**, **e** The uniform distribution of BODIPY monomer molecules is seen in fluorescence micrographs taken at the emission wavelength 510–562 nm. **c**, **f** The distribution of BODYPY (panel **c**) and BODIPY-C5 (panel **f**) in and around the –1/2 defect line in nematic liquid crystal 5CB as seen in the fluorescent images taken at the emission wavelength 606–684 nm. Scale bars 20 μm. **g** Super-resolution fluorescence image showing details of distribution of PC-C12 inside –1/2 disclination line in 5CB. Scale bar 100 nm. Reprinted by permission from Macmillan Publishers Ltd: Nature Materials **15**, 106 (2015), Xiaoguang Wang, Daniel S. Miller, Emre Bukusoglu, Juan J. de Pablo and Nicholas L. Abbott, copyright (2015)

Quite recently, Wang et al. [121] presented interesting images of self-assembled molecular structures, which were infiltrated into the core of the −1/2 disclination lines, including the Saturn rings. They studied the aggregation of fluorescent molecules of BODIPY, conjugated to different fatty acids and dispersed in the nematic liquid crystal 5CB. Additional colloidal particles were introduced into the 5CB with a chiral additive, which created clearly visbile disclination lines, similar to knotted colloidal structures.

Because BODIPY is a fluorescent molecule that changes the initial spectrum when forming self-associated structures of the amphiphiles, it was possible to trace and detect any significant trapping and collective assembly of these molecules into the −1/2 disclination lines. For BODIPY alone, dispersed in 5CB, there was no fluorescent contrast between the disclination lines and the nematic surroundings, as illustrated in Fig. 2.49a–c. However, when BODIPY and the conjugated amphiphile were added above the critical concentration of 95 μm, fluorescent contrast was clearly resolved from the −1/2 line as shown in panel (f) of Fig. 2.49. Using the

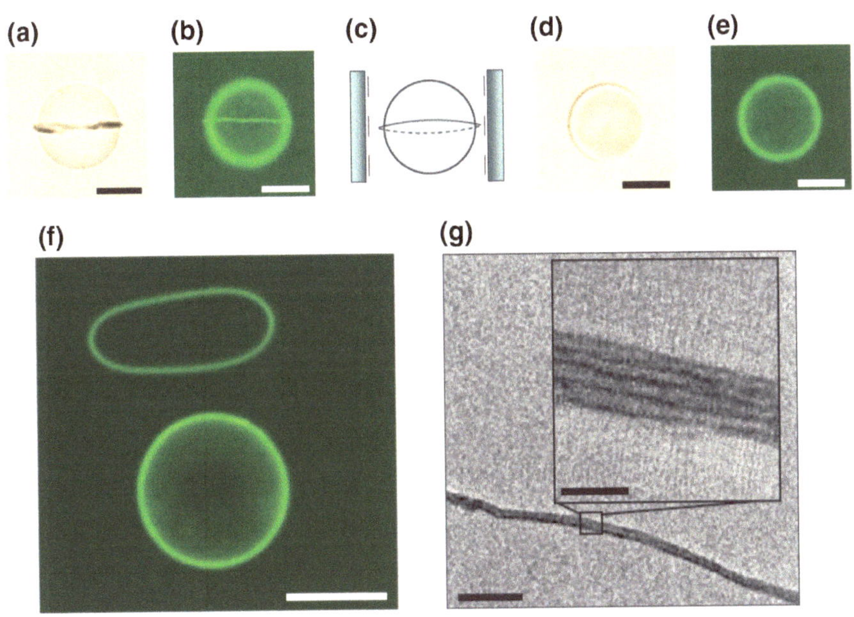

Fig. 2.50 Formation and cross-linking of self-assembled amphiphilic molecules infiltrating a −1/2 Saturn ring. **a** Bright field of a Saturn ring. **b** Fluorescence image of a distribution of lipid DIYNE PC doped with BODIPY-C5 in nematic liquid crystal 5CB. **c** Schematic picture of a Saturn ring encircling the spherical particle. **d**, **e** The distribution of diyne PC in isotropic nematic 5CB as seen in bright field (**d**) and fluorescence images (**e**). **f** Fluorescence images of a polymerised Saturn ring core with diameter of 20 nm that was released from the equator of the microparticle. Scale bars 50 μm. **g** TEM image of the polymerised core of the Saturn ring shown in **f**. Scale bars 100 nm (main image); 20 nm (*inset*). Reprinted by permission from Macmillan Publishers Ltd: Nature Materials **15**, 106 (2015), Xiaoguang Wang, Daniel S. Miller, Emre Bukusoglu, Juan J. de Pablo and Nicholas L. Abbott, copyright (2015)

super-resolution optical microscopy (STORM), it was actually possible to image the fluorescence of another fluorescent molecule (ATTO) which were trapped into the core of the −1/2 defect, as shown in Fig. 2.49g. The width of this region was of the order of 80 nm. Images with better resolution were obtained from BODIPY-labelled amphiphiles assembled in −1/2 defect lines by using cryogenic transmission electron microscopy. This showed ordered tubular-like structures of BODIPY-labelled amphiphiles in the cores of disclination lines. The assembly of individual molecules into the cores of the −1/2 disclination lines is therefore possible because these molecules spontaneously form a kind of a collectively ordered tubular structure (phase) and interact among themselves.

Topological defect lines were used as a template for directing the BODIPY-C5 self assembly, as shown in Fig. 2.50. Here, the Saturn ring around a colloidal particle was filled with BODIPY-C5 amphiphiles (Fig. 2.50a–c), which could reversibly be dissolved back to the nematic liquid crystal by heating it into the isotropic phase. By using photoreactive lipids, the self-assembled amphiphile structures could be polymerized in the core of the disclination loop, which could be diss-assembled from the colloidal particle by heating it into the isotropic phase, as shown in Fig. 2.50f. TEM images in Fig. 2.50g clearly prove the lamellar structure of amphiphiles, self assembled in the cores of the −1/2 disclination loops.

2.7 Vortices and Nematic Colloids

So far, we have considered topological defects, which are singularities of the order parameter tensor, appearing as points and closed loops in nematic colloids. Yet, there is another, less studied class of nonsingular defects in nematics, which was first observed by Williams et al. [122] in their studies of nematic topological defects in capillaries. They noticed that the free energy, which is stored in a singular defect, could be released by escaping of the director into the third dimension. In the case of nematic liquid crystals in a capillary, the point defects situated on the axis of the capillary could escape along this axis, leaving behind a smooth, but splayed nematic liquid crystal. This escape strongly reduces spatial variations of the nematic order and prevents local melting of the nematic due to high elastic deformation. This class of defects is rarely observed in Nature.

Interestingly, the first signs of nonsingular defects binding colloidal particles were reported by Poulin et al. [72], who observed a tubular-like, smoothly deformed region between two tightly bound colloidal particles in an achiral nematic liquid crystal. They called this kind of binding the "bubble-gum" defect, because it could be stretched continuously like a bubble-gum, connecting the two colloidal particles. Whereas this kind of binding is rarely observed in achiral nematic colloids, it turns out that chirality strongly promotes this kind of topological defects. Somehow, chiral liquid crystal environment triggers and stimulates a transformation of singular point defects into nonsingular structures, as will be discussed further on.

When silica particles with perpendicular surface anchoring are introduced in a chiral nematic liquid crystal with rather low chirality, interesting phenomenon of a spontaneous formation of a "bubble-gum" colloidal binding is observed [123]. An example is shown in Fig. 2.51 where 4.72 μm silica microspheres with perpendicular surface alignment were introduced into a 5CB-filled glass cell of 8 μm thickness and a total twist of 90°. This twist was imposed by rubbing both surfaces of the cell in perpendicular directions. One can see from Fig. 2.51a that colloidal particles appear in a form of dipoles, with their axes tilted at 45° with respect to both rubbing directions, which are in this case horizontal and vertical on either surface. This orientation of the dipole is not surprising, as the point hedgehog chooses its orientation in the middle of the cell, where it is located and the director is oriented there at 45 degrees. By bringing the two particles at a separation of approximately ~10 μm, the particles are strongly attracted to each other, as shown in the sequence of frames in Fig. 2.51a. Whereas in a usual case of achiral nematic such two dipoles would bind together with their dipoles antiparallel, another mechanism is obviously acting in the chiral nematic liquid crystal. As seen from the last panel of Fig. 2.51a, both point hedgehogs

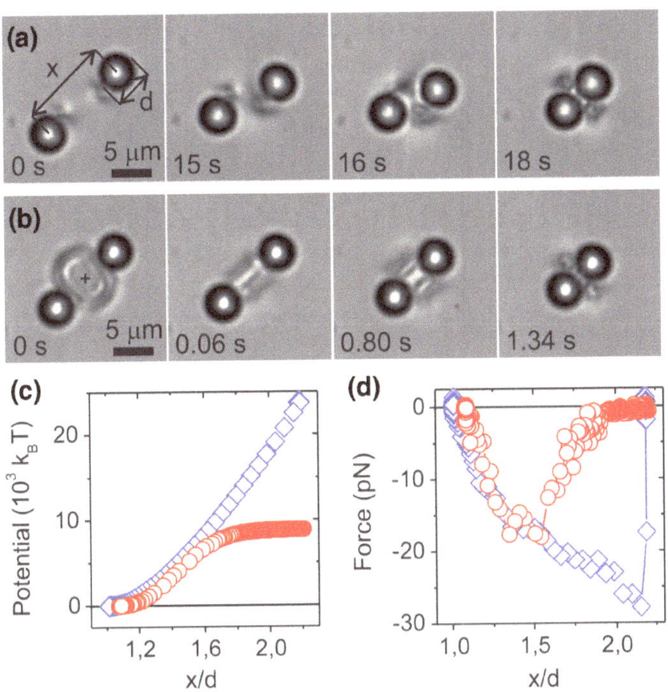

Fig. 2.51 **a** In a 90° TN cell, two antiparallel dipolar colloidal particles spontaneously fuse into a "bubble gum" dimer. **b** An isotropic droplet formed by the laser tweezers separates the particles. After switching off the light, the particles are attracted by two thin birefringent lines. Note the difference between final states in **a** and **b**. **c** Separation dependence of the attractive potential determined from **a** (*circles*) and **b** (*squares*). **d** The corresponding forces. Cell thickness $h = 8\,\mu$m

somehow fuse together and form an axially symmetric, tubular-like structure, which is connecting both particles. The line orienting their centres of gravity is now oriented at 45°, i.e. along the director in the middle of the twisted cell.

By using the laser tweezers, one is able to melt the liquid crystal forming this tubular bond, pushing the particles apart, as shown in Fig. 2.51b. After the light is switched off, the tubular-like bond is reconstructed in a fraction of a second, forming a string-like elongated region, pulling both particles together. When reaching the equilibrium, the original binding is recovered, as shown in the last panel of Fig. 2.51b. Using the particle tracking method, one could reconstruct the separation dependence of the force (Fig. 2.51d) and the attractive pair potential. The structural force due to this nonsingular binding is typically ten times bigger compared to singular binding and so is the corresponding potential. It reaches values of $\sim$10.000 $k_B T$, which is the highest value observed for binding of nematic colloids.

These strings, observed in Fig. 2.51b are characteristic of the escaped nonsingular ring disclination, which must have been created from two singular hedgehog defects, which were present at the beginning of the experiment. R.B. Meyer proposed the "director-field escape" mechanism [124] which transforms two hedgehog point defects belonging to the antiparallel pair of dipoles into a toroidal escaped and nonsingular disclination loop, encircling the line linking two interacting colloidal particles. Topologically, such a smooth loop carries a topological charge of –2 with the local winding number of –1. These escaped hyperbolic defect rings are extremely rare in planar cells but are ubiquitous in chiral nematic liquid crystals, which indicates that the chiral environment increases the relative stability of nonsingular topological structures. This chiral environment embedding colloidal particles therefore has a similar role as the chiral spin-orbit interaction in magnetic systems, which generates magnetic vortices and skyrmions. A detailed analysis of the optical appearance of nonsingularly bound colloidal dimers in chiral nematic cells clearly shows that there are actually two types of dimers, which are optically different, as illustrated in Fig. 2.52a–f.

Because there are two different colloidal dimers observed in the experiments in chiral nematic cells, one immediately conjectures that these two types of colloids should differ in their chirality. This means that somehow the nonsingular, elastically distorted field connecting both particles is intrinsically chiral, and there are two possibilities for the chirality of this region: a left-handed or a right-handed structure. So how could this chiral region be generated from two hyperbolic hedgehog defects? The answer is the way how these two point defects could escape and form a nonsingular structure. The mechanism is illustrated in Fig. 2.52g–j. It shows the cross section of a singular hyperbolic line, where the order parameter of the defect core is low due to strong splay deformation. This stress is released if the director escapes into the third dimension, i.e. out of the plane of the Fig. 2.52g–i. Because of this escape, the singularity of the director field in the centre of the originally singular defect line disappears and a nonsingular escape director field is established with minor variations of the order parameter and lower free energy. By connecting the escaped defect line into a loop, a chiral object is created, with the symmetry of a belt-like object shown in Fig. 2.52j. As the director field could escape in two different

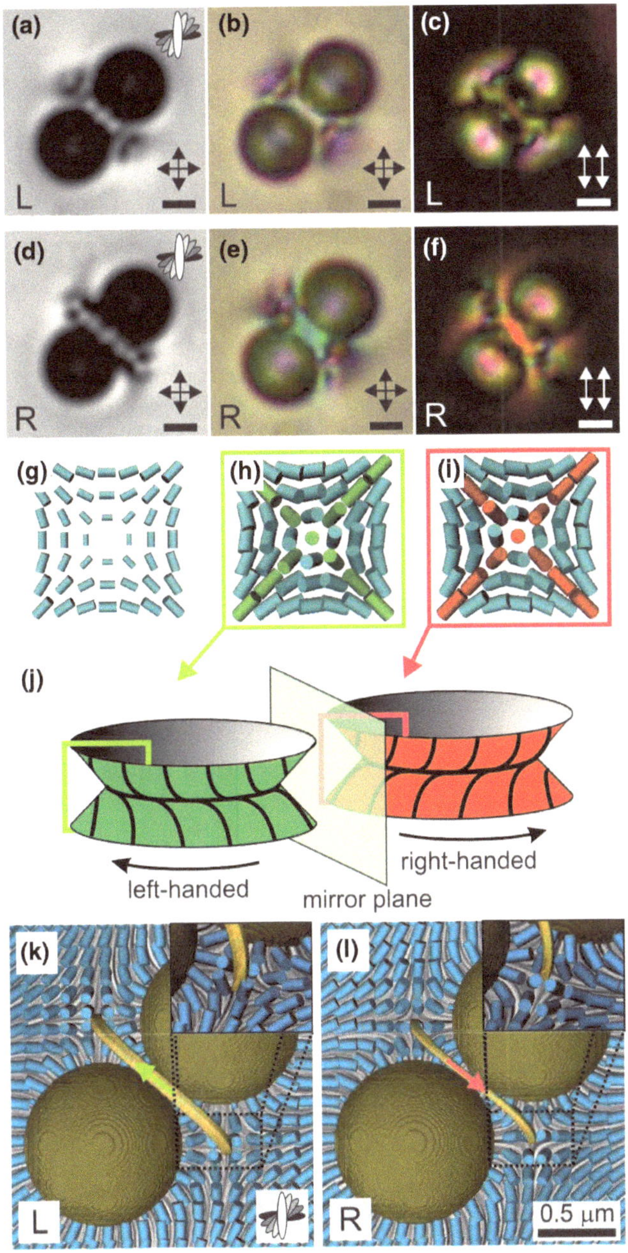

Fig. 2.52 **a** Two colloidal dimers of different chirality in a 90° right-twisted cell (R). Energetically preferable left twisted dimer (L-dimer) between crossed (**a**, **b**) and parallel (**c**) polarisers. R dimer between crossed (**d**, **e**) and parallel polarisers (**f**). The R-handed nematic twist is indicated in **a** and **d**. Scale bar in **a**–**f** is 2 μm. **g** Cross section of a non-escaped hyperbolic line. **h**, **i** Escaping of (**g**) into the third dimension. **j** The escaped hyperbolic rings are chiral. *Dark solid lines* show the local director. **k** Left-handed and **l** right-handed escaped colloidal dimer, calculated from LdG theory. *Colour arrows* indicate handedness of rings

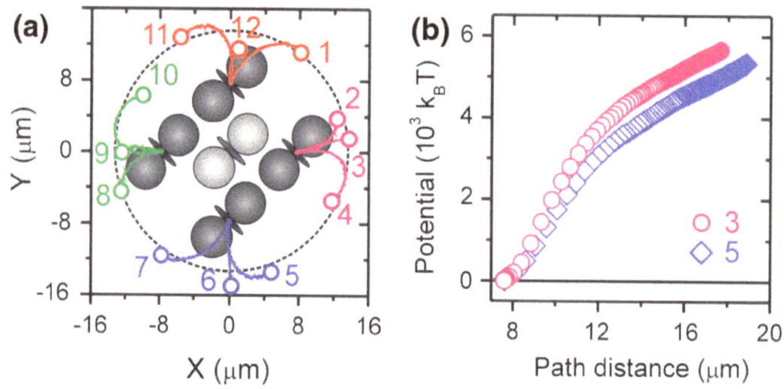

Fig. 2.53 **a** Anisotropy of attraction between two L-colloidal dimers in a 90° R-twisted cell. Starting positions are labelled 1–12, ending positions of the approaching dimer are drawn. **b** Separation dependence of the attractive potential between two colloidal dimers of opposite chirality, determined for trajectories starting from positions No. 3 and 5, presented in **a**. Cell thickness $h = 8\,\mu$m

directions, two different chiral belt-like objects could therefore be created, as shown in the cross sections in Fig. 2.52h and i. The 3D illustration of these chiral belt-like objects is shown in Fig. 2.52j, clearly illustrating their chirality. Here the left-handed belt is viewed in the middle and one also sees the right-handed belt, therefore we have two different and non-superposable chiral objects. This hypothesis of the escape of the hyperbolic line defect was supported also by an extended LdG analysis of the colloidal pair interaction in twisted nematic cells. The results of the modelling are shown in Fig. 2.52k and l. There are indeed two metastable dimer configurations with opposite handedness of the elastically distorted region between the two particles.

It was not surprising to find that colloidal dimers interacted strongly with each other, leading to the formation of dimer clusters, chains, and 2D colloidal crystals in as-prepared samples [123]. The reason for this interaction is quite simple, as each colloidal dimer represents a new colloidal entity, which creates significant elastic distortion of the liquid crystal surroundings. Two colloidal dimers are therefore expected to interact elastically, and this interaction is analysed using the laser tweezers to manipulate a selected chiral colloidal dimer. Figure 2.53a shows the schematics of the experimental arrangement, where one colloidal dimer was kept in the centre of the field of view, and the other colloidal dimer of the same or opposite chirality was positioned close to the first dimer. After being released, dimer trajectories in Fig. 2.53a are clearly attracted from any of the starting positions 1–12, and the dimers spontaneously pack into a square 2D lattice in short time.

The strength and the range of the dimer pair potential were measured using videomicroscopy and particle tracking technique described in Chap. 3 and are shown in Fig. 2.53b. The pair interaction is quite strong and is of the order of 5.000 $k_B T$ per 5 μm diameter particle. Dimers of opposite chirality exhibit roughly the same binding strength as the dimers of the same chirality. This means that using this dimer pair colloidal interaction, one is able to assemble rather large 2D colloidal crystals with pure chirality of dimers (either left (L) or right (R)) or mixed chirality.

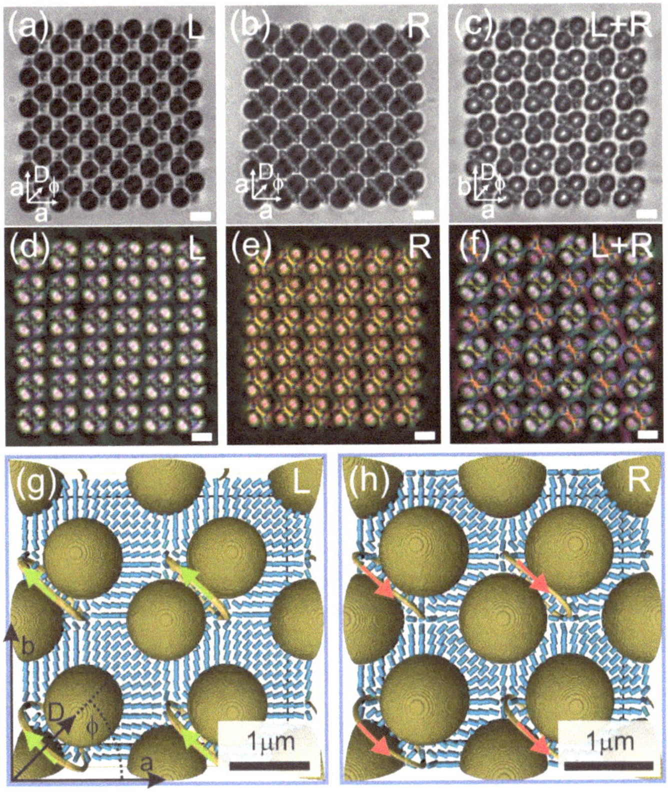

Fig. 2.54 2D colloidal crystals assembled from chiral dimers by laser tweezers. **a** L dimers in R-TN cell between crossed and **d** parallel polarisers. R dimers in R-TN cell between crossed **b** and parallel polarisers **e**. **c** 2D mixed colloidal crystal from equal number of R and L dimers between crossed and **f** parallel polarisers. Note the colour difference between R and L dimers. Scale bar in **a**–**f** is 5 μm. **g** L and **h** R colloidal crystals, as obtained from numerical modeling in R-handed nematic. The iso-surface corresponds to the geometrical escape parameter $S = 0.8$. *Colour arrows* indicate handedness of the ring

Figure 2.54 shows three selected 2D colloidal crystals, assembled from chiral colloidal dimers. Figure 2.54a and b present L-handed and R-handed colloidal crystals, respectively, whereas Fig. 2.54c presents mixed (L+R) chirality colloidal crystals. One can see from the images taken between parallel polarisers that there is a clear optical difference between L- and R-handed crystals; this becomes even more evident when crystals are observed between crossed polarisers. Colloidal interaction in chiral nematic crystal cells therefore provides for an easy assembly of chiral colloidal crystals, which might be interesting for optical applications. In all cases shown in Fig. 2.54, the unit cell is square with basis, which is quite a rare example in colloidal crystals and is most likely due to the very specific nature of the angular dependence of the pair colloidal interaction.

2.8 Ferromagnetism in Dispersions of Magnetic Platelets in Nematic Liquid Crystals

It was predicted in 1970 by Brochard and de Gennes [51] that a truly fluid ferromagnetic phase could be stable in dispersions of magnetic nanoparticles in a nematic liquid crystal. In a ferromagnetic nematic, the orientational order of the nematic liquid crystal imposes spontaneous and collective orientational ordering of anisotropic magnetic nanoparticles and a collective magnetic ordering appears. This spontaneous magnetization of a nematic is quite different than the magnetization in ferro-fluids, which is induced only by the presence of an external magnetic field. This prediction inspired many experiments, in which magnetic nanoparticles of different shapes and sizes were dispersed in the nematic liquid crystal. No ferromagnetism was observed in suspensions of spherical or ill-defined magnetic particles until the experiments were reported by A. Mertelj et al. in 2013 [125, 126] in nematic dispersions of thin magnetic platelets.

When monodomain magnetic nanoparticles are dispersed and entropically stabilised in isotropic liquids, one obtains a ferrofluid. This is a stable colloid dispersion of magnetic nanoparticles that behaves as a paramagnetic or superparamagnetic fluid. Because individual nanoparticles are so small, they are repealed either by the entropic stabilisation (i.e. their vivid Brownian motion) or the electric double-layer repulsion. Due to thermal movement, there is no macroscopic magnetic moment of such an ensemble of magnetic nanoparticles. However, when an external magnetic field is turned on, individual magnetic moments tend to align along the field direction, which causes aggregation of nanoparticles in nanocolloidal chains. The chaining of particles along the lines of the external magnetic field is somehow similar to the chaining of colloidal particles in the nematic liquid crystal, discussed in Sect. 2.3.1. This chaining has two important consequences: *(i)* because of the magnetic chains, the viscosity of a ferrofluid increases with increasing magnetic field, and *(ii)* because of the magnetic ordering, the magnetic susceptibility is increased and the fluid behaves as a (super) paramagnet.

Previous experiments with magnetic nanoparticle dispersions in nematic liquid crystals either used rod-shaped magnetic nanoparticles such as ferrite, or the shape of the nanoparticles was not well defined. They all resulted in the aggregation of nanoparticles introduced in the nematic liquid crystal. This aggregation could be avoided by using a very diluted suspension of elongated ferromagnetic particles dispersed in the nematic liquid crystal. Because of the strong dilution, the interactions between the magnetic particles are very small and no macroscopic magnetisation was observed in the experiments. Typical response of such a diluted magnetic dispersion was paramagnetic, which contributed to the overall response of the nematic director to an external magnetic field, but no true ferromagnetism was observed.

We have discussed in Sects. 2.6.1 and 2.6.2 the interactions and self-diffusion of nanoparticles of different shapes in the nematic liquid crystal. These kinds of nanoparticles are typically made of silica glass or other isotropic material, and in nematic dispersion they experience pair-interaction forces, which are the nematic colloidal forces. These forces are of elastic origin and are strong enough to induce spontaneous aggregation of nanoparticles on a longer term. The nanoparticles experience a rapid Brownian motion and, at a given concentration, spontaneously form aggregates, which act as bigger trapping sites for other nanoparticles until practically all nanoparticles are aggregated and expelled to the confining surfaces. This mechanism of aggregation is effective down to the 20 nm diameter of nanoparticles. It is therefore clear that another, repulsive, force is needed, which could compensate the attractive nematic colloidal force of elastic origin. If the colloids are made of an isotropic material, it would be possible, in principle, to stabilise nanocolloidal dispersions by surface charging of nanoparticles. This electric repulsion between silica nanoparticles coated with DMOAP was in fact observed in the experiments. However, the spontaneous charging of silica nanoparticles in the nematic liquid crystal does not assure enough electrical charge on nanoparticles to counterbalance the attractive elastic colloidal force.

In contrast to the experiments using rod-like magnetic particles, Mertelj et al. [125] used ferromagnetic platelets made of barium hexaferrite. The platelets used were quite thin, around 5 nm, and the distribution of the platelet diameter was around 70 nm with a rather large standard deviation of 3.8 nm. TEM image of barium hexaferrite magnetic platelets is shown in Fig. 2.55b.

Barium hexaferrite has high magnetocrystalline anisotropy with spontaneous magnetisation perpendicular to the plane of the platelets. When such platelets are dispersed in the nematic liquid crystal, the liquid crystal molecules are anchored perpendicular to the surface of the platelets. Each platelet therefore acts as a pancake-like colloidal particle with homeotropic surface anchoring. Such a platelet is therefore an elastic quadrupolar particle interacting elastically with neighbouring platelets, which are also elastic quadrupoles. We know from Sect. 2.3.2 that elastic quadrupoles interact in a very anisotropic way. In a planar nematic cell, the quadrupoles tend to form kinked chains with their direction of attraction at an angle of approximately 50° with respect to the director. For a different mutual orientation of the line joining a pair of colloidal particles, the elastic interaction is purely repulsive. However, our platelets are not only pancake-like colloidal particles, they also carry a rather large spontaneous magnetic moment, which is oriented perpendicular to the platelet, as illustrated with a red arrow in Fig. 2.55a. Now, the magnetic interaction between two ferromagnetic platelets is also important and this interaction is repulsive at the angle where the platelets are elastically attracted. There is therefore a competition between the elastic colloidal interaction and magnetic interaction. This competition is of course non-trivial in nature, but the experiments show that it results in a spontaneous and macroscopic ordering of magnetisation of dispersed nanoplatelets.

When a suspension of magnetic platelets with number concentration of $10^{+13} - 10^{+14}\,\mathrm{cm}^{-3}$ in the nematic liquid crystal 5CB is introduced into a planar glass cell, the suspension is first kept at a temperature above the nematic-isotropic transition of 5CB. If such a sample is slowly cooled down, one observes spontaneous segregation of magnetic nanoplatelets in elongated aggregates. This aggregation is due to the appearance of many domain interfaces between the isotropic and nematic phase during slow cooling. These interfaces act as extremely efficient trapping sites for dispersed nanoplatelets and collect practically all nanoparticles during

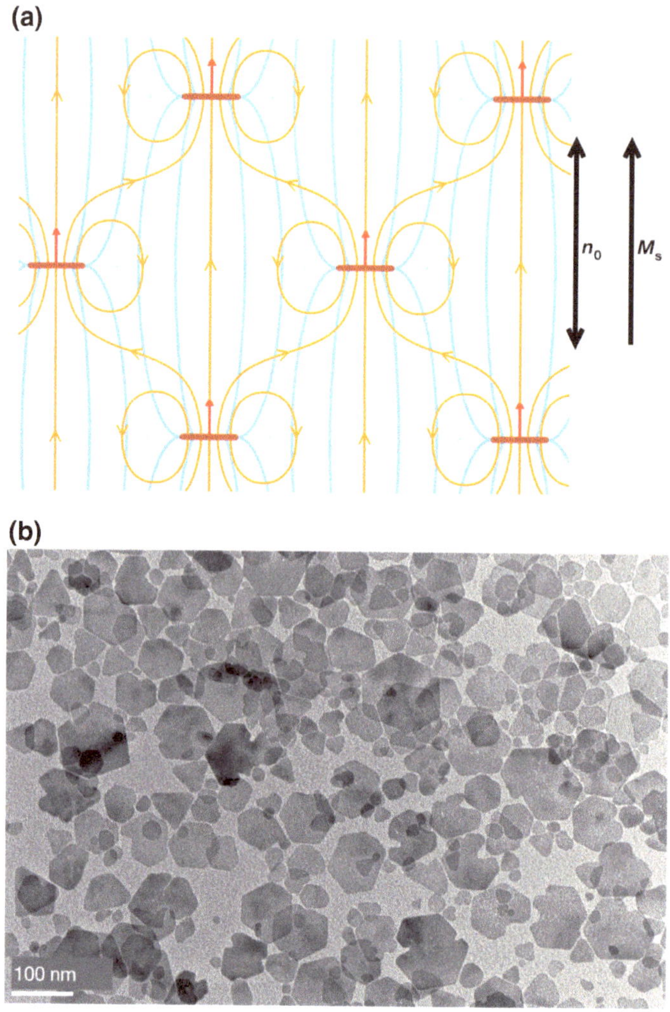

Fig. 2.55 Organisation of magnetic platelets in a nematic liquid crystal matrix. **a** A schematic drawing of the platelets (*red discs*), the nematic director field (*blue lines*) and the lines of magnetic field (*orange lines*). Small *blue dots* represent the cross-sections of the tiny Saturn rings around each platelet. The magnetic moments of the platelets are indicated by *red arrows*. **b** TEM image of barium hexaferrite platelets lying flat on a surface. Reprinted by permission from Macmillan Publishers Ltd: Nature **504**, 237 (2013), Alenka Mertelj, Darja Lisjak, Miha Drofenik and Martin Čopič, copyright (2013)

slow cooling down. According to the experiments, the situation is quite different if the sample of magnetic nematic suspension is quenched from the isotropic phase into the nematic phase in a small magnetic field of approximately 5–10 mT. In this case, there is no formation of well-developed domain walls and much less trapping of nanoparticles as well as much less extraction of nanoparticles from the nematic liquid

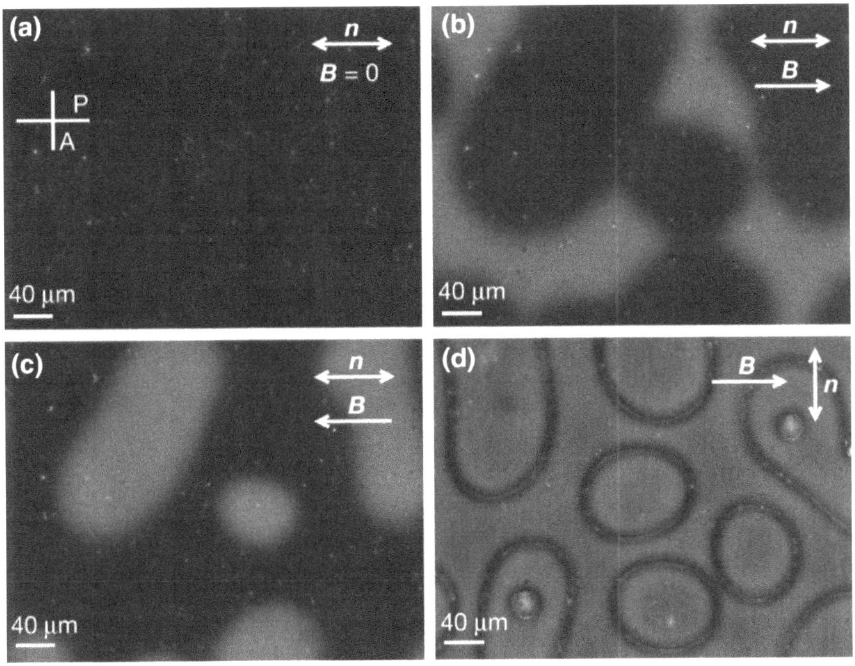

Fig. 2.56 Ferromagnetic domains in the planar nematic can be switched by an external magnetic field. All images are taken under crossed-polariser P and analyser A as shown in **a**. **a** In the absence of magnetic field, the ferromagnetic nematic is uniformly ordered. **b**, **c** An external magnetic field of 3.2 mT is applied either along the director (pointing to the *right*), or in the opposite direction, pointing to the *left* in **c**. **d** Magnetic field of 16 mT is applied perpendicular to the nematic director, which induces the formation of domain walls, visible as *dark lines*. Platelet concentration in 5CB is 0.16 wt%. Reprinted by permission from Macmillan Publishers Ltd: Nature **504**, 237 (2013), Alenka Mertelj, Darja Lisjak, Miha Drofenik and Martin Copic, copyright (2013)

crystal to the defect cores. Initially, small domains of several micrometer diameter are formed, which grow up to macroscopic domains of several hundred micrometers. If the external magnetic field is applied in-plane of such a rapidly quenched ferromagnetic nematic suspension, the magnetic response is linear in the field, and therefore depends on the direction of the field as illustrated in Fig. 2.56.

The first panel in Fig. 2.56a shows the ferromagnetic dispersion of barium hexaferrite in 5CB between crossed polarisers, with the director pointing horizontally. If an in-plane magnetic field is applied along the director and is pointing to the right side, one observes the appearance of domains. These domains are brighter, which indicates that the nematic liquid crystal is distorted by the external magnetic field. If the direction of the magnetic field is reversed, the contrast of the domains is also reversed. The originally dark domains become brighter and the originally bright ones become darker. Therefore, we have two types of spontaneously formed domains, which differ in their response to the external magnetic field. Only one kind of domains responds to the field which is parallel to the nematic director, and the other type responds when the direction of the field is reversed. If the magnetic field is applied in the direction perpendicular to the director, one observes switching of both types of domains and the domain walls become clearly visible, as shown in Fig. 2.56d. This behaviour clearly shows that the spontaneous magnetisation of the ferromagnetic-nematic dispersion is along the director and that two types of magnetic domains are formed, with magnetisations in opposite directions. This is a clear proof of spontaneously appearing ferromagnetism in a dispersion of barium hexaferrite nanoplatelets in the nematic liquid crystal 5CB.

Having two types of magnetically ordered domains in the nematic matrix, one should be able to observe the ferromagnetic hysteresis loop when such a sample is exposed to an external magnetic field varying in magnitude and direction. These magnetisation curves were indeed observed in ferromagnetic-nematic suspension using vibrating-sample magnetometer and are shown in Fig. 2.57. The hysteresis loop indeed shows the presence of a spontaneously formed ferromagnetic liquid state. It clearly shows magnetisation saturation and the coercitive magnetic field, which is of the order of several mT at a platelet concentration of 0.3 wt%.

Ferromagnetic response of a nematic liquid crystal is possible because the colloidal interactions between ferromagnetic solid platelets prevent aggregation of platelets and produce macroscopic spontaneous magnetisation. The shape of the dispersed solid particles is very important for the onset of spontaneous polar order, as well as the quadrupolar symmetry of the elastic deformation of the nematic liquid crystal around such a colloidal inclusion. With rod-like particles, dipolar

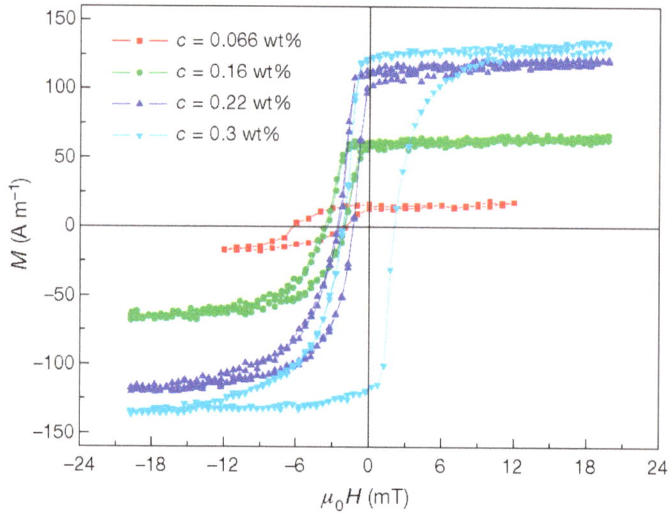

Fig. 2.57 Hysteresis loops of the measured magnetisation of monodomain ferromagnetic nematic samples. The magnetisation curves are different for different concentrations of the particles. When the concentration of the platelets is small, the magnetisation returns reversibly to its initial value. For larger concentrations, the magnetisation curve shows a well-developed hysteresis, which is centred around zero. Reprinted by permission from Macmillan Publishers Ltd: Nature **504**, 237 (2013), Alenka Mertelj, Darja Lisjak, Miha Drofenik and Martin Copic, copyright (2013)

symmetry would be induced, which would promote dipolar chaining of colloidal particles along the nematic director. At the same time, this colloidal chaining would also be favourable for stacking of magnetic dipole moments in a registry. This would result in overall chaining of rod-like magnetic inclusions, which is in favour of both interactions. On the other hand, quadrupolar symmetry of the elastic distortion around the magnetic platelet with magnetic moment perpendicular to it causes repulsion between two platelets placed collinearly and along the direction of magnetisation. This quadrupolar interaction becomes attractive at an angle of around 50° between the two platelets, but then the magnetic interaction becomes repulsive. Overall, the competition of both interactions reduces the total interaction energy to less than kT, but is still sufficient to induce ferromagnetic ordering by the magnetic interaction.

A ferromagnetic nematic liquid crystal is therefore a liquid, which shows all the common properties of a magnet. Monodomain samples of uniform magnetisation can be prepared by cooling in an external magnetic field. By reversing the magnetic field, the magnetisation can be flipped and the material shows domain walls and their motion. Such a material is therefore a combination of an ordered fluid and an ordered magnetisation field, which mutually interact. It can therefore be considered as a multi-ferroic fluid, in which the application of an external electric field couples to the dielectric anisotropy of a liquid crystal and induces change of the magnetic properties due to the coupling of two fields. Vice versa, the external magnetic field couples to the magnetisation and thus promotes change in the dielectric constant of the nematic field, which is coupled to the magnetisation field. Because of its fluidity and softness, a ferromagnetic nematic liquid crystal is an intriguing new material that responds to very small magnetic fields and might find use in new devices.

Whereas ferromagnetism in an ordered fluid is by itself a fascinating phenomenon, it opens up new and interesting questions, such as the mutual interplay of nematic ordering and magnetisation, long range orientational order and the nature of topological defects [127, 128]. These are separately well-explored in nematic liquid crystals and in magnetism. By merging both fields, a new class of topological defects should emerge, where the singularities of the tensorial ordering field and magnetisation field are coupled.

2.9 Forces Between Particles in Chiral Nematic Liquid Crystals

Chirality of the medium in which the colloidal particles are dispersed has a strong influence on the pair-interaction forces. It will be shown in Chaps. 5 and 6 that chirality promotes colloidal entanglement up to the most complex form of colloidal knotting and linking. This makes the topology of chiral nematic colloids very rich in various topological phenomena. In this section we shall discuss the effect of chirality of the nematic liquid crystal on the long-range interaction of colloidal particles.

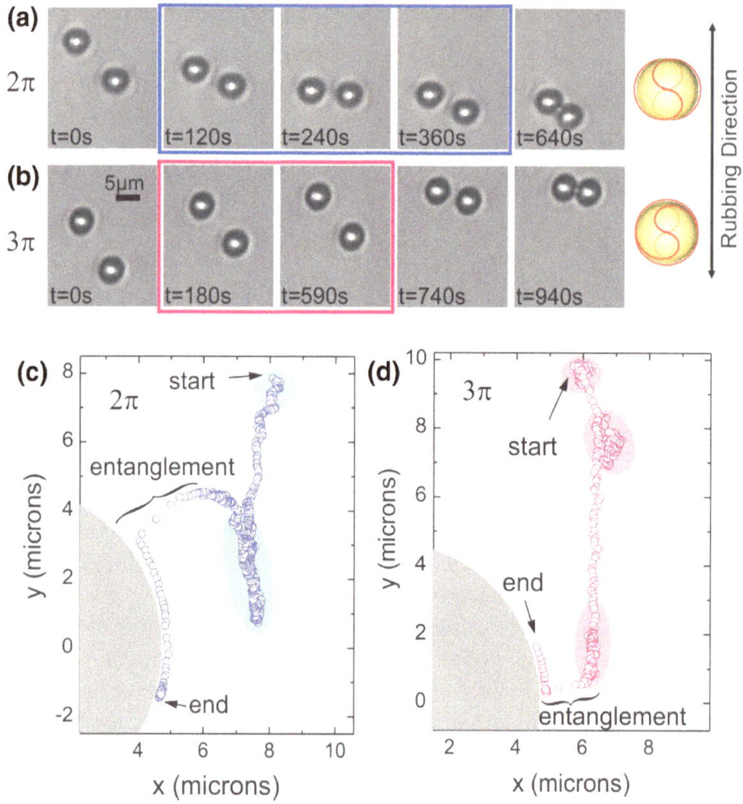

Fig. 2.58 Colloidal pair-attraction in chiral nematic cells. **a** Snapshots of a colloidal pair in a 2π cell demonstrate metastable state in the pair-interaction. Between 120 and 360 s the particles fluctuate at a practically fixed separation. **b** In 3π cell, the particles are temporarily bound between 180 and 600 s. **c** Trajectory of a particle approaching a second particle in a 2π cell. Metastable state is shaded. **d** Colloidal trajectory in a 3π cell reveals three metastable regions

In view of the spatial periodicity of the chiral nematics, it is natural to expect that the pair-interaction force between two colloidal particles reflects the signature of periodicity of the chiral nematic liquid crystal. Indeed, the experiments of Jampani et al. [129, 130] revealed the oscillatory nature of the pair-interaction force in chiral nematic colloids and screening at large separations. These are new phenomena that are not observable in non-chiral nematic colloids and are attributed solely to the chirality of the system. Jampani et al. [129, 130] have used silica particles with a diameter of $d = 4.7\,\mu$m treated with DMOAP to obtain strong perpendicular surface anchoring. The particles were dispersed in right-handed nematic cell of 5CB doped with various amounts of chiral dopant CB15 to tune the helical pitch p. The thickness of the liquid crystal was held constant to $h = 5 - 6\,\mu$m and the amount of chiral dopant varied between the experiments to match a multiple of half-pitch $h = N(p/2)$ in glass cells, which are denoted as Nπ cells.

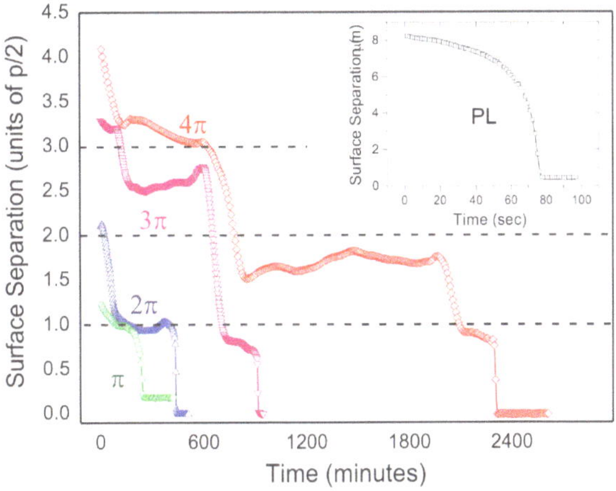

Fig. 2.59 Metastable states and the pair-interaction potential for chiral nematic colloids. The time evolution of the surface-to-surface separation of a pair of colloidal particles with a diameter of 4.7 μm in a chiral nematic liquid crystal with variable chirality. The *inset* shows the pair-attraction in planar, homogeneous liquid crystal cell

The forces between colloidal particles were measured by positioning a selected pair at the starting centre to centre separation of several μm. After the particles were released form the optical trap, their movement was video-monitored, which allowed for the reconstruction of their trajectories and calculation of pair interaction forces and pair interaction elastic potentials. Figure 2.58a shows snapshots of the colloidal pair in the 2π cell during their interaction, indicating that in the time interval 120–360 s, their separation does not change substantially. The pair exhibits Brownian motion around their centres of gravity, as if they experienced a local energy minimum. Similar behaviour of temporal arrest in a local energy minimum is also observed for higher chirality, such as in the 3π cell illustrated in Fig. 2.58b.

The reconstructed trajectories from Fig. 2.58a, b are shown in Fig. 2.58c, d and clearly show the existence of a single minimum for a 2π cell and two local minima for a 3π cell. In both cases, the particles spent some time in the local energy minimum, typically several hundreds of seconds for 2π and 3π cells. The particles would eventually be kicked out of this metastable state by thermal fluctuations and start a rapid approach, which always ends with a spontaneous entanglement of the two colloidal particles.

The experiments clearly indicate that the number of metastable states in chiral nematic colloids depends on the chirality of the surrounding liquid crystal. By conducting experiments in chiral nematics with a higher twist, the number of metastable states is increased, as shown in Fig. 2.59. We can also see that the strength of the pair-interaction decreases with increasing chirality, because the time the two colloids need to come into the final state is longer for higher chirality. Whereas it takes only 80 s in a planar cell, 4 min are needed in a π cell and 45 min in a 4π cell. The second conclusion is that the number of metastable states increases with increased chirality.

The energies of these metastable states in chiral nematic colloids are analysed by calculating the work done by the attractive colloidal force mediated by the chiral nematic liquid crystal and the results are shown in Fig. 2.60a. Here, the pair interaction potential is shown as a function of time for various chirality. The metastable states

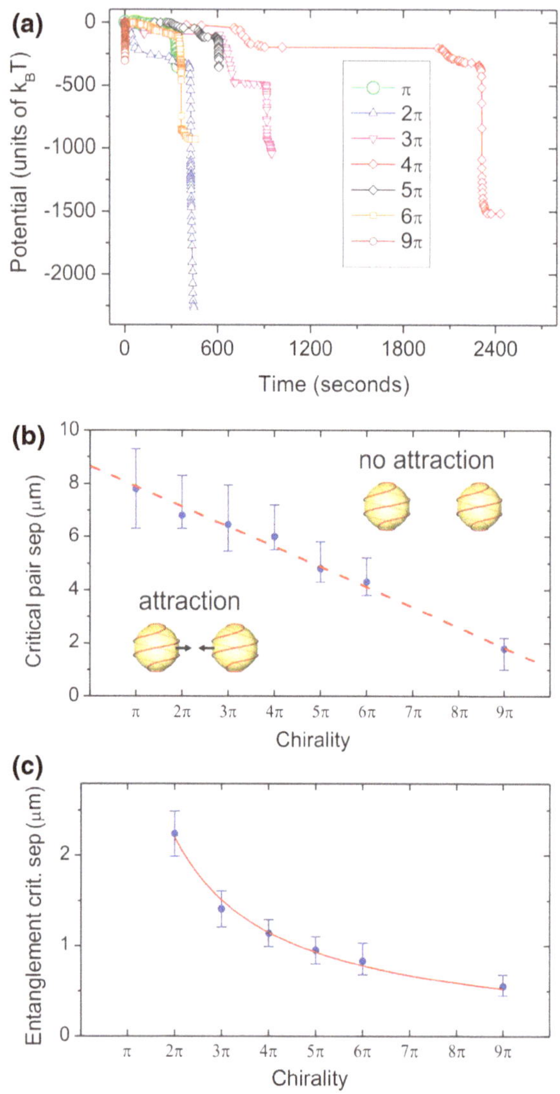

Fig. 2.60 **a** Elastic energy of a 4.7 μm colloidal particle as a function of time, as it approaches the second particle, shown for cells with different chirality. **b** Critical starting separation for the colloidal pair-attraction as a function of chirality. **c** Critical separation for the onset of spontaneous entanglement as a function of chirality

are clearly observable as nearly flat parts of the energy-vs-time curve. When the Brownian motion of colloidal particles in these metastable states is analysed, the depth of this local potential well is several k_BT.

The experiments indicate that the colloidal pair interaction is strongest when the helical period p is equal to the diameter of the colloid $p = 2R$, as indicated by the blue triangles in Fig. 2.60a. If the chirality is increased, the colloidal pair interaction becomes weaker. The analysis shows that the colloidal interaction can be separated into the far field and near field sections and the range decreases with increasing chirality. To characterise this decrease of the pair interaction force with increasing chirality, the critical pair separation was determined for each chirality, as shown in Fig. 2.60b. If the starting separation is larger than the critical, the colloids do not attract within a reasonable time of measurement, which is several hours. If the starting separation is smaller than the critical, the pair attracts and finally entangles. Figure 2.60b clearly shows that the critical pair separation decreases linearly with increasing chirality. This could be explained by considering the spatial dependence of the interaction which is carrying a perturbation. At $N\pi$ turns of the director, the perturbation is a N-times scaled down version of the field at π turns. This scaling of all distances with the pitch is also observed in the positions of metastable states (Fig. 2.59), which makes the range of interactions shorter with increasing chirality. Similar screening of the pair interaction is observed for the near field where the entanglement takes place, as shown in Fig. 2.60c.

The structure of the Saturn ring defects around silica microspheres with perpendicular surface anchoring was also studied in cholesteric liquid crystals with a variable pitch by Gvozdovskyy et al. [131]. The pitch was varied by illuminating the chiral nematic mixture, which responded to the light by a change in the pitch. Spontaneous winding of the Saturn ring around each colloidal particle and the interaction with the Gradnjean-Cano lines was observed. An interesting experiment with the colloidal particles in the chromonic liquid crystals revealed spontaneous twisting of the chromonic liquid crystal structure around the inserted particle Nych et al. [132]. There is a number of theoretical (Lintuvuori et al. [133, 134]) and experimental [113, 134, 135] studies of the spontaneous aggregation and interaction of colloidal particles in chiral nematics.

Chapter 3
Optical Trapping and Manipulation of Nematic Colloids

Abstract The chapter starts with discussion and analysis of optical trapping of particles in isotropic fluids and then continues with a question why optical trapping in nematic colloids is so different compared to the isotropic liquids. Different mechanisms of this anomalous trapping in birefringent fluids are discussed and optical techniques of measuring the forces between nematic liquid crystals are presented in details.

3.1 Optical Tweezing of Particles in Isotropic Media

Small particles can be trapped and moved by a strongly focused laser beam if they are transparent and their index of refraction is larger than the index of refraction of the surrounding medium. This phenomenon was discovered in 1970s with the advent of the first commercially available lasers and triggered an intense research of laser optics. In 1966, Rawson and May [136] and soon later May et al. [137] reported that small dust particles inside a cavity of the He-Ne laser were trapped and accelerated by light to move along straight or kinked trajectories. These phenomena are caused by an intense and spatially non-uniform electromagnetic field of a Gaussian laser beam. They were explained using a simple physical picture in terms of radiation pressure of photons refracting on the surfaces of particles.

These pioneering experiments were followed in 1970 by a seminal experiment of A. Ashkin [138], who was able to show that micron-sized transparent particles could be accelerated and trapped in stable optical-potential wells, using only the force of radiation pressure from a 1 W of continuous Ar^+ laser light at the wavelength of 514.5 nm. He was using transparent latex spheres of 0.59–2.68 μm diameter, which were freely floating in water. A TEM_{00}-mode beam of an argon laser with radius of the waist $w = 6.2\,\mu$m was focused horizontally through a 120 μm-thick glass cell and focused on a single particle. If a beam with milli-Watts of power hits a 2.68 μm sphere off-centre, the sphere is immediately drawn into the beam axis and accelerated in the direction of the light. The sphere moves with a constant velocity of microns per second, until it hits the glass wall, where it remains trapped in the beam.

I. Muševič, *Liquid Crystal Colloids*, Soft and Biological Matter,
DOI 10.1007/978-3-319-54916-3_3

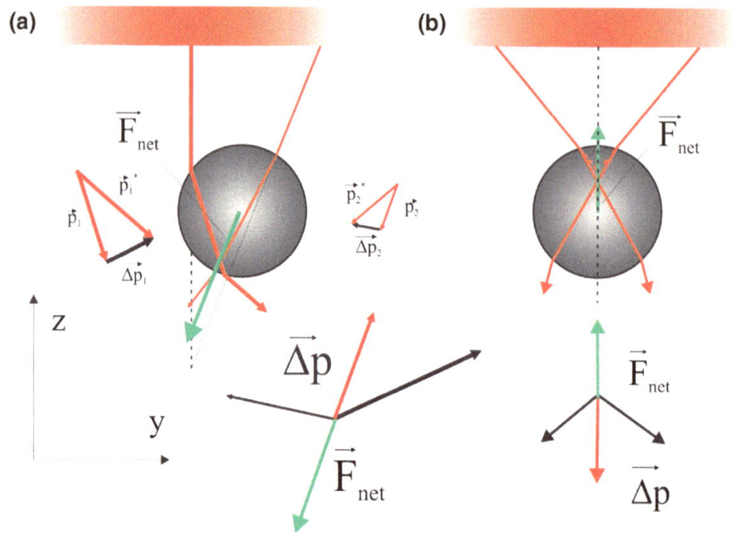

Fig. 3.1 Gradient optical force in the regime of ray optics. The gradually shaded *red boxes* represent the gradient profiles of the incident light, the dark *red* color indicates high intensity. **a** Lateral component of the trapping force. The particle is located on the *right* side of focal point, so the intensity of the *left* ray is bigger than the *right* ray. Each ray is refracted and transmitted through the particle, so they have different momenta ($\mathbf{p}_1^*$ and $\mathbf{p}_2^*$). The momentum changes of both rays $\Delta\mathbf{p}_1$ and $\Delta\mathbf{p}_2$ causes an equal and opposite reaction force, $\mathbf{F}_{net}$, which pushes the particle to the left and towards the focal point. **b** Axial component of trapping force. Two incident rays with same intensity and different direction are passing through the particle, which is placed below the focal point. The resultant momentum on the particle produces an opposite reaction force, which is directed to the focal point. Image courtesy of M. Nikkhou

In a series of subsequent experiments, Ashkin et al. [139–146] demonstrated many different variants of laser trapping and manipulation of micrometer dielectric particles, including a method with two counter-propagating and focused argon beams in a cell with water dispersion of latex spheres. The particles were observed to move into the brightest part of the beam, where they decelerated rapidly and stopped at a stable equilibrium point.

Ashkin explained this phenomenon on the basis of ray-optics and considered the Fresnel refraction of rays of the Gaussian beam on a sphere with a higher refractive index than the surroundings, as shown in Fig. 3.1. Let us first consider the situation when the sphere is displaced laterally from the axis of a Gaussian laser beam, as shown in Fig. 3.1a. We shall consider two rays impinging on the curved surface of the sphere. The less-intense ray from the edge of the Gaussian beam is refracted twice on the curved surface of the sphere, and we do not consider the reflected ray, which gives rise to the *scattering force*. After exiting the sphere, the direction of this weaker ray is towards the optical axis. If we now consider the stronger ray in the centre of the Gaussian beam, we can see that it will be refracted on the sphere in a direction pointing away from the optical axis. After passing through the sphere, the beams will be highly divergent.

We know that each photon carries a momentum, $\mathbf{p} = \hbar\mathbf{k}$. Because the direction of refracted light is changed, so is the momentum of the corresponding photons. The photons from the more intense part, which carry a larger momentum, are deflected away from the optical axis and the change of their momentum $\mathbf{p}_1^*$ is directed away from the optical axis. The photons from the less intensive part of the beam are deflected towards the optical axis, with the momentum change of $\mathbf{p}_2^*$. In overall the momentum of light after refraction on the sphere $\Delta\mathbf{p}$ is directed away from the optical axis.

Because of the conservation of the momentum, this change of momentum of light $\Delta\mathbf{p}$ is accompanied by a reaction force $\mathbf{F}_{net}$, exerted by light on the sphere. This force $\mathbf{F}_{net}$ is opposite to the direction of the change of the momentum of light and therefore points towards the optical axis and towards the focal point. This means that the light refraction on a sphere forces the sphere to move towards the region of more intense light. In the end, the sphere will find equilibrium exactly in the centre of the Gaussian beam, where the light-scattering forces will be totally symmetric and will cancel each other totally, as shown in Fig. 3.1b. The force due to the change of the momentum of light caused by refraction is also called the *gradient force*, or the *trapping force*. It is understood that the situation is quite different if the refractive index of the sphere is smaller than the surroundings, because in that case, the rays will be refracted differently. This will cause a gradient force, which will be directed out of the most intense light field, and the low-refractive-index particle will be expelled from the beam.

The gradient optical force on a dielectric microsphere can be calculated analytically in the ray-optics regime (also named Mie regime), as discussed in Refs. [145, 147–149]. We first consider a single ray of power P hitting a dielectric sphere at an angle of incidence θ, having an incident momentum $n_1 P/c$. The net force on a sphere is given by [148]:

$$F_z = F_{scattering} = \frac{n_1 P}{c} \cdot \left\{ 1 + R\cos 2\theta - \frac{T^2[\cos(2\theta - 2\varphi) + R\sin 2\theta]}{1 + R^2 + 2R\cos 2\varphi} \right\} \tag{3.1}$$

$$F_y = F_{gradient} = \frac{n_1 P}{c} \cdot \left\{ R\sin 2\theta - R\cos 2\theta - \frac{T^2[\cos(2\theta - 2\varphi) + R\sin 2\theta]}{1 + R^2 + 2R\cos 2\varphi} \right\} \tag{3.2}$$

In the above expressions, θ and φ are the angles of incidence and refraction, c is the speed of light in vacuum, n_1 is the refractive index of the surrounding medium. The two angles, θ and φ are connected via Snell's law of refraction, R and T are the Fresnel reflection and transmission coefficients at incident angle θ, which depend on light polarization.

The F_z component in Eq. 3.1. is called the scattering force and acts in the direction of the incident light. The F_y in Eq. 3.2. is the gradient force and is directed towards the optical axis and the focal point of the focusing lens. The total force on the sphere due to reflection and refraction of that ray is the sum of contributions of all reflected and refracted rays, and are therefore exactly calculable. For beams of complex shape, such as highly convergent TEM_{00} laser beam, used in the single-beam gradient trap, the overall scattering and trapping forces are the vector sums of the scattering and gradient force contributions of the individual rays of the beam and are therefore calculated numerically. It can be shown that the gradient force is conservative and the work of this force along a closed path equals zero. More details can be found in Ashkin [148].

If the diameter of the microsphere is reduced, we leave the ray-optics regime and enter into the Rayleigh regime, where the size of the microsphere is much smaller than the wavelength of the incident light. In this case, we can use induced point dipole picture, where the scattering and gradient forces are given by Ashkin [145]:

$$F_z = F_{scattering} = \frac{I_\circ}{c} \cdot \frac{128\pi^5 r^6}{3\lambda^4}(\frac{m^2-1}{m^2+2})n_{medium} \tag{3.3}$$

In the above expression, **E** is the electric field of light, $I_\circ$ is the intensity of light, r is the radius of the particle and $m = n_2/n_1$ is the effective index, equal to the index of the particle divided by the index of the surrounding medium n_{medium}. The scattering force is always directed perpendicularly to the wavefronts of the laser light and therefore pushes the particle in the direction of light propagation.

The gradient force, which is responsible for the stable trapping of dielectric particles with the dielectric constant larger than the dielectric constant of the surrounding material, is directed along the direction of the intensity gradient and is given by:

$$F_y = F_{gradient} = -\frac{n_b^3 r^3}{2}(\frac{m^2-1}{m^2-2})\nabla E^2 \tag{3.4}$$

The gradient trapping force is therefore directly related to the gradient of the electric energy, which is therefore the mechanism, responsible for trapping high refractive index particles. This force is directed towards the regions of highest light intensity. It is clear that the high gradient of the electric energy of the light field and the highest trapping force is obtained when the laser light is focused to the smallest possible dimension, which is limited by the diffraction of light.

If we consider the total electric energy of a dielectric sphere and the surrounding dielectric medium in the presence of a uniform Gaussian beam, we can conclude that the electric energy is at its lowest value when the high-dielectric sphere is positioned in the most intense region of the Gaussian beam. This is because the electric energy of the sphere in the electric field of the light ray will be the lowest. If the sphere is moved away from the brightest region, it will be replaced by the material with lower dielectric constant, which will increase the total electric energy of the system (it is negative in sign). As a result, the gradient force will appear, which will move the sphere with the higher dielectric constant into the centre of the Gaussian beam.

The direction of the gradient force will be opposite if the dielectric constant of the sphere is lower than the surrounding medium. In such case, it is energetically more favourable to move the sphere away from the bright region and fill the emptied space with a higher-dielectric-constant surrounding medium. The low-dielectric-constant sphere will be therefore expelled from the brightest region of the Gaussian beam.

The gradient force is crucial for the operation of the optical tweezers and provides for a restoring force, which forces the particle to always move towards the regions of highest intensity, provided the index refraction of the particle is bigger than the refractive index of the surrounding medium. The focused beam can therefor be considered as a parabolic potential pot in the first order approximation. The light gradient force acts as a spring with a spring constant k, which forces the colloidal particle to move towards the regions of strong light. In the absence of damping, the particle will therefore freely oscillate in the optical trap with its natural resonant frequency (Molloy and Padgett [150]):

$$f_{res} = \frac{1}{2\pi} \cdot \sqrt{\frac{k}{m}} \tag{3.5}$$

Here k is the elastic constant or stiffness of the optical trap and m is the mass of the trapped particle. For typical dimensions of the trapped object of $R = 1\,\mu\text{m}$ and the stiffness of the optical trap of the order of 0.05 pN nm^{-1}, the resonant frequencies of trapped microparticles is typically ~50 kHz. If we consider trapping in a viscous fluid, the movement of the trapped particle is hindered by the Stokes drag force $6\pi R\eta \text{v}$, which provides damping to the harmonic oscillations in a parabolic potential.

Trapping experiments in colloidal science and also biology are performed around room temperature (i.e. 300 K), which means that our trapped particle is exposed to the rapid bombardment of the molecules in the fluid around the trapped particle. This gives rise to a random fluctuating force of thermal origin, which produces a mean-squared deviation of the position of the particle along a selected axis:

$$\frac{1}{2} k_B T = \frac{1}{2} k \langle x^2 \rangle \tag{3.6}$$

For a typical micrometer colloidal particle and tweezers stiffness, the *rms* deviation is of the order of 10 nm and it is very unlikely that the particle will spontaneously diffuse out of the optical trap that has a typical range of capture of 300 nm. This *rms* displacement becomes important when trapping small particles of the molecular scale.

In practice, Gaussian single-mode beams from a high-power laser sources are strongly focused by using high numerical aperture microscope objectives. Prior to entering the objective aperture, the laser beam is expanded to completely fill the available aperture D of the microscope lens with a focal length f. Having assured this, the Gaussian beam will be focused to a diffraction-limited diameter $d_\circ$ [151]

$$d_\circ = \frac{2f\lambda}{D} = \frac{\lambda}{NA} \tag{3.7}$$

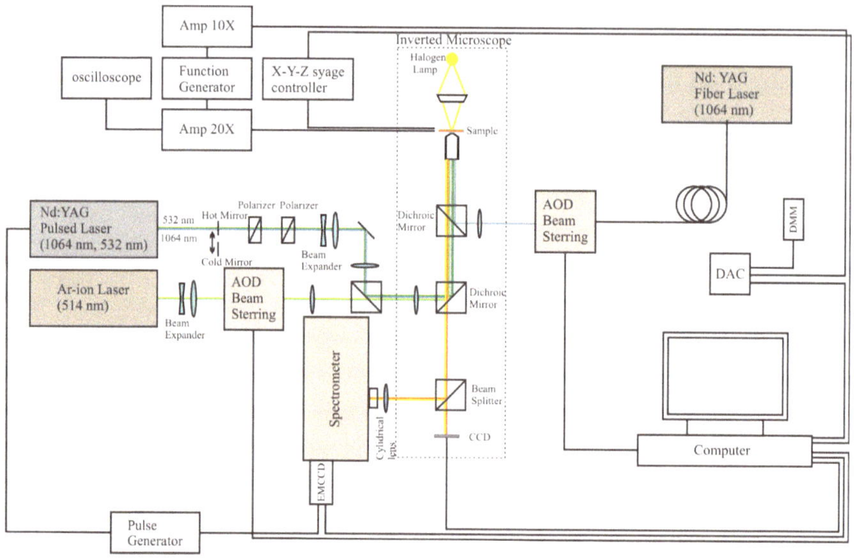

Fig. 3.2 Schematic of the optical set-up consisting of two different laser-tweezers operating at 1064 nm (Nd:YAG laser) and 514.5 nm (Ar^+ laser). For details see article by N. Osterman [152]. Image courtesy of M. Nikkhou

where NA is the numerical aperture of the focusing lens. The experimental realisation of typical laser tweezers is shown in Fig. 3.2.

In reality, colloidal particles of typical micrometre diameters are used in experiments and the electric field strengths of a typical laser beam are of the order of $10^4 - 10^5$ V/m. Upon focusing, these field strengths could be increased by two orders of magnitude, reaching field strengths of the order of 10^7 V/m or 10 V/μm. One can immediately see that the strength of the electric field on a focused laser beam is strong enough to influence considerably the collective order of liquid-crystal molecules in the nematic phase. These effects of the influence of a focused laser beam on a nematic liquid crystal are actually the basis of laser-trapping phenomena in nematic liquid crystals and will be explained in the Sect. 3.2.

We conclude this discussion of the principles of laser tweezing of colloidal particles in isotropic fluids by emphasising that nowadays, laser tweezers are commonly used in many scientific disciplines besides physics, including chemistry, biophysics and medicine [153–160]. The reason for this is simple: one is able to grab from a distance an object under an optical microscope, simply by moving the laser trap or a multitude of traps with a slight movement of a computer joystick. Complex tweezers, including multi-colour tweezers, are nowadays available, which allow a number of traps to be generated holographically or by using acousto-optic beam steering and multiplexing.

3.2 Trapping and Manipulation of Particles in Nematic Liquid Crystals

Optical trapping of dielectric particles in nematic liquid crystals is substantially different from optical trapping in isotropic fluids, and the physics involved is substantially different from the physics of trapping in isotropic liquids.

The first experiment on optical trapping of silica microspheres with diameter 0.97 μm in a nematic liquid crystal was performed by Muševič et al. in 2004 [161]. The refractive index of silica, $n = 1.37$, is substantially smaller than both refractive indices of the nematic liquid crystal used in the experiments E12 and 5CB. The ordinary refractive index of the nematic E12 is $n = 1.52$ and the extraordinary index is $n = 1.74$; and both values are higher than the refractive index of silica. The surfaces of silica particles were coated with DMOAP silane to induce strong perpendicular anchoring of liquid crystal molecules and the silica microsphere dispersion was put inside a homeotropic nematic cell with a thickness much larger than the colloidal diameter.

Contrary to expectations, silica particles were attracted into the focus of the laser tweezers. This attraction was of long range and very strong. The question that arises is how can one trap dielectric particles with lower refractive index compared to the indices of the surrounding fluids? The answer is given by a careful observation of video images taken during the trapping of silica microsphere by rather strong and linearly polarised laser tweezers. When the laser is focused at one point in the nematic liquid crystal, the strength of the electric field of the light field of the laser tweezers is extremely large for laser power (typically 100 mW at the focal point) used in the experiments, reaching up to $E = 10^6$ V/m = 10 V/μm. This electric field is linearly polarised inside the nematic liquid crystal because, inside the cell, it propagates along the optical axis of the nematic liquid crystal in the homeotropic cell. However, the linearly polarised electric field of light exerts mechanical torque on collectively aligned liquid-crystal molecules due to the coupling of the strong electric field of light $\mathbf{E}$ and dielectric anisotropy for optical frequencies (i.e. birefringence). When the field exceeds a certain value, the light may induce a local Fredericksz transition by turning the illuminated liquid-crystal molecules with their long axes directed along the light polarisation for positive birefringence.

This optically induced Fredericksz transition is observable as a change in the local transmittance of the nematic liquid-crystal cell and is presented in a sequence of images in Fig. 3.3. This sequence of video images was taken during the approach of the colloidal particle, which can be seen as a bright circle with a dark Maltese cross in Fig. 3.3a, which was taken between crossed polarisers. The region of the nematic liquid crystal illuminated by the focused laser tweezers appears as as brighter diffuse area above the colloidal particle in Fig. 3.3a. This region is called a "ghost" because it cannot be seen without crossed polarisers. It corresponds to a locally elastically distorted liquid crystal, forced by the torque of linearly polarised light to turn from the regionally out-of-plane position to in-plane position. This localised and distorted

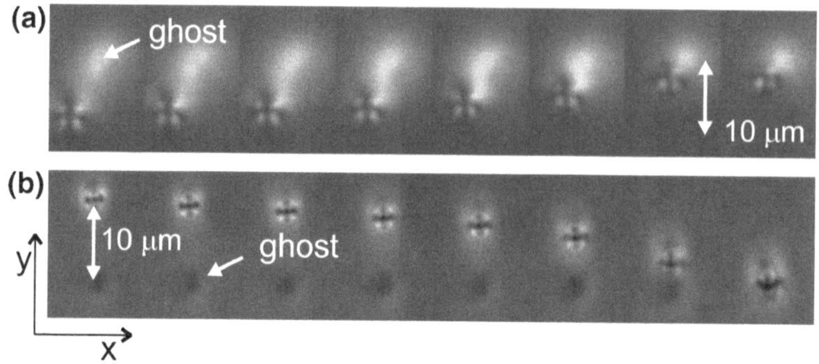

Fig. 3.3 **a** Sequence of micrograph images under crossed polarisers, showing the trapping of a 0.97 μm silica colloidal particles by 64 mW laser line at 1064 nm. Note the apparent size of a 0.97 μm colloid and the bright spot of the ghost at the laser focus. This ghost is formed due to the local birefringent area created by the laser-light-induced distortion of the director field. **b** If 5CB is used instead of E12, the laser power (64 mW) is strong enough to locally melt the 5CB into the isotropic phase. This isotropic island is now seen as a *black* ghost because of crossed polarisers

region of a nematic liquid crystal therefore acts as an artificially created colloidal particle, which interacts with the real silica colloidal particle.

If the laser power is further increased or if one uses a nematic liquid crystal with a lower clearing point, such as the clearing point of 5CB compared to E12, one is able to locally melt the liquid crystal into the isotropic phase. This is illustrated in a sequence of images in Fig. 3.3b, which were taken in 5CB with the same colloidal particles dispersed in it. The molten circular island is optically isotropic and appears completely dark between crossed polarisers in Fig. 3.3b. Because the nematic liquid crystal is homeotropic, it also appears dark between crossed polarisers in Fig. 3.3b. The colloidal particle is now positioned on "top" of the black ghost in Fig. 3.3b and is attracted towards the molten volume of the nematic liquid crystal.

This simple experiment clearly demonstrates that optical trapping of particles in nematic liquid crystals is fundamentally different than optical trapping of particles in isotropic fluids [161–164]. Of course, the scattering and the gradient force, which we have mentioned in the Sect. 3.1 are still present, but are much less important than these new forces, which are obviously present when the optical trap is positioned into a birefringent and long-range-ordered fluid, such as a nematic liquid crystal. The ordinary isotropic trapping forces are observable if the nematic colloidal dispersion is heated up to the isotropic phase. In this case, silica particles are clearly and strongly repelled from the optical trap because their refractive index is smaller than the refractive index of the isotropic liquid crystal.

It was found that there are four different mechanism are responsible for this "forbidden trapping" of low-refractive-index particles in high-refractive-index liquid crystals. In order of increasing laser power, these mechanism are:

(i) At low laser power, i.e. below the threshold for the optical Fredericksz transition, the linearly polarised light of the laser tweezers couples dielectrically to the

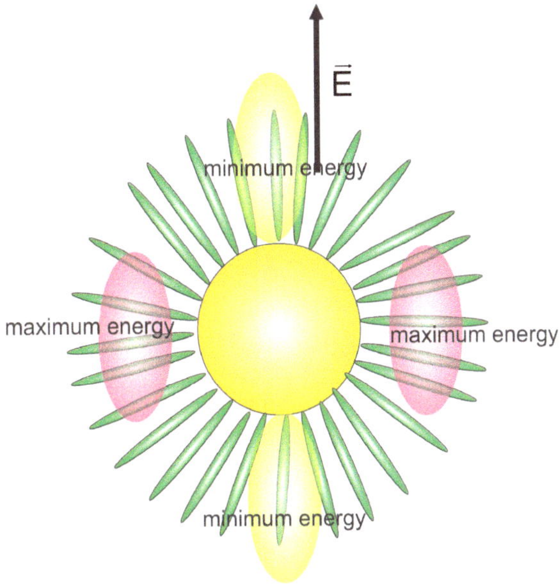

Fig. 3.4 Illustration of the *top* view of a homeotropic silica microsphere in a nematic liquid crystal. *Green* ellipsoids are liquid crystal molecules around the equator of the microsphere. Note the inhomogeneous region of distorted nematic liquid crystal around the microsphere. A linearly polarised light will act on this colloidal particle in such a manner as to minimise the electric energy. This means that local director will be aligned along the polarisation of the beam in the equilibrium position

anisotropic dielectric constant of the liquid crystal. In the equilibrium position the director is aligned along the electric field of light, resulting in the lowest possible electric energy. A spherical colloidal particle that induces perpendicular surface anchoring of liquid crystal molecules is surrounded by a strongly anisotropic region of a liquid crystal, as illustrated in Fig. 3.4. The electric energy of the liquid crystal in a linearly polarised light field will be minimum if the local director is parallel to the light polarisation. This will generate a structural force which will position the colloidal particle in such a way as to obtain the minimum energy, as illustrated in Fig. 3.4. The experiments with homeotropic colloidal particles and a linearly polarised laser trap indeed show that below the threshold for optical Fredericksz transition, the colloidal particle is trapped exactly at its surface. The orientation of the particle is such that the local normal to the particle surface is directed along the polarisation of the beam. Roughly speaking, this physical mechanism of laser trapping is based on the minimisation of the electric energy of the elastically distorted region around the colloidal particles in the laser trap.

(ii) When the laser power of the optical trap exceeds the threshold for the optical Fredericksz transition, the nematic liquid crystal is elastically distorted by polarized light. This distortion is not visible without polarised light of observation, but becomes clearly visible as a bright diffuse region above the threshold. In this case, we can imagine that a colloidal particle "sees" an artificial colloidal particle created by the strong laser tweezers. The force of interaction between the real particle and the ghost colloid depends in a complex manner on the mutual position of the particles, and the resulting trajectories during trapping are "spider-like", as illustrated in Fig. 3.5.

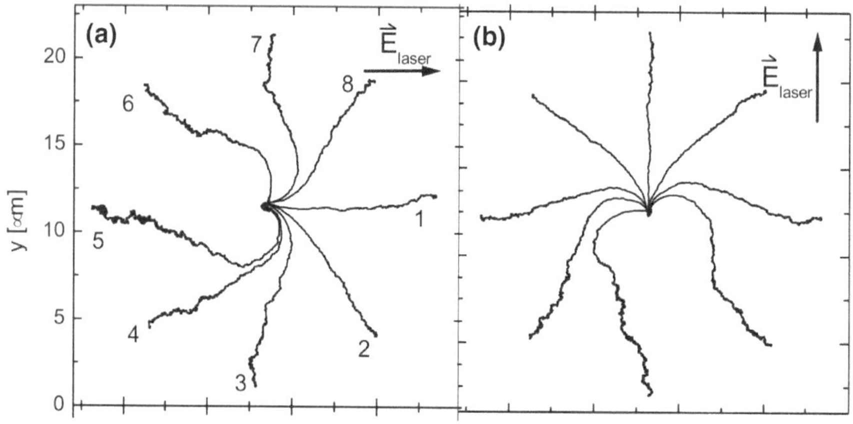

Fig. 3.5 **a** Spider-like trajectories during the optical trapping of a 0.97 μm silica colloid in a 15 μm homeotropic 5CB cell. The laser power is 35 mW, which is well above the optical Fredericksz transition. The starting positions are labeled 1–8. **b** The set of spider-like trajectories rotates as a whole as the polarisation of the laser light is rotated for 90°

The strength of confinement of silica particle in such a light trap was determined from the recorded time series of the coordinates $\mathbf{r} = (x(t), y(t))$ of the particle during its Brownian motion in the plane of the cell. The probability density $p(r)$ to find a particle at a given distance r from its equilibrium position is

$$p(r) = A \cdot r \cdot \exp(-\frac{U(r)}{k_B \cdot T}) \tag{3.8}$$

In the vicinity of the equilibrium point, the potential well $U(r)$ can be treated as a parabolic one, $U(r) = \frac{1}{2}k_{trap} \cdot r^2$. The spring constant of the optical trap k_{trap} is therefore obtained by fitting the experimentally determined probability distribution $p(r)$.

The results are shown in Fig. 3.6 for a laser power of 8 and 32 mW, respectively. Figure 3.6a shows the positions (approximately 5000 points out of 15.000 recorded) of a colloid in the $x-y$ plane, recorded during the Brownian motion, whereas Fig. 3.6b shows the corresponding histograms for each coordinate. The probability distribution $p(r)$ is presented in Fig. 3.6c. It is symmetric with respect to x in y coordinates and shows perfect agreement to a Gaussian fit. Finally, the reconstructed profile of the laser trapping potential is presented in Fig. 3.6d, together with a parabolic fit. The corresponding spring constant of the trap is $k_{trap} = 9.2(1 \pm 0.1)$ pN/μm for 32 mW laser power and is reduced to $k_{trap} = 2.0(1 \pm 0.1)$ pN/μm for 8 mW power, as expected.

(iii) By further increasing the laser power, the liquid crystal is locally heated because of the light absorption, which creates a temperature gradient around the optical trap. In a practical way, this can be realised by using a glass cell with a thin layer of

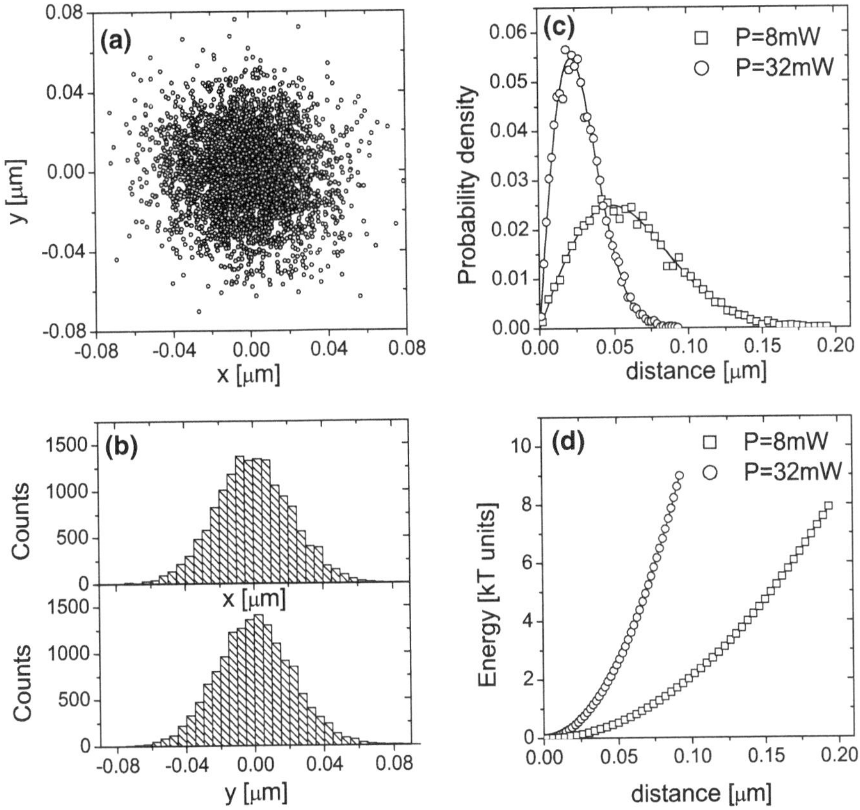

Fig. 3.6 **a** The positions of 5000 points out of 15.000 recorded during the Brownian motion of a 0.97 μm silica particle in an optical trap in homeotropic E12. **b** The corresponding histogram for each coordinate. **c** The probability density p(r) for the Brownian cloud at two power levels. The solid lines are the best fits to Eq. 3.8. **d** The reconstructed profile of the optical potential. The parabolic fit yields spring constants of the trap $k_{trap} = 9.2(1 \pm 0.05)\ pN/\mu m$ for 32 mW laser power and $k_{trap} = 2.0(1 \pm 0.05)\ pN/\mu m$ for 8 mW

indium-tin-oxide (ITO) deposited on the inner surfaces of glass. The light of the tweezers is locally absorbed by the ITO, which creates a temperature gradient with a hotspot in the optical trap and decreasing temperature in the liquid crystal. As a result of local heating, the degree of order of liquid crystal molecules (S) depends on the position and this spatial variation of the liquid-crystalline order generates forces on the particles, as first proposed by Samitsu et al. [165]. The orientational order is depressed in the hottest spot and less depressed as we move away from this spot. An example of the temperature-induced inhomogeneities of the order parameter in the nematic liquid crystal cell is shown in Fig. 3.7.

One can see from Fig. 3.7a that the illuminated part of the nematic liquid crystal shows different birefringent colours compared to the rest of the cell. Because this crystal is viewed between crossed polarisers positioned at 45° with respect to the

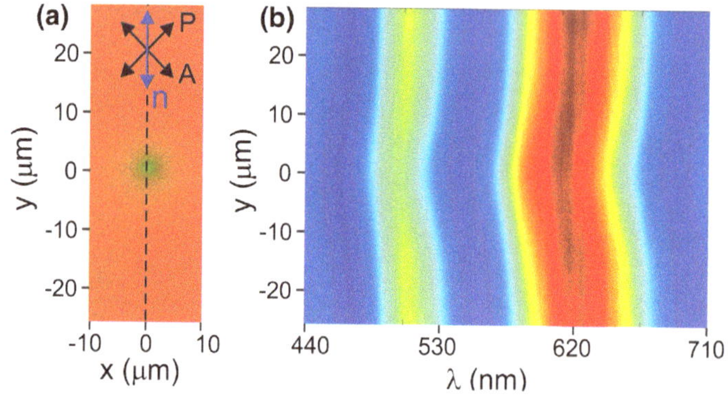

Fig. 3.7 **a** Micrograph of a 5 μm-thick layer of E12 under crossed polarisers, locally heated by the laser tweezers with a power of 30 mW. The rubbing direction is at 45° with respect to the polarisers. The hot spot becomes visible as a *green* area due to different local change of the birefringence of the heated nematic liquid crystal. **b** Wavelength dependence of the intensity of the transmitted line measured along the dashed line shown in (**a**)

rubbing direction of the cell, we can explain this colour change as a local change in the birefringence of the liquid crystal. This change of birefringence is caused by locally increased temperature due to heat dissipation of the absorbed light. The change in the birefringence must be accompanied by the change of local degree of order S, which can be measured by taking the line spectrometry, as illustrated in Fig. 3.7b. Finally, one can calculate the change of order from the measured change in birefringence. It turns out that in the hottest spot in Fig. 3.7, the order is depressed for ∼4% as compared to the surroundings. If the colloidal particle is positioned close to this "hotspot", it is attracted directly into the spot, as illustrated in Fig. 3.8.

The analysis of this trapping mechanism [166] shows that particles are attracted to the hotspot because the degree of the orientational order S is lower in the hot region, which means that the elastic constants of the liquid crystal at the hotspot are lower than in the surroundings. Let us remind that the elastic constants of a nematic liquid crystals are proportional to the square of the order parameter and therefore decrease with increasing temperature. Having this in mind, we can see that the elastic free energy of the colloidal particle is lowered if the particle moves into the region with lower elastic constants. This generates a very efficient trapping force due to the gradient of the order parameter, which is so efficient that even individual fluorescent molecules could be trapped in the hot region [166]. Similar transport of particles due to the gradient of the order parameter is reviewed by O.D. Lavrentovich [167]. We should note that the classical Soret effect, which is responsible for the transport of particles in a temperature gradient, is here excluded by the experimental evidence.

(iv) At an ultimate power level of the optical trap, the liquid crystal is locally molten and an isotropic island is created within the surrounding nematic phase of the liquid crystal, as shown in Fig. 3.9.

The isotropic island clearly shows a distinct interface with the nematic surrounding, visible as a dark grey line in Fig. 3.9a. This isotropic-nematic interface

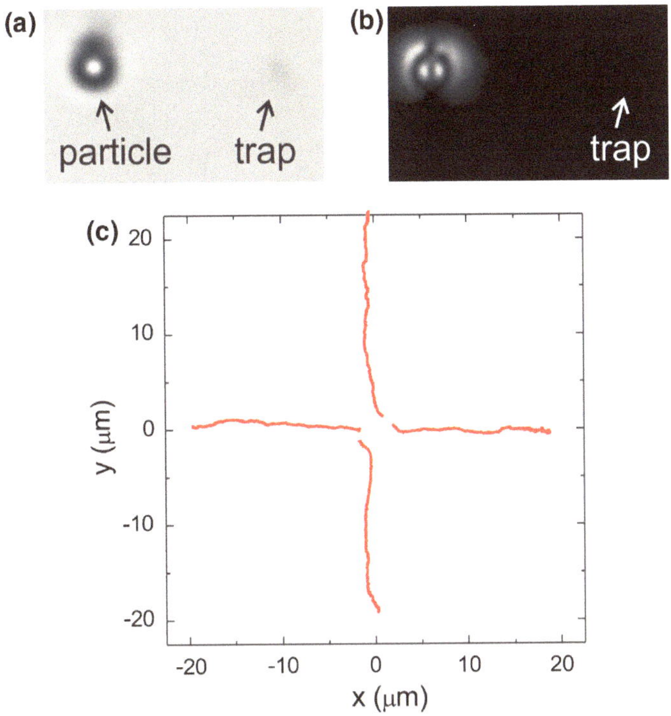

Fig. 3.8 **a** A dipolar colloidal particle is positioned close to the optical trap. Due to light absorption, the trap locally increases the temperature and decreases the order parameter. **b** The same as in (**a**), but between crossed polarizers. **c** Trajectories of the dipolar colloidal particle, as it is attracted by the S-gradient force into the trap "hot-spot"

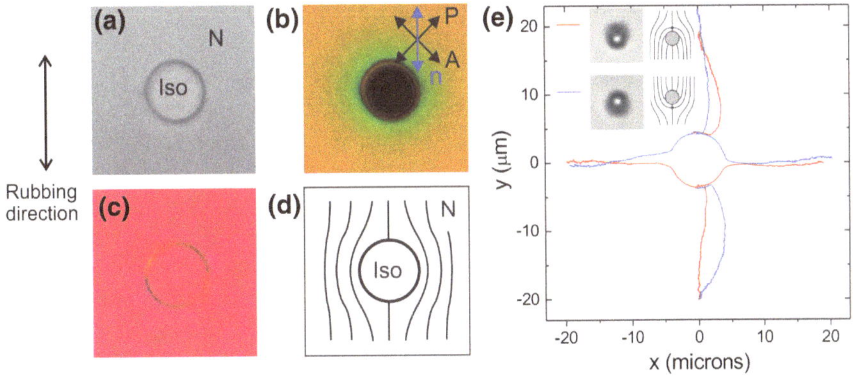

Fig. 3.9 Images of a locally molten 5CB **a** with no polarisers, **b** between crossed polarisers, and **c** with a lambda-plate added. **d** Schematic sketch of the director around the isotropic molten region. **e** Trapping trajectories for two different orientations of a dipolar, 2.32 µm colloidal particle

is optically sharp, since the isotropic-nematic transition is of first order and therefore discontinuous. At the interface, we have the coexistence of the isotropic and nematic phase, which have different refractive indices. This makes the interface visible. When observed between crossed polarisers oriented at 45° to the rubbing direction (Fig. 3.9b), the isotropic island appears black, whereas the surrounding nematic is coloured. The colour is different in a narrow rim surrounding the island, indicating a region of gradually changing birefringence from the interface out.

The anchoring of the nematic surrounding of this isotropic island can be determined with the use of the lambda-plate, as shown in Fig. 3.9c. Yellow and blue quadrants of the rim surrounding the island clearly prove the parallel anchoring of the nematic phase on the interface to the isotropic phase. This is illustrated in Fig. 3.9d, where the isotropic island appears as a kind of colloid with tangential surface anchoring, which should exhibit two boojums on each side of the island.

It is now clear from the above that this isotropic island, created by the laser trap, is acting as an artificial two-dimensional colloid with tangential surface anchoring, which of course interacts with other colloidal particles. The corresponding mechanism is quite strong, as expected for any colloidal interaction in the nematic liquid crystals.

In addition to "forbidden" optical trapping and manipulation of colloidal particles with low refractive index in high-refractive-index nematic liquid crystal, normal trapping of high-refractive-index colloids attracted a lot of attention. One of the earliest observations of the interaction of laser tweezers with liquid-crystal textures was reported by Hotta et al. [168] and Iwashita et al. [169], who manipulated defects in the lyotropic phase. Soon after that, Joudkazis et al. [170] reported the first laser tweezers manipulation and switching of nematic microdroplets, using optical tweezers. By using a circularly polarised beam, they were able to spin the nematic microdroplets at a sub-millisecond rotation time [171]. This phenomenon is due to the dielectric torque exerted by a circulating electric field and the transfer of the spin-angular momentum of circularly polarised light to the birefringent droplets. The mechanism of the angular momentum transfer was studied in detail by Gleeson and Wood [172]. Recently, similar experiments with circularly polarised light were performed on chiral nematic liquid-crystal droplets [172–174]. In these droplets, planar surface anchoring induces an onion-like spherical Bragg structure of the cholesteric droplet, which is either left- or right-handed [175]. Because chiral nematic structure is a Bragg mirror, which selectively reflects only the light with the same handedness as the chiral nematic structure, interesting opto-mechanical effects are observed. They include a selective trapping of chiral microdroplets by circularly polarised trap, pushing and stabilising chiral microdroplets by a single circular beams or a pair of counter-propagating circular beams [176].

Laser tweezers were used in low-refractive-index nematic liquid crystals to trap and measure forces between high-refractive-index colloids made of polystyrene or melamine resin. Smalyuk et al. had used laser tweezers extensively [177] and in and combination with other optical methods, such as the Fluorescence Confocal Polarising Microscopy (FCPM) and multimodal nonlinear microscopy [178, 179], to study long range order and nematic defect structures [180, 181]. Recently,

Lucchetti et al. [182] have performed a detailed analysis of colloidal trapping in nematic liquid crystals. The influence of an external electric field on the optoelastic interaction between the optical trap and colloidal particle was considered by Criante et al. [183]. Large area optoelastic manipulation of colloidal particles in lyotropic nematics using photoresponsive surface monolayers was studied by Martinez et al. [184]. Whereas in all these studies, a Gaussian beam profile was used for trapping, the group of I. Smalyuk have used Laguerre-Gaussian beam profile to optically create 3D structures in chiral nematic liquid crystal [29, 178, 185]. These chiral structures, also called "torons", carry a topological charge of −2. Multistable optical switching of triple-twisted-particles was also demonstrated. Torons will be discussed in Sect. 8.3.

3.3 Measuring Forces Between Nematic Colloids by Video-Microscopy and Particle Tracking

Although measuring the structural forces between colloidal particles in nematic liquid crystals might seem like an impossible task, it is in fact very simple and straightforward. The method for measuring forces in nematic colloids was developed in the colloidal physics and is based on the simple video-microscopy technique [186]. Consider, for simplicity, two colloidal particles in a fluid, which are attracted to each other by a force generated by the fluid itself. The two particles are first set in motion and accelerated towards each other according to the Newton's law. Since the particles are moving in a viscous fluid, the viscous drag force starts to increase as it is proportional to the particle velocity in the linear regime:

$$\mathbf{F}_{Stokes} = 6\pi R\eta \mathbf{v} \tag{3.9}$$

This relation, which is also known as a Stokes' drag force, tells us that the net force on a particle will diminish with time, until the viscous drag force due to increased velocity will be equal in strength but opposite in direction to the structural force on the particle. Finally, the velocity of the particle will be constant in time and there will be zero total force and therefore zero acceleration on the particle:

$$\mathbf{F}_{Stokes} + \mathbf{F}_{structural} = 0 \tag{3.10}$$

Looking at the Eq. 3.10, one can immediately notice that it is possible to determine the unknown structural force $\mathbf{F}_{structural}$ at a given time, if one is able to determine the Stokes' drag force $\mathbf{F}_{Stokes}$ at that time. Let us recall that the Stokes' drag force on the particle at a given time is determined by the velocity of the particle $\mathbf{v}(t)$ at that time. If one can therefore measure the velocity of the particle, it is possible to determine the unknown force at the same given time. The instantaneous velocity of the particle can be calculated if one has recorded the trajectory of the particle, because the velocity is the time-derivative of the trajectory:

$$\mathbf{v} = \frac{d\mathbf{r}}{dt} \tag{3.11}$$

The problem is therefore reduced to measuring and recording the trajectory $\mathbf{r}(t)$ of a small particle, driven by an external force, which can be done simply by taking video images of the motion of the particle. By using an appropriately high frame rate, one takes images separated in time by a very small time interval. In reality, the frame rate is between a hundred Hz and thousands of Hz and the time interval between consecutive images ranges from 10 ms to tens of µs. In a typical experiment, one records hundreds of thousands of video frames, each showing the selected particle at slightly different position.

The next question is how to extract the position of the particle in a given frame. If one is able to do this, one can obtain the complete trajectory of the particle driven by external force. The position of the particle in a video frame is determined by overlapping a graphical object, such as small circle, over the image of the particle. For example, if the particle appears as a bright circular spot on a dark background, one creates a circular object of approximately the same size as the colloid and lets the computer program find the best overlapping of that object with the bright image of the colloidal particle. As a result, one can obtain the position of the particle in a plane of the image with a surprising accuracy of the order of several nm. This accuracy depends only on the image quality and can be obtained with sufficient illumination and vibration isolation of the experimental set-up.

Let us note that the nm precision in the measurement of the position of the particle has nothing to do with diffraction-limited resolution of an optical microscope. Due to diffraction effects, even a sub-micron and nanometer sized colloidal particle will be observed as a diffraction-limited spot, determined by the optical properties of the microscope (such as the numerical aperture of the objective) and the wavelength of the light used to illuminate the object. However, the centre of gravity of this diffraction-limited image can be determined with a precision, which is much lower than the diffraction limitation value. This is typically 200–400 nm for a good optical microscope and blue-light illumination.

Once the trajectory of a particle is numerically reconstructed from the recorded video images, the velocity of the particle is determined as a ratio between the two positions and the corresponding time interval, i.e. the trajectory is numerically derived. As we now have the instantaneous velocity of the particle, we can also determine the instantaneous force, provided that we know the Stokes' drag coefficient of the particle $6\pi R\eta$. The Stokes' drag coefficient for a given particle can be determined by monitoring the Brownian motion of the same particle, when it is free to randomly move in the same fluid. This was first proposed and used in nematic liquid-crystal colloids by Loudet et al. [187], who monitored the Brownian motion of a silicone oil droplet with quadrupolar distortion of the nematic liquid crystal around it. Brownian trajectory of a 1 µm diameter droplet was video imaged by using the time interval of 80 ms and thus sampled a trajectory with 18 000 Brownian steps. After the trajectory was recorded, the length of each step was measured and the histogram of particle displacements was constructed both for the movement along and perpendicular to the director. The histograms could be well fitted to the Gaussian function, where the width of the Gaussian distribution determines the self-diffusion coefficient D for each direction. The probability P that the colloid would diffuse a certain distance δ in the time τ is:

$$P(\delta) = P_{\circ}(\tau) \cdot \exp\left(\frac{-\delta^2}{\Delta^2(\tau)}\right) \tag{3.12}$$

here, $P_{\circ}(\tau)$ is the normalisation constant and $\Delta(\tau)$ is the width of the Gaussian distribution, which is related to the self-diffusion constant D of the particle exhibiting random walk

$$\Delta^2 = 4D\tau \tag{3.13}$$

The self-diffusion constant D can therefore be determined experimentally by fitting the histograms of particle displacements to Gaussian function. On the other hand, D is given by the Stokes-Einstein relation for the random walk of spherical particles with radius R in a fluid with viscosity η:

$$D = \frac{k_B T}{6\pi\eta R} \tag{3.14}$$

It is clear from this expression that the diffusion of particles in the nematic liquid crystal is very anisotropic because of the anisotropy of liquid crystal viscosity η.

Figure 3.10 shows an example of a typical thermal motion experiment, where around 3000 positions of a selected 125 nm dipolar particle were recorded in a time-interval of 300 s. Because the nematic phase is orientationally ordered and has a rotation symmetry axis along the director, the random motion of colloidal particles is characterised by two independent diffusion coefficients $D_{\|/\perp} = k_B T/6\pi\eta_{\|/\perp} R$, with two different viscosity coefficients for motion along or perpendicular to the

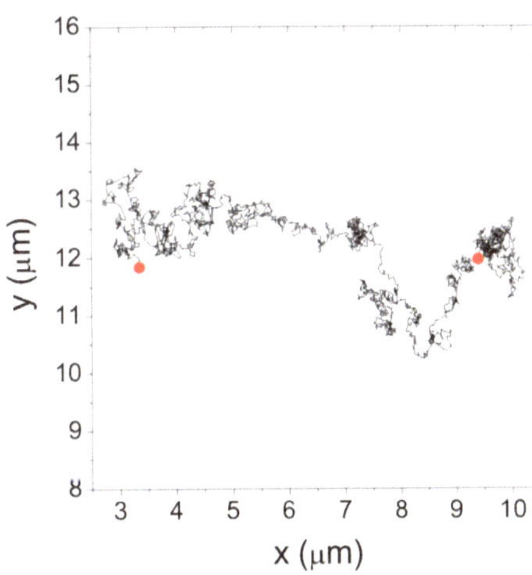

Fig. 3.10 Trajectory of a 125 nm dipolar particle in a time interval of 300 s in a planar nematic liquid crystal cell of 5 μm thickness, filled with E12

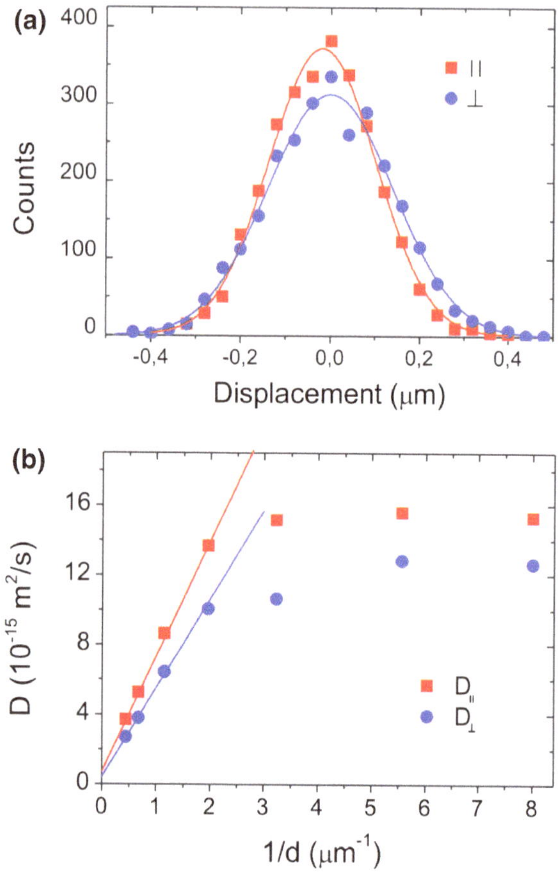

Fig. 3.11 **a** histograms of the displacements of 125 nm dipolar nanospheres between two positions separated by the time of $\tau = 640$ ms. **b** Diffusion constants of nanospheres for two perpendicular directions of motion are proportionate to the inverse colloidal diameter down to d = ~500 nm. Below this size, the diffusion coefficients of nanocolloids are quite constant

director **n**. In a Brownian motion, the mean square displacement (MSD) of a particle grows linearly with time τ. It was recently demonstrated that at short times [188], the anisotropic diffusion in a nematic becomes anomalous with the MSD growing slower or faster with time τ. The anomalous diffusion occurs at time scales comparable to the relaxation times of director fluctuations around the diffusing microsphere.

One can see from Fig. 3.11 that the diffusion coefficients for a colloidal particle in a nematic liquid crystal are different for the motion along and perpendicular to the director. In general, diffusion coefficient for dipolar colloids is larger for motion along the director (Fig. 3.11b) than in the perpendicular direction. For dipolar colloids, this difference in the effective viscosities is of the order of 10%. For quadrupolar particles, it is much lower, whereas it is much higher for microrods with an aspect ration of 4:1, as discussed in Sect. 2.4. (Tkalec et al. [99]).

When the diffusion coefficients $D_{||/\perp}$ are determined from the histograms, one can determine the two Stokes' drag coefficients for that particle:

$$6\pi\eta_{||/\perp}R = \frac{k_B T}{D_{||/\perp}} \tag{3.15}$$

Once the two Stokes drag coefficients are determined, the structural force $\mathbf{F}_{str}$ can be calculated at each point on the trajectory of the colloidal particle. From the known force, one can calculate the work done by this force along the trajectory of the particle by numerical integration:

$$A_{str} = \int_{path} \mathbf{F}_{str} \cdot d\mathbf{s} \tag{3.16}$$

The work of the structural force done along the trajectory of the particle is equal to the change of the elastic (binding) energy of the two colloidal particles. An example of the binding potential is shown in Fig. 3.12 for two dipolar nanoparticles of 125 nm diameter, which are spontaneously attracted to each other.

In nematic colloids, one is able to measure the pair-interaction potential (or forces) for very different sizes of particles, ranging from several micrometer down to tens of

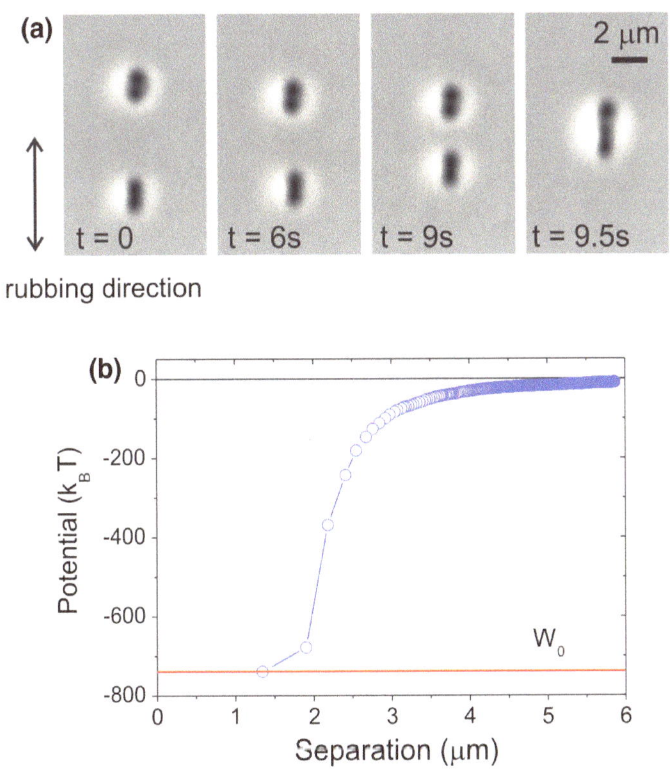

Fig. 3.12 **a** Two dipolar nanoparticles of 125 nm diameter are attracted along the nematic director into a dipolar pair. **b** The pair-potential energy as a function of separation of two 125 nm dipolar nanoparticles

nanometers. Particularly interesting video monitoring experiments were performed by Ryzhkova et al. [118, 119] on a series of silica nanospheres with the diameter from 2 μm to 22 nm. The Brownian motion of such small colloidal particles in a nematic liquid crystal is very fast, and it is an experimental challenge to track such small particles emitting very low light and moving very fast. This can be done with the dark-field video microscopy and a fast and ultra-sensitive CMOS camera. Using the state of the art technique, it was possible to determine the pair-interaction and binding energy of only tens-of-nm diameter silica particles in 5CB.

Chapter 4
2D and 3D Colloidal Crystals and Superstructures

Abstract This chapter is devoted to colloidal crystals assembled by the nematic structural forces. We give an overview the structure and properties of photonic crystals, which is followed by examples of realisation of photonic crystals form nematic colloidal crystals: 2D dipolar and quadrupolar nematic colloidal crystals and binary nematic colloidal crystals. The assembly of 3D nematic photonic crystals is described, as well as assembly of colloidal particles of very different sizes in nematic liquid crystals.

4.1 Photonic Crystals

In Chap. 2, we have seen that small spherical particles dispersed in a nematic liquid crystal exhibit new types of forces, which are quite different from the forces which are common for ordinary colloids. Whereas in water-based colloids, the forces which are responsible for the interaction of colloidal particles are mainly of electric origin, the colloidal forces in nematic colloids are caused by the elastic distortion of long-range-ordered nematic fluid. The differences in the nature of forces result in differences in their strength. Whereas in water-based colloids, the strength of electrostatic and van der Waals force depends on the surface charge and the differences in refractive indices of colloids and the fluid, the forces between nematic colloids depend on the size of inclusions, elasticity of the liquid crystal and the surface anchoring strength.

Unlike completely isotropic electrostatic and van der Waals force, structural forces in nematic colloids are strongly anisotropic. They also are accompanied by topological features, including topological defects in the nematic solvent, which are absent in classical colloidal systems. The strength of anisotropy and the topological nature of the particle forces in nematic colloids raise an interesting question about what kind of colloidal crystals do this forces form. This question is not only interesting per se, but also because of its implications for technology. One wonders whether the forces in nematic colloids could be used for self-assembly or directed-assembly of regular colloidal crystals, which could be used as photonic crystals.

I. Muševič, *Liquid Crystal Colloids*, Soft and Biological Matter,
DOI 10.1007/978-3-319-54916-3_4

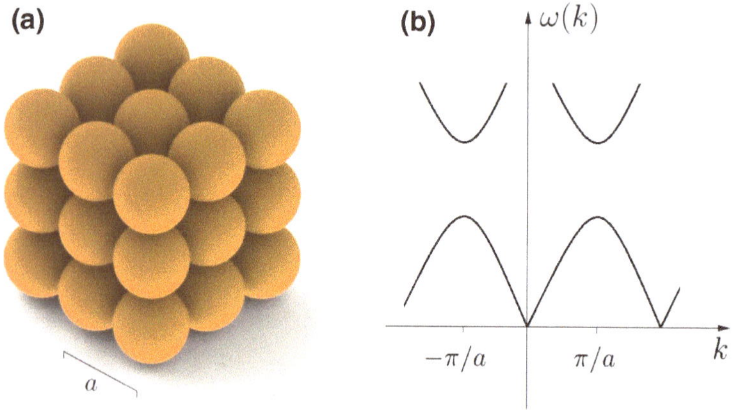

Fig. 4.1 **a** Schematic 3D view of a cubic photonic crystal, which is formed of regularly spaced dielectric microspheres with a crystal lattice a of the order of the wavelength of visible light. **b** Dispersion relation for electromagnetic waves propagating through a photonic crystal shows periodic structure and forbidden frequency bands. These bands are due to the Bragg scattering and interference of light. Image courtesy E. Zupanič

Photonic crystals were invented in 1989 when Yablonovitch and Gmitter [189] were considering the optical properties of regular crystal structures formed of dielectric and transparent objects of the size of the wavelength of light. An example of a photonic crystal is shown in Fig. 4.1.

Optical properties of a photonic crystal can be understood considering the analogy with the motion of electrons and the corresponding electron energy spectrum in a regular crystal structure. We know that the dispersion relation for electrons in periodic potential is periodic in the reciprocal, **k**-space and exhibits forbidden energy levels between the valence and conductive bands. These forbidden energy levels are due to the interference of electron wave-function and Bragg-scattering of electrons on a periodic potential formed by a crystalline lattice.

An electromagnetic wave experiences the same interference effects when propagating through a regular crystal formed of transparent dielectric objects. In analogy to the energy levels of electrons in solid crystals we expect that the dispersion relation for electromagnetic waves propagating along different direction in a photonic crystal $\omega(\mathbf{k})$ will have the same structure. There will be a "valence and conductive" bands for light in a photonic crystal, separated by a forbidden frequency band. This means that no light can propagate in a photonic crystal if its frequency is within this forbidden frequency gap. If not propagating, such light will be reflected from a photonic crystal, which will act as a perfect and lossless Bragg mirror.

Bragg mirroring of photonic crystals is nowadays used in design and manufacturing of photonic microcircuits to confine the light and allow its flow through "conducting channels" confined between Bragg mirrors. Fascinating photonic structures can be assembled simply by etching thin dielectric materials, as the one shown in Fig. 4.2.

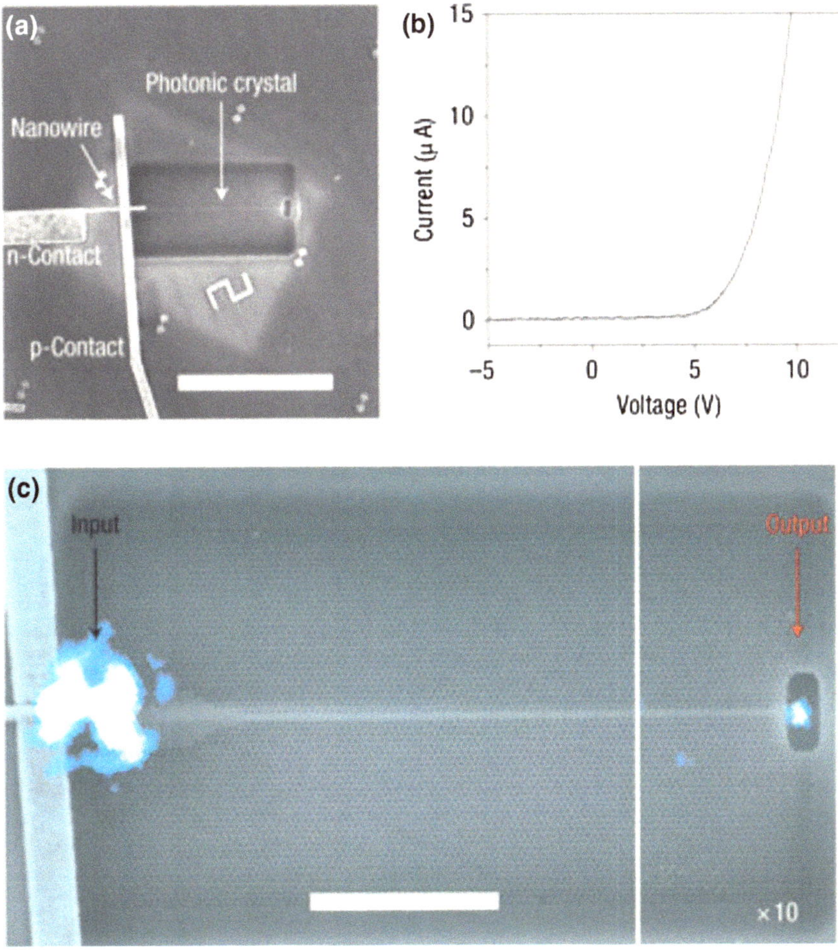

Fig. 4.2 The light from a nanowire light source is coupled into a photonic crystal waveguide. **a** SEM image of the nanowire light-emitting diode, which faces the photonic crystal waveguide. The waveguide terminates with a large hole, which serves as the output scattering site, more clearly visible in (**c**). Scale bar 20 μm. **b** Current-voltage relation for the nanowire light-emitting diode. **c** Superposition of the emitted light and the SEM image of the waveguide. The white irregular object on the left side of the waveguide is where the light is coupled (and partially scattered) into the waveguide (i.e. Input). The small bright object in the hole on the left side is the light transmitted along the photonic crystal waveguide (i.e. Output). Scale bar 5 μm. Reprinted by permission from Macmillan Publishers Ltd: Hong-Gyu Park, Carl J. Barrelet, Yongning Wu, Bozhi Tian, Fang Qian and Charles M. Lieber, Nature Photonics **2**, 622–626 (2008); copyright (2008)

In this chapter, we shall overview the assembly of 2D and 3D colloidal crystals made of spherical microparticles in a nematic liquid crystal, which could potentially be used as photonic crystals. For perpendicular surface anchoring, these particles appear in two different topological forms: as elastic dipoles and elastic quadrupoles.

It is possible to obtain both types of colloids, provided they are confined in a precisely thickness-tuned planar nematic cells. For thickness below the critical thickness, the colloidal particles appear as quadrupoles, whereas in thicker parts, they appear as elastic dipoles. In the intermediate interval of thicknesses around the critical thickness, we have both types of particles available. In this interval of thickness, it is therefore possible to assemble purely dipolar colloidal crystals, purely quadrupolar colloidal crystals or a mixed type of colloidal crystals, consisting of elastic dipoles and quadrupoles. Of course all these crystals will be of a planar type, i.e. consisting only of a single layer of colloids, confined in a nematic crystal between two tightly spaced glass plates. The last subsection of this chapter describes the assembly of nematic colloidal crystals in 3D. As expected, this is a much more difficult experimental task. In thick nematic cells, one needs to control the z-position (i.e. perpendicular to the glass surface) of colloidal particles in addition to viewing and presenting 3D distribution of colloidal particles. This is possible by using confocal optical imaging, which allows for reconstruction of 3D colloidal crystals in the nematic liquid crystal.

When discussing possible 2D colloidal structures, consisting of dipolar, quadrupolar or both types of colloidal particles in the nematic liquid crystal, it is useful to summarise the anisotropy of pair interactions for each type of colloidal particles. This is illustrated in Fig. 4.3, which shows the angular dependence of colloidal pair interactions for dipole-dipole, dipole-quadrupole and quadrupole-quadrupole interactions.

It is evident from Fig. 4.3 that dipole-dipole interaction results primarily in the formation of linear chains. On the other hand, the pure quadrupole-quadrupole interactions allows more freedom of colloidal assembly, which results in kinked chains or

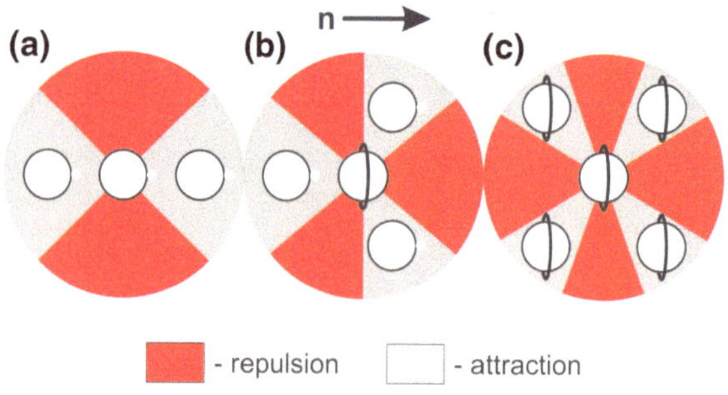

Fig. 4.3 Schematic representation of angular dependence of colloidal pair interactions. **a** Dipole-dipole interaction, **b** dipole-quadrupole interaction and **c** quadrupole-quadrupole interaction. The *white dots* are hyperbolic hedgehog defects and the *black rings* are Saturn ring defects. The central particle is the referential one and the remaining particles are placed along the direction of the strongest attraction to the referential particle. In the *red-shaded* sectors, pair-wise interaction is repulsive. Image courtesy A. Nych

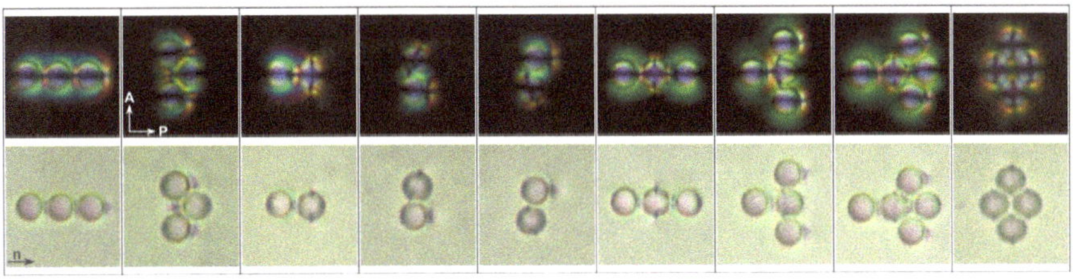

Fig. 4.4 Experimental images of several basic arrangements of 4 μm colloids in a planar nematic cell between crossed polariser (*top row*) and in unpolarised light (*second row*). Image courtesy A. Nych

small quadrupolar clusters. Finally, one can see from the mixed interaction in Fig. 4.3 that binary nematic colloids, consisting of dipoles and quadrupoles, should exhibit a huge variety of 2D colloidal motives.

Some examples of colloidal clusters assembled from pure dipoles, pure quadru poles or mixed dipole-quadrupole structures are shown in Fig. 4.4. It is clear already from this very simple presentation that the variety of the mixed type colloidal crystals is huge.

4.2 Two-Dimensional Dipolar Colloidal Crystals in Nematic Liquid Crystals

2D dipolar and quadrupolar colloidal crystals are assembled by using laser-tweezers-assisted manipulation of colloidal particles in a planar cell filled with a nematic liquid crystal [73, 84, 190]. If the thickness of the cell is much larger than the diameter of colloidal particle with perpendicular surface anchoring, dipolar colloidal particles are spontaneously formed. In a cell, we observe spontaneous formation of chains of dipolar colloidal particles or colloidal aggregates with spontaneously ordered parts, as shown in Fig. 4.5.

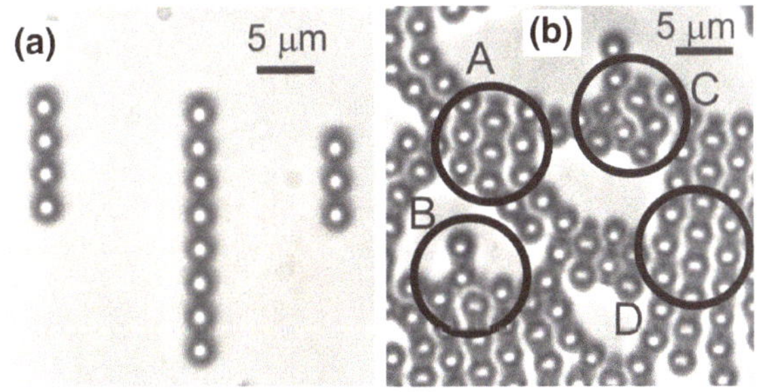

Fig. 4.5 **a** Dipolar colloidal chains are spontaneously formed in nematic colloids. **b** In a dense sample, dipolar colloidal clusters are formed, which exhibit several different kinds of local colloidal binding

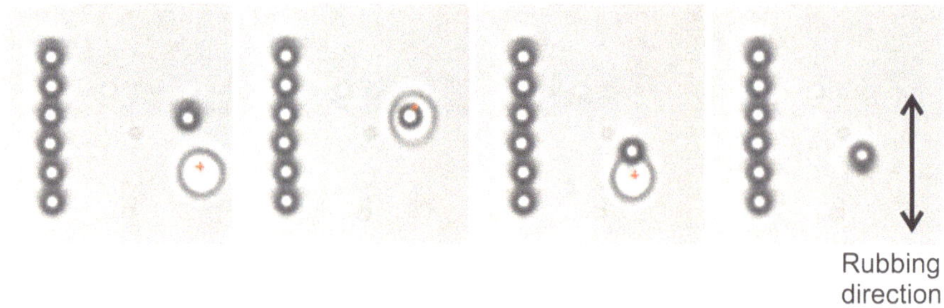

Fig. 4.6 The orientation of the topological dipole can be reversed by locally melting the nematic liquid crystal and dragging the molten crystal in a preferred direction for the new orientation of the dipole

In order to control the assembly of dipolar colloids into chains and 2D clusters, one should be able to control the position of the hedgehog point defect on a selected particle. This can be done with the help of laser tweezers and is shown in Fig. 4.6, where the direction of the topological dipole is reversed. This is done by increasing the power of the laser tweezers, which results in local melting of the nematic liquid crystal. The tweezers focus is positioned on the colloidal particle so that the isotropic phase symmetrically surrounds the particle. After the isotropic island is dragged "downwards" in Fig. 4.6 with the use of tweezers, the topological defect appears on the opposite side of the particle after the tweezers is switched off.

Reversal of the dipole orientation with the melting-and-dragging technique can be understood if we remember that there are two energetically equivalent positions of the hyperbolic hedgehog defect. This is because of the invariance of the director sign, which leaves two possible and equivalent positions of the point defect close to the particle. By melting the liquid crystal and dragging the isotropic melt, the point defect is preferably deposited on the same side of the colloid where the isotropic melt was positioned last. Having control over the orientation of the dipolar colloidal particle, we are able to selectively add colloidal particles to the pre-existing colloidal clusters, manipulate the particles, and measure the forces between deliberately positioned pairs of particles. An example is shown in Fig. 4.7, where the formation of colloidal chains is shown. The colloidal particles were prepared with their dipoles oriented in the same direction and were added one by one to form a chain of dipoles oriented along the far-field director.

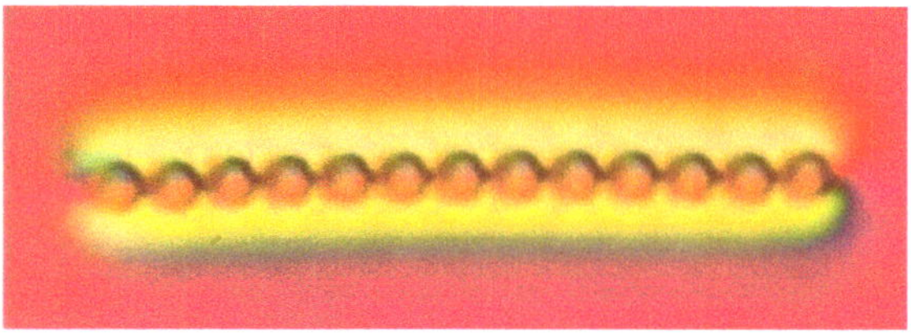

Fig. 4.7 Dipolar chains of colloidal particles are formed by laser-tweezers assisted manipulation of individual colloidal particles and their incorporation into colloidal chains

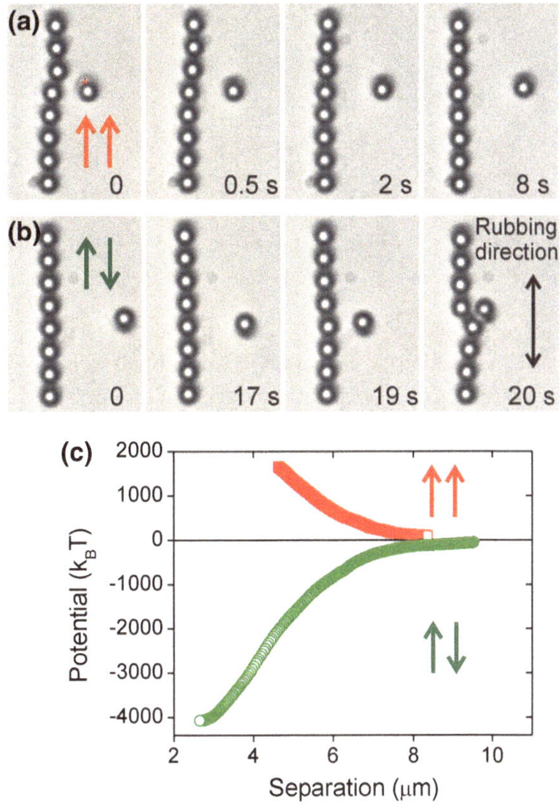

Fig. 4.8 The interaction between a dipolar colloidal chain and an isolated dipolar colloidal particle. The *arrows* indicated the direction of the topological dipoles. **a** For parallel orientation of the topological dipoles and side position, the colloidal particle is repelled from chain. **b** If the direction of the topological dipole is reversed, the colloidal particle is strongly attracted to the chain. **c** Using the particle tracking software, the energies of repulsion (*red squares*) and attraction (*green circles*) of a single colloidal particle from the dipolar chain are calculated. Note the energy scale and the strength of interaction

After formation of colloidal chains where all topological dipoles of particles in the chain are pointing in the same direction, one is wondering how regular 2D clusters could be formed from such a chain? To this aim, we measure the interaction of an additional colloidal particle, which is positioned close to the chain, as shown in Fig. 4.8. There are two possible directions of its dipole. In Fig. 4.8a the dipole of the neighbouring particle is pointing along the direction of the dipoles, in the chain. One can clearly see that this particle is repelled from the chain, after it is released from the optical trap. If the direction of the topological dipole of this extra particle is reversed, it is clearly attracted to the chain, as demonstrated in Fig. 4.8b. Using the technique of video microscopy, described in Chap. 3, one can determine the interaction force and the interaction potential between an isolated dipolar colloidal particle and a colloidal chain of 8 dipolar particles, as shown in Fig. 4.8c. The interaction potential is extremely strong, reaching more than 4000 k_BT for antiparallel orientation of dipoles and strong repulsion of 1000 k_BT at one colloidal diameter separation.

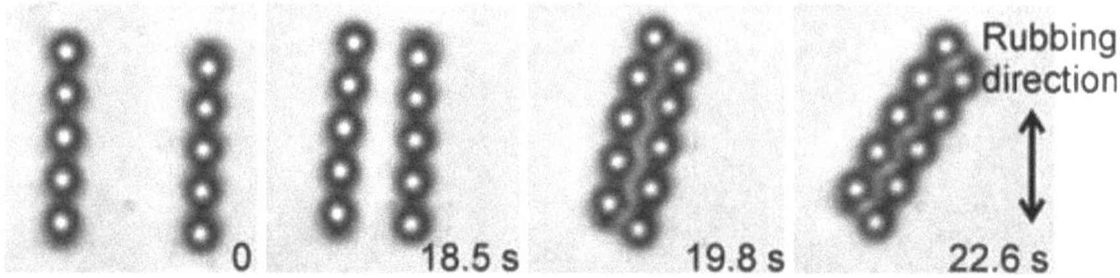

Fig. 4.9 Attraction between a pair of dipolar nematic colloidal chains with reversed directions of their dipoles

The observation that a singular dipolar colloidal particle is strongly attracted to an already assembled dipolar colloidal chain is an important step towards directed assembly of macroscopic dipolar colloidal crystals in 2D. It is natural to anticipate that two dipolar chains with antiparallel topological dipoles should attract each other, whereas two chains with parallel dipole moments should repel. This is indeed observed in the experiments where two dipolar colloidal chains are brought close together using the laser tweezers and then released, as shown in Fig. 4.9.

In a real-time experiment, one can see clear attraction between the two chains under the microscope. The chains start moving towards each other from a separation of tens of μm, as shown in Fig. 4.9. Upon approaching, they start to tilt away from the rubbing direction until they firmly attach to each other and form a robust, elongated and tilted crystallite. Analysing the forces between two colloidal chains of five particles each, one finds the binding energy exceeding 10 000 k_BT for 2.32 μm diameter silica spheres. The binding energy is extremely large and is of the order of 1000 k_BT per colloidal particle.

There is an interesting geometric feature observed during the assembly of colloidal chains - an odd - even effect, illustrated in Fig. 4.10. A crystallite formed of an even number of colloidal chains is always tilted with respect to the far-field nematic director; its tilt is symmetric and as high as 30° for two colloidal chains. The magnitude of the tilt decreases with increasing number of chains and is absent for an odd number of colloidal chains. The origin of this odd-even effect in the tilt of the colloidal chains is understood by considering the elastic distortion of the nematic liquid crystal around colloidal chains, as illustrated in Fig. 4.10e.

In case of an odd number of chains, there is a region of strong elastic distortion located at both ends of the chains. It is energetically more favourable to relax the strong distortion by a slight rotation of the crystallite. In case of an odd number of chains, the distortion is symmetric, which leaves the crystallite directed exactly along the far-field director. This understanding of the mechanism of dipolar colloidal assembly allows for directed assembly of large-scale dipolar colloidal crystals. It could be done either by preparing 2D blocks of particles by using the laser tweezers or assembling the 2D crystal particle by particle. The process results in large-size 2D dipolar colloidal crystals, which are formed of pairs of antiparallel dipolar chains, as illustrated in Fig. 4.11. The 2D dipolar colloidal crystal has a general parallelogram unit cell with the lattice constants $a = (2.95 \pm 0.03)$ μm, $b = (2.84 \pm 0.02)$ μm,

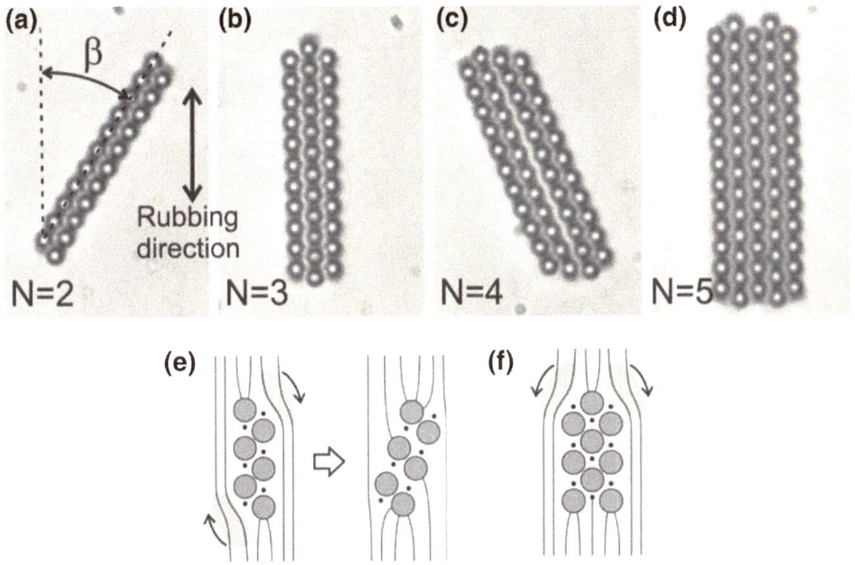

Fig. 4.10 **a–d** Odd-even effect in the tilt of 2D dipolar nematic colloidal assemblies. **e** For an even number of dipolar colloidal chains, regions of strong elastic deformation are present when the assembly is oriented along the rubbing direction. This strain is released by slight rotation of the chains. **f** For an odd number of dipolar colloidal chains, the elastic distortion is symmetric and the overall elastic torque is balanced. Any rotation does not decrease the elastic energy

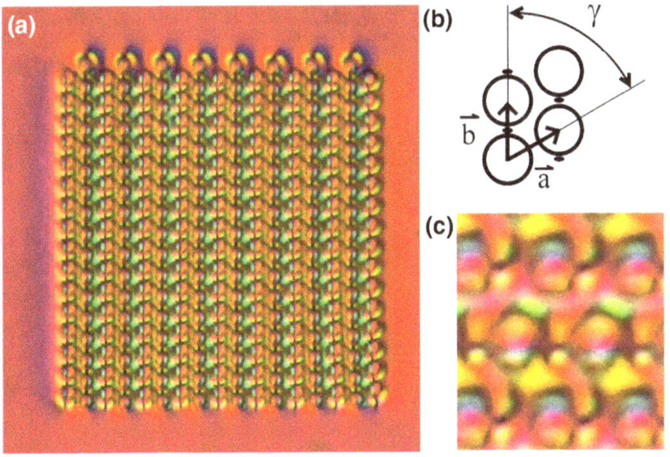

Fig. 4.11 **a** An example of a 2D dipolar nematic colloidal crystal assembled by the laser-tweezers manipulation. The crystal consists of 280 silica spheres of 2.32 μm diameter and is stable for more than a year at laboratory conditions. **b** The lattice vectors and the unit cell of the dipolar colloidal crystal shown in **a**. **c** Enlarged detail of the colloidal lattice

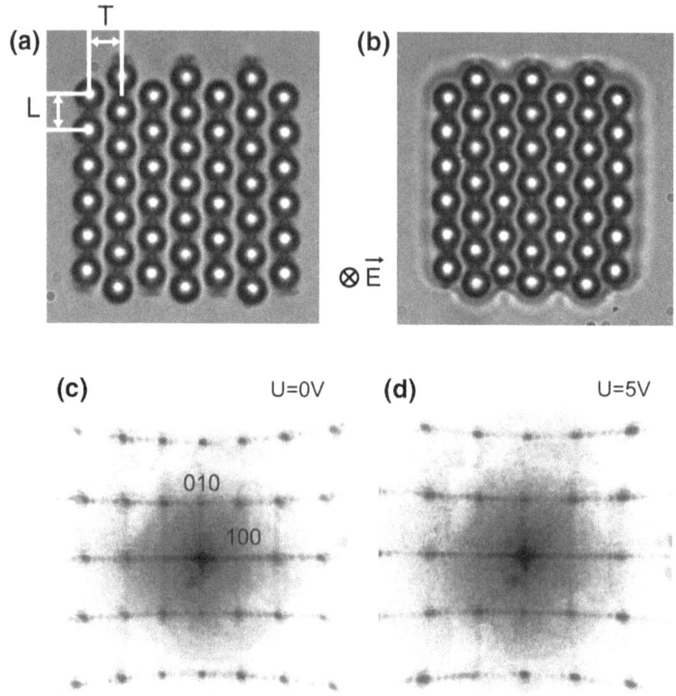

Fig. 4.12 **a**, **b** 2D dipolar colloidal crystal of 4.7 μm diameter silica microspheres shrinks upon the application of 5 V across the 9.9 μm thick planar nematic cell of 5CB. **c**, **d** Diffraction of the 532 nm laser light from 2D dipolar colloidal crystal shown in (**a**), (**b**) at 0 V (**c**) and 5 V (**d**)

and $\gamma = 61^\circ \pm 1^\circ$, as indicated in Fig. 4.11b. This 2D colloidal crystalline structure is extremely robust against external perturbations and remains untouched for several months. The crystal can also be grabbed by laser tweezers and slowly moved to a new position as a single whole, which further illustrates its extreme robustness.

Such a 2D dipolar nematic colloidal crystal is not only very robust, but also responds to an external electric field applied perpendicularly to the plane of the crystal [191]. The 2D dipolar crystal is inherently anisotropic, as it is formed of strongly bound colloids forming a colloidal chain, whereas the chain-to-chain interaction is somewhat weaker and the crystal is correspondingly softer in this lateral direction. This is indeed reflected in the elastic response of the crystal to an external electric field, as illustrated in Fig. 4.12a, b. The crystal shrinks as a whole already at an electric field of 0.2 V/μm. The shrinking is stronger in a transverse direction, indicating softer elasticity between two neighbouring chains. At approximately 0.5 V/μm, the crystal shrinks for 20% in this transverse direction, whereas it shrinks for only 2% along the chains. This is because there are topological defects in between the colloidal particles forming the chains and because they are very resistant to external field. This electric-field response of the 2D dipolar colloidal crystal can be nicely observed via the diffraction of light in such a crystal in Fig. 4.12c, d. The square-like diffraction pattern changes upon the application of the electric field and the crystal acts as a tunable diffraction grating.

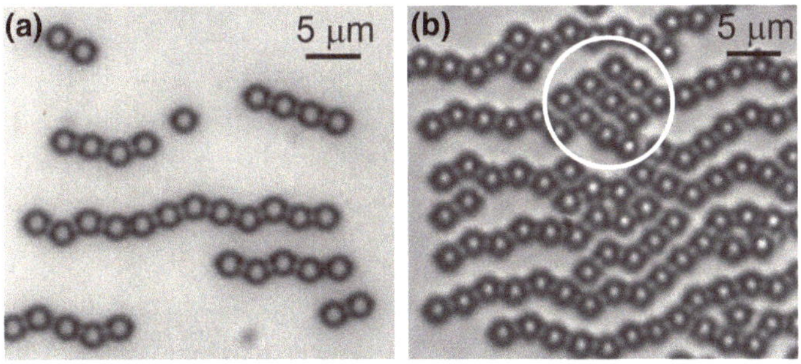

Fig. 4.13 A majority of self-assembled quadrupolar colloidal clusters are just kinked chains, shown in (**a**). However, in some parts of the sample, spontaneously ordered 2D quadrupolar crystallites are observed, such as the one in the encircled region

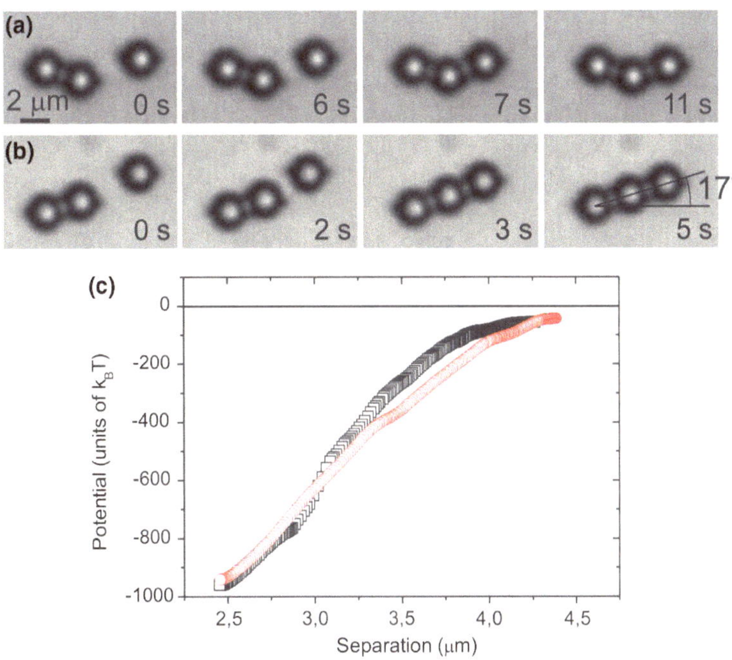

Fig. 4.14 **a**, **b** Quadrupolar colloidal chains can grow in a form of kinked (**a**) or straight chains (**b**). **c** The quadrupole–quadrupole interaction as a function of separation between the particles. The *black squares* correspond to (**a**), the *red circles* correspond to (**b**)

4.3 Two-Dimensional Quadrupolar Colloidal Crystals in Nematic Liquid Crystals

The interaction of quadrupolar nematic colloids [84] are quite different and much weaker compared to dipolar nematic colloids [73]. Roughly speaking, the quadrupole-quadrupole binding energy is an order of magnitude smaller than the dipole-dipole binding energy. As a consequence, quadrupolar colloidal crystals are quite fragile and easily perturbed by external forces caused by, for example, liquid crystal flow and temperature change. When the sample of a quadrupolar colloidal crystal is prepared with the same particles that form dipolar colloids in thicker cells, one can observe clusters of quadrupolar particles, which are spontaneously formed in the thinner parts of the cell, as shown in Fig. 4.13.

In order to be able to assemble 2D quadrupolar colloidal crystals, one first analyses the rules of attraction and repulsion of an individual quadrupole with a quadrupolar chain or cluster. Several example of the interaction of quadrupolar colloids, resulting in the growth of kinked or straight chains or smaller crystallites, is shown in Fig. 4.14. Figure 4.14a, b shows a spontaneous assembly of a quadrupolar colloidal particle to

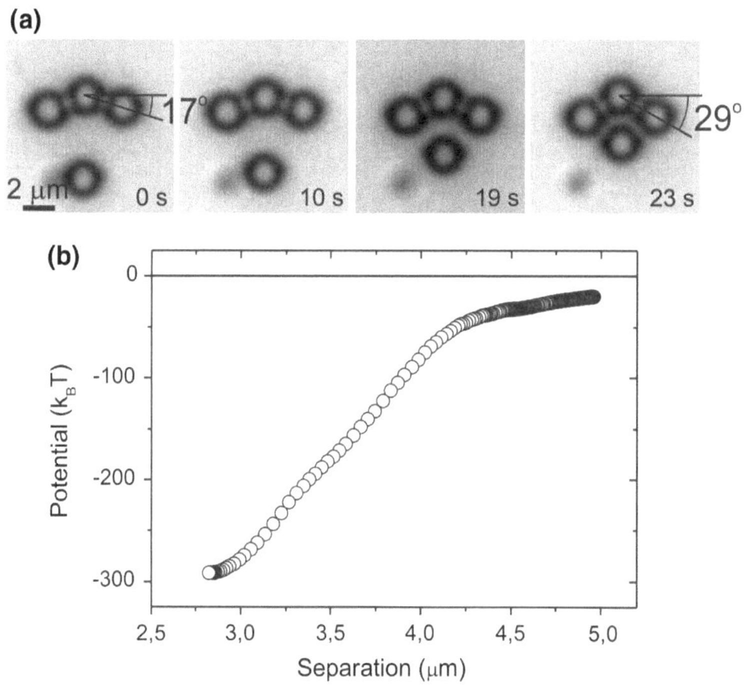

Fig. 4.15 **a** An isolated quadrupolar particle is attracted laterally to an already formed quadrupolar chain. **b** The lateral quadrupolar interaction is weaker and amounts to 1/3 of the head-to-head quadrupolar interaction

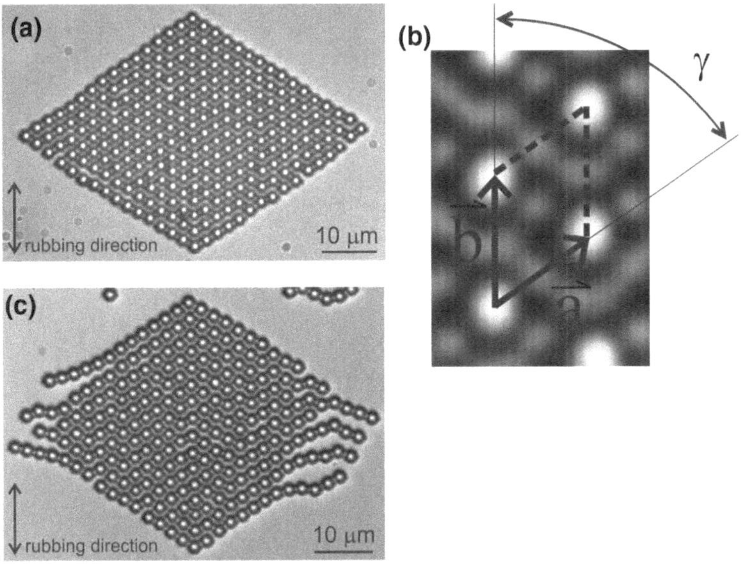

Fig. 4.16 **a** Large-scale quadrupolar nematic colloidal crystal, assembled by using the laser tweezers. **b** The lattice vectors for 2.32 μm colloidal particles in a 2.7 μm planar nematic cell are $a = (2.69 \pm 0.04)$ μm, $b = (3.01 \pm 0.05)$ μm and $\gamma = (56 \pm 1)°$. **c** Unlike the dipolar colloidal crystal, the quadrupolar crystal is fragile and sometimes spontaneously disrupts over a period of a month

an already assembled pair of quadrupolar colloidal particles. The structural force is the strongest for a slightly off-axis attraction of the particle, which results in the formation of either kinked chains or tilted straight chains.

One can see from Figs. 4.14 and 4.15 that the quadrupolar interaction is also anisotropic. Quadrupoles have the preference of forming straight and tilted, or kinked chains, whereas any lateral interaction of an additional particle is much weaker. This is reflected in the appearance of macroscopic 2D quadrupolar colloidal crystal, assembled from hundreds of particles, shown in Fig. 4.16a. Small irregularities are clearly visible, also reflecting the history of formation of such a crystal, which was assembled "by hand" using the laser tweezers manipulation of individual particles. The unit cell of a quadrupolar colloidal crystal is an oblique one, corresponding to a slightly distorted hexagonal order, as shown in Fig. 4.16b.

4.4 Numerical Simulations of Two-Dimensional Quadrupolar Colloidal Crystals in Nematic Liquid Crystals

In line with experimental investigations of nematic colloidal crystals in 2D, theoretical analyses were performed by M. Ravnik and S. Žumer to understand the nature of the liquid-crystal ordering in nematic colloidal systems [5, 192]. The Landau-de

Gennes (LdG) formalism with full tensorial description had to be used in these cases because of the smallness of the colloidal particles and the presence of a multitude of topological defects in these systems. The degree of nematic ordering depends strongly on the position, which means that the director field has to be considered as a fully tensorial field. The order parameter tensor is a 3×3 symmetric traceless matrix whose invariants are used to construct the free energy F of the nematic liquid crystal, constrained by the surfaces of colloidal particles and the surfaces of the cell:

$$
\begin{aligned}
F = &+ \frac{1}{2} L \int_{LC} \left(\frac{\partial Q_{ij}}{\partial x_k}\right)\left(\frac{\partial Q_{ij}}{\partial x_k}\right) dV \\
&+ \int_{LC} \left(\frac{1}{2} A(T) Q_{ij} Q_{ji} + \frac{1}{3} B Q_{ij} Q_{jk} Q_{ki} + \frac{1}{4} C (Q_{ij} Q_{ji})^2\right) dV \\
&+ \frac{1}{2} W \int_{Surf.Col.} (Q_{ij} - Q_{ij}^0)(Q_{ji} - Q_{ji}^0) dS \quad .
\end{aligned}
\tag{4.1}
$$

The first term in Eq. 4.1 describes the increase of the free energy due to spatial inhomogeneities of the nematic orientation and order. The second term represents the contribution of the bulk nematic free energy due to the magnitude of the nematic order. The third term represents the interaction of the nematic liquid crystal with the surfaces of the colloidal particles and is of the standard Rapini-Papoular type. For the sake of numerical simplification, the elastic constants are taken to be equal (L), which is the single elastic constant approximation. $A(T)$, B, and C are the conventional nematic material constants, which describe the isotropic-nematic phase transition. W is the strength of the surface anchoring and Q_{ij}^0 is the order parameter preferred by the surface. The confining surfaces are taken into account by prescribing the average orientation of the nematic molecules parallel to each other on both surfaces. The order parameter at the confining surfaces is equal to the bulk order parameter. These three energy terms therefore cover all three fundamental liquid-crystal phenomena relevant to the experiments: elasticity, possible formation of topological defects, and finite interaction of a liquid crystal with the surfaces of the colloids.

We shall not enter further into the discussion of the numerical techniques necessary to obtain the equilibrium order parameter tensor profile, meaning that the tensor should be fully defined at all points of a 3D mesh describing the sample. In short, the free energy F is minimised according to the Euler–Lagrange formalism, which ends with two separate equations describing the bulk behaviour of the order parameter and the surface or boundary conditions. One obtains a set of 6 coupled non-linear partial differential equations that need to be solved together with a constraint that the Q-tensor has to remain traceless. One usually applies numerical algorithms based on explicit Euler finite difference relaxation algorithm on a cubic mesh with suitable numerical parameter values. An additional requirement is that the maximum cell thickness can be set to 2 μm and the largest feasible diameter of the colloidal particles is 1 μm. Larger dimensions substantially increase the time needed for the minimisation of the free energy. Furthermore, one usually starts with a chosen colloidal configuration, such as an orthorhombic unit cell, with a suitable number of

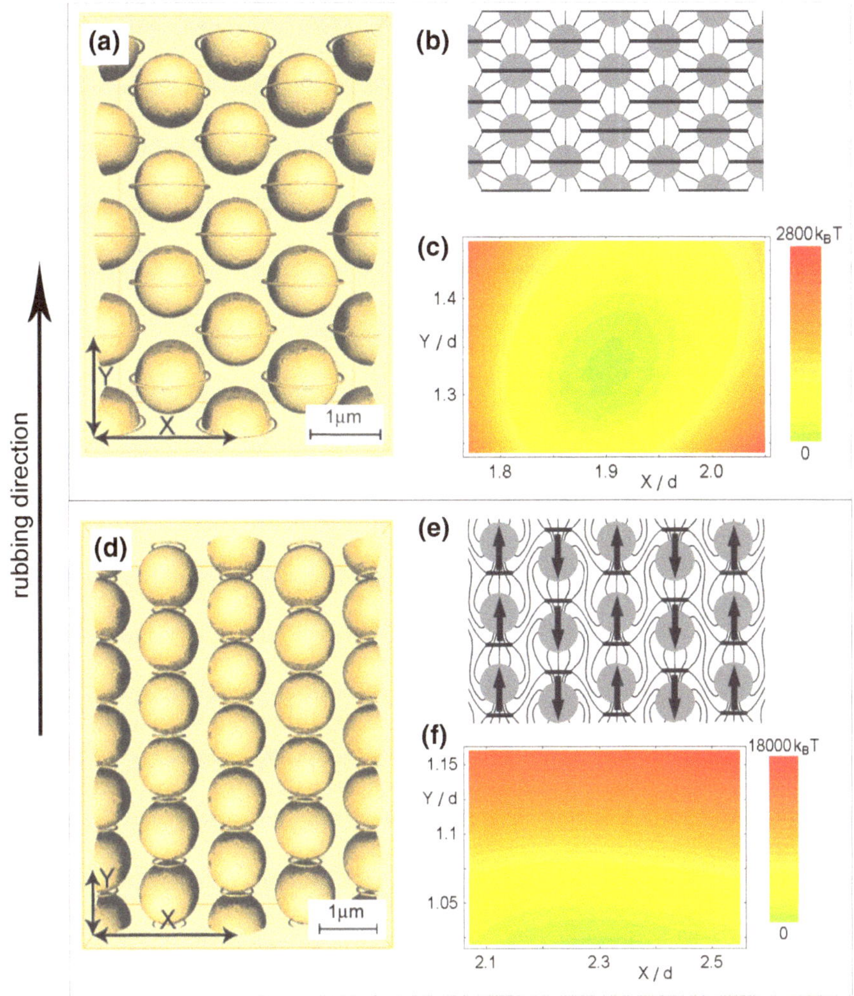

Fig. 4.17 Stable quadrupolar and dipolar nematic colloidal crystals in two dimensions. **a** Crystal with quadrupolar symmetry. The Saturn defects are localised and surround the colloids. The image represents iso-surfaces, where the magnitude of order parameter is $S = 0.45$. **b** Schematic view of the quadrupolar director field, with the heavy line representing the Saturn defect. **c** Free energy landscape of F, calculated for a unit cell with a total of two quadrupolar colloids bound by Saturn defects. **d** Stable 2D crystal with dipolar symmetry. The hyperbolic hedgehog defects have opened into small rings because of the strong confinement. Again, a surface with $S = 0.45$ is shown. **e** Schematic view of a 2D stable crystal of dipolar nematic colloids. The *arrows* indicate the direction of the topological dipole; note the antiferroelectric arrangement of dipoles. *Thin lines* represent the local direction of nematic molecules. **f** Calculated F for a unit cell with a total of two dipolar colloids

colloidal particles. At such parameter values of the starting geometry, several equilibrium order parameter tensor profiles and defect configurations are found as a solution of the Euler–Lagrange equations, and they depend primarily on the initial conditions, chosen at the start of simulations. The starting tensor profile is usually expanded in multipole series and the corresponding results, giving stable configurations of a 2D dipolar colloidal crystal are shown, in Fig. 4.17 for elastic quadrupoles and dipoles.

Figure 4.17a, b show one of the stable solutions for the 2D nematic quadrupolar crystal, where the director field with local quadrupolar symmetry is periodic in two dimensions. The Saturn rings of each colloidal particle are clearly localised around each particle, and together they form a hexagonal lattice. The binding force between the colloids comes from sharing of the elastically distorted regions around individual colloids. The calculated lattice constants are in good agreement with the experiments. This numerically calculated crystal could also artificially be stretched in the x and y directions and the corresponding free energy is constructed for these two distortions. One can clearly see the minimum of the free energy (green area in Fig. 4.17c), which proves that quadrupolar colloids are bound collectively in two dimensions by liquid-crystal Frank elasticity.

A stable 2D colloidal structure with local dipolar symmetry of the director field is presented in Fig. 4.17d. Because of the smallness of the numerical system, which results in strong confinement of colloids, the hyperbolic hedgehog defect does not remain intact but "opens" into a small loop with the same topological properties. Because of this opening, the separation along the dipolar chains is reduced and the anisotropy of the lattice is enhanced. The calculated lattice constants of the oblique 2D dipolar lattice are in reasonable agreement with experiments. The correspondence between the numerical predictions and experimental results are better for the quadrupolar 2D crystal, which is due to the numerical limitations preventing calculations for thick samples with bigger colloidal particles.

4.5 Binary Colloidal Crystals in Nematic Liquid Crystal: Mixture of Dipoles and Quadrupoles

When considering the assembly of dipolar and quadrupolar colloidal particles, one faces a huge variety of possible combinations of binding dipoles and quadrupoles [87, 88]. Besides choosing either a dipole or a quadrupole, one is free to chose the direction of the dipolar particles where two orientations of dipolar particles are possible. Furthermore, dipoles can be assembled not only in straight chains of parallel dipoles, but also in kinked chains of anti-parallel dipoles which grow in a direction perpendicular to the rubbing direction. Figure 4.18 shows schematic representation of clusters of dipolar, quadrupolar or mixed-type colloids together with corresponding "chemical" formulas which help to organise the schemes of all possible bindings of dipoles and quadrupoles.

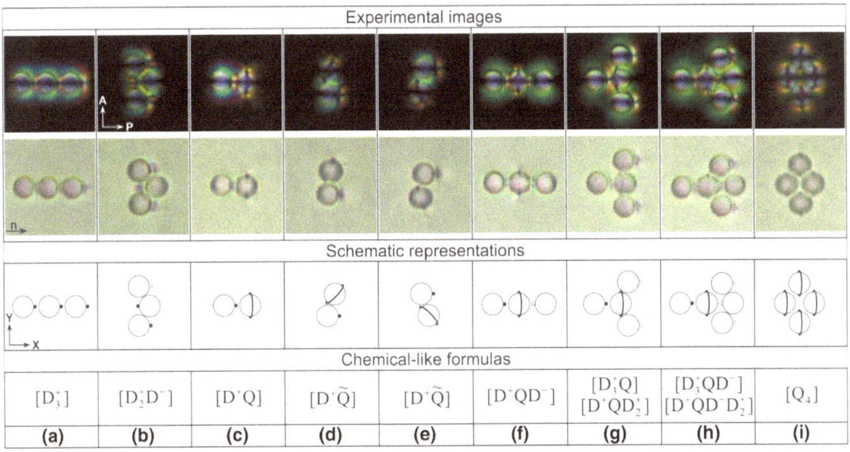

Fig. 4.18 Experimental images and schematic representation of several basic arrangements of dipoles and quadrupoles together with the corresponding "chemical-like" formula. We take the dipolar moment of the particle as positive, when the hyperbolic hedgehog defect (*black dot*) points along the x axis, i.e. to the right on all images

"Chemical-like" formulas describing the structure of colloidal clusters consisting of dipoles and quadrupoles were introduced in order to organise the classification of all possible colloidal 2D motifs. For notational purposes, we assign positive sign to a dipolar colloidal particle when its hyperbolic hedgehog defect points along the x axis, as shown in Fig. 4.18. In these terms, the configuration of three collinear dipolar particles in Fig. 4.18a is represented as D_3^+ because it consist of three dipolar particles lying along the nematic director. We ascribe a formula $D_2^+D^-$ to the configuration in Fig. 4.18b, which consists of two positively oriented and one negatively oriented dipolar particle. Configuration in Fig. 4.18c is represented as D^+Q because it is formed of a positive dipolar and a quadrupolar particle. Configurations in Fig. 4.18d, e are represented as $D^+\tilde{Q}$, where $\tilde{Q}$ implies that the quadrupolar ring is distorted. The rest of configurations in Fig. 4.18d–i are uniquely described by the corresponding formulas.

This chemical-like notation can successfully be applied to describe periodic 2D colloidal structures of binary nematic colloidal mixtures. In this case, one formula is ascribed to one particular structure, and the formula tells the structure of the elementary crystallographic cell of that structure. Of course, simple lattices will have simple "chemical" structure of the unit cells and more complex lattices are represented by sophisticate formulas.

The simplest example of a 2D binary colloidal structure composed of one dipole and one quadrupole is a unit cell is presented in Fig. 4.19. Although it possesses all possible kinds of bonds between the dipolar and the quadrupolar colloid, the structure is stabilised solely by dipole-quadrupole interaction. The structure is not dense and contains regularly spaced voids of the same shape, unlike the densely packed 2D crystalline lattices composed of pure dipolar or pure quadrupolar colloids.

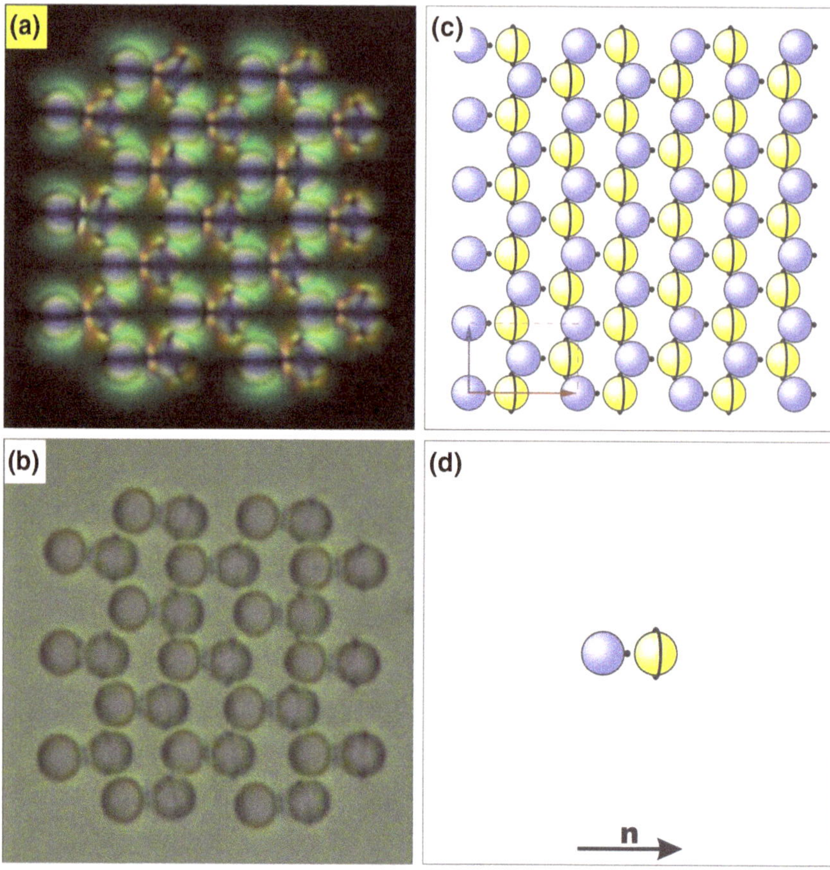

Fig. 4.19 Structure $D^{+}Q$ made of 4 μm colloidal particles under crossed polarisers (**a**) and unpolarised light (**b**). The schematic view of the lattice is shown in (**c**), whereas **d** shows the elementary building block

In this simple lattice in Fig. 4.19, all dipoles are aligned in the same direction and this crystal is "ferroelectric" in the topological sense. It was quite interesting to find that the voids in the lattice in Fig. 4.19 could be occupied by additional dipolar colloids, which would point into the opposite direction, therefore compensating the topological dipoles of the original lattice, as shown in Fig. 4.20. There is a large variety of 2D colloidal motifs of dipolar and quadrupolar particles, which are summarised in Fig. 4.21. Altogether 12 different colloidal motifs have been identified and assembled.

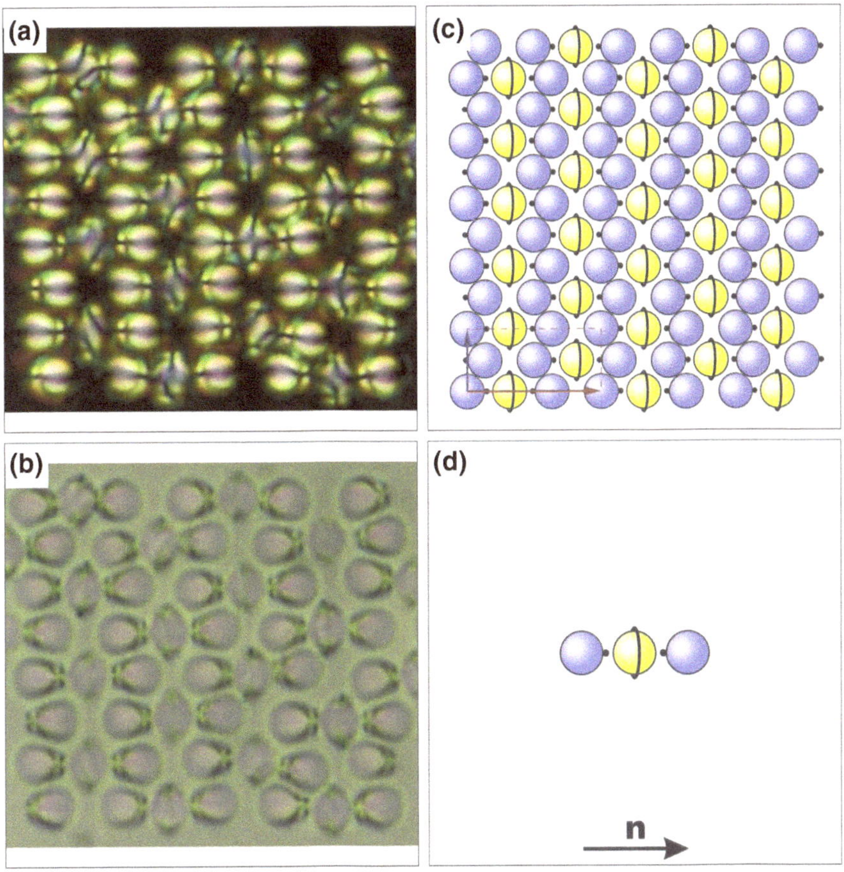

Fig. 4.20 Structure *DQD* under crossed polarisers (**a**), and unpolarised light (**b**). Together with schematic representation (**c**) and the elementary building block (**d**)

4.6 Three-Dimensional Nematic Colloidal Crystals

Optical observation and manipulation of nematic colloidal crystals in thin nematic cells is quite simple, as only a single layer of colloidal particles is usually assembled in such cells. This layer is of thickness of several μm and can easily be analysed using an ordinary transmission optical microscope. On the other hand, observation and manipulation of colloidal clusters of large volume and dimensions is quite a difficult task because of the limited depth of the field of view along the thickness of the sample. This requires use of confocal optical microscopes, which are able to image and reconstruct optical objects of large dimensions in 3D. However, confocal microscopes are not fast enough to allow real-time imaging of objects under study. This makes assembly and study of 3D nematic colloidal crystals quite difficult.

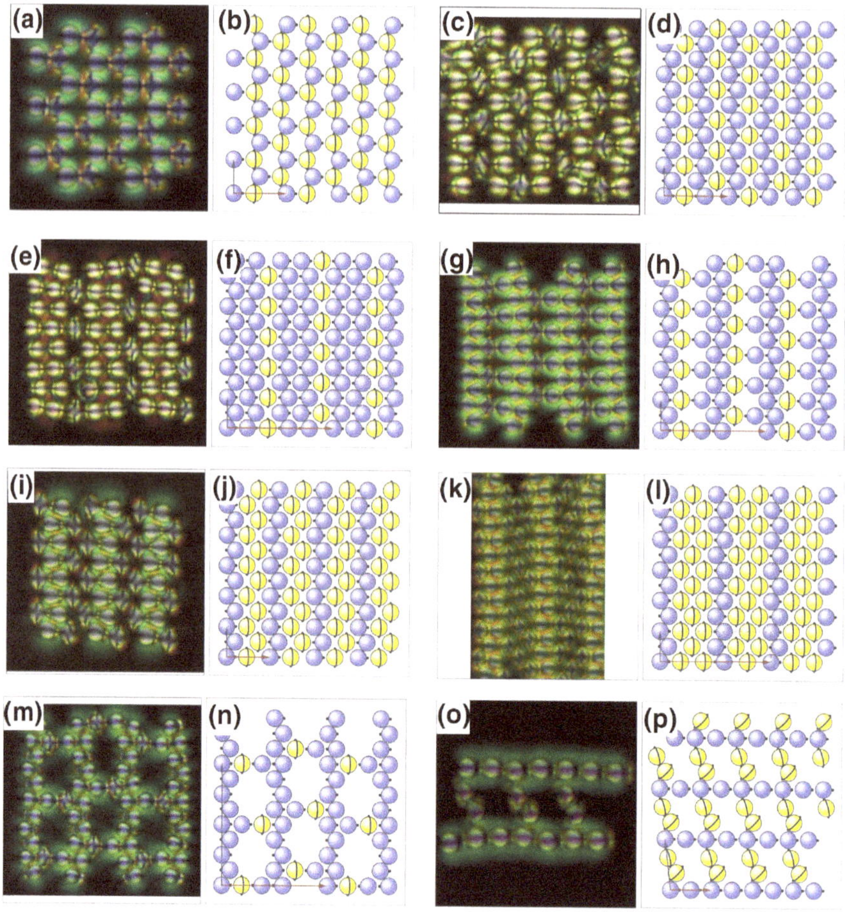

Fig. 4.21 **a–p** Overview of some of 2D colloidal motifs, assembled from elastic dipoles and quadrupoles. Images are organized in pairs, the *left panel* shows crossed polarisers images, *right panel* shows schematic drawings of the crystal structure

Nevertheless, Nych et al. [193] succeeded in assembling, imaging and controlling 3D nematic dipolar colloidal crystals with more than a hundred of colloidal particles assembled in colloidal blocks of 3-colloidal layer thickness. This kind of crystal was prepared by putting dipolar colloidal particles of 4 μm diameter in a rather thick homeotropic nematic cell with thickness of at least 25 μm. As a starting point, one observes colloidal arrangement in such a thick homeotropic cell when the density of colloids is low and allows for the spontaneous assembly of a single colloidal layer in a homeotropic nematic cell. The resulting structure is shown in Fig. 4.22, where a checkerboard colloidal pattern can be seen (Fig. 4.22c).

This checkerboard appearance of a colloidal layer can be understood in terms of dipole orientation, which is different when the topological dipole points towards the

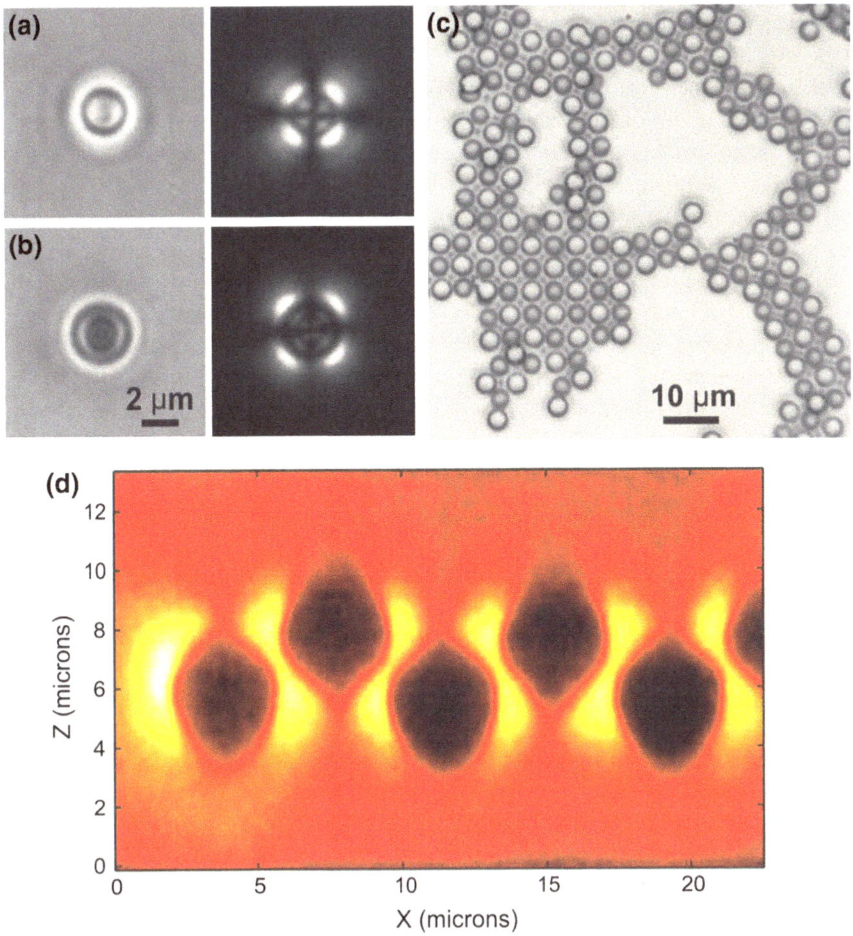

Fig. 4.22 Colloidal particles with dipolar symmetry in a thin homeotropic layer of a nematic liquid crystal. **a**, **b** 2.32 μm dipolar silica microspheres in 5CB homeotropic cell. In **a** the topological defect is above the microsphere, in **b** the defect is below the microsphere. Scale bar, 2 μm. **c** Quasi-2D checkerboard structure formed by 4.32 μm silica particles in 10 μm thick homeotropic cell. **d** A vertical cross-section of a checkerboard colloidal crystal shown in (**c**)

observer or when it points in the opposite direction, as shown in Fig. 4.22a, b. When let free, the dipolar colloids in a homeotropic cell will spontaneously self-assemble with neighbouring dipoles alternating in their orientation. This means that a particle with its dipole pointing up will be surrounded by nearest neighbours with their dipoles pointing down. This alternation in the orientation of the directions of dipoles is indeed observable when measuring the cross-section of such a checkerboard 2D colloidal crystal. The cross-section is shown in Fig. 4.22d and one can clearly see that the positions of neighbouring colloidal particles are alternating in height and the direction of the dipoles in neighbours is also opposite.

Proceeding from one layer to several layers of colloidal particles is quite an elaborate task, since one has to prepare blocks of colloidal particles, where each block consists of several colloidal layers. The term "colloidal layer" means that several layers of colloidal particles are arranged in neighbouring planes, which are parallel to the surfaces of the confining cell. It turns out the successful strategy is to prepare colloidal chains of several particles, which are directed along the depth of the cell. The assembly of such a chain of dipolar particles in a thick homeotropic nematic cell is shown in Fig. 4.23a. The first frame is a crossed-polarisers image of 3 colloidal particles, floating freely in the nematic liquid crystals and obviously positioned at the same height, measured from the surface. The small red cross indicates the position of the laser tweezers. The colloidal particles are shown as bright circles, transmitting light between crossed polarisers, with an additional Maltese cross due to crossed polarisers and the direction of observation along the optical axis of the nematic liquid crystal.

With the tweezers switched on, one is able to grab and drag the selected colloidal particle (lowest particle) next to the red cross in Fig. 4.23a. When this particle approaches the central one, it clearly goes out of the focal plane, i.e. it sinks below the central colloidal particle, and obviously attaches to it. This attachment results in a chain of two colloidal particles, which is oriented along the optical axis, and the visible particle clearly becomes brighter. The reason for this is the increased distortion of the nematic liquid crystal due to the presence of two particles in the chain and larger elastic distortion. In the next step, the third particle is grabbed and brought to the vicinity of the chain of two colloidal particles. Again, the third particle sinks out of the focal plane and obviously attaches as the third particle in the colloidal chain directed along the optical axis. This chain again becomes brighter than before.

After that, several chains are prepared, each one consisting of three colloidal particles. The dipoles of each chain are all aligned, but different chains have different orientation of their dipoles. The orientation can again be distinguished, similar to the case of a single particle presented in Fig. 4.22a. Figure 4.23b shows snapshots from a video of spontaneous assembly of 3 dipolar chains, which were brought close to each other by the laser tweezers. After they are attracted, the three chains form a distorted trio of vertical colloidal chains. It looks like these three bodies are topologically frustrated, which is similar to the orientational frustration in spin systems, such as the Kagome lattice. In that case, a pair of spins has the lowest energy in the antiparallel orientation, but when the third spin is added, it is frustrated in the orientation. It cannot be simultaneously up, to satisfy one of the neighbours, and down, to satisfy the other. The formation of such a frustrated trio is actually the most critical part of the 3D nematic colloidal crystals assembly. If the fourth chain is added, the orientational frustration is released and a final colloidal block of four chains of microspheres is formed, as shown in the first panel of Fig. 4.23c.

After building several separated and well-ordered blocks of $2 \times 2 \times 3$ colloidal particles, a bigger, 3D, dipolar colloidal crystal can be easily assembled by guiding and directing colloidal blocks using the laser tweezers. Figure 4.23c shows such an example, where a larger colloidal block of $2 \times 4 \times 3$ particles is assembled from two smaller blocks. These blocks are spontaneously attracted to each other

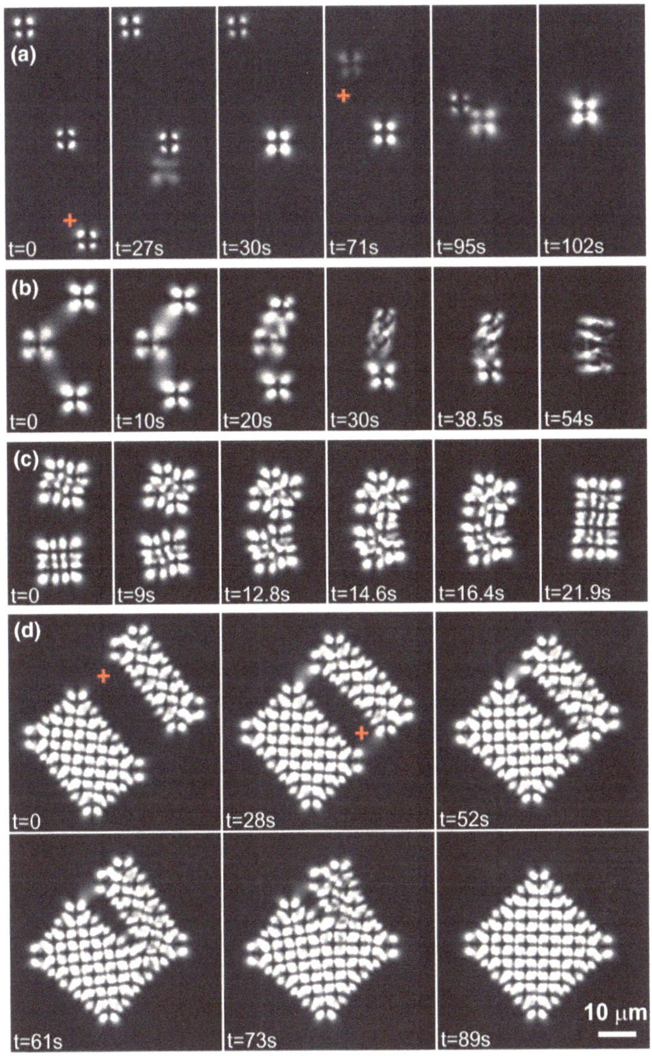

Fig. 4.23 Laser-tweezers assembly of a 3D dipolar colloidal crystal observed under crossed polarisers. **a** Three isolated colloidal particles of 4 μm diameter in the ZLI-2806-filled homeotropic cell of 25 μm thickness appear as bright objects with a dark cross in the centre. Using laser tweezers, one particle is brought close the other and they spontaneously form a chain of two particles in a direction perpendicular to the plane of the image. The pair appears like a single but larger and brighter particle (3rd image from the *left*). The third particle is brought to the couple and it spontaneously forms a dipolar colloidal chain of three particles on top of each other. **b** Three chains, each made of three dipolar particles, are brought close to each other and they start to assemble into a frustrated colloidal trio. Note the tilting of the chains. **c** Two colloidal blocks of 2 × 2 × 3 particles self-assemble into 2 × 4 × 3 blocks. **d** Colloidal blocks of 2 × 2 × 3 and 2 × 4 × 3 particles assemble in the final 6 × 6 × 3 dipolar colloidal crystal. The assembly at the initial stage was guided by laser tweezers until blocks started to attract themselves. Scale bar, 10 μm. In all images, the small *red cross* is the optical trap, used to direct the colloidal assembly

by the elastically distorted liquid crystal in between them. The elastic distortion is minimised by spontaneous attraction and assembly of colloidal blocks. Figure 4.23d shows another example of assembly of a 6 × 6 × 3 dipolar colloidal crystal from two smaller colloidal blocks. The tweezers are operated at such a power that they cause elastic attraction of colloidal blocks to the focus of the tweezers. In this way, the tweezers act as a kind of welding machine which fuses colloidal blocks into a perfectly ordered final colloidal crystal.

The arrangement of individual colloidal particles in such a 3D dipolar colloidal crystal can be determined by using Fluorescence Confocal Polarising Microscopy (FCMP). In this case, a fluorescent dye is added to the liquid crystal and the dye molecules align with their radiative dipole moment along the local director. By selecting a polarisation during the imaging, one is able to reconstruct the director pattern around colloidal particles. One FCPM cross-section of a 3D dipolar colloidal crystalis shown in Fig. 4.24a for two different positions of the focal plane.

The colloidal particles appear dark because they do not fluoresce. In Fig. 4.24a, one can clearly see the regular arrangement of colloidal particles in a square lattice. By taking subsequent FCPM cross-sections along the depth of the cell, one finds that the colloidal particles are arranged in a regular tetragonal Bravais lattice with basis.

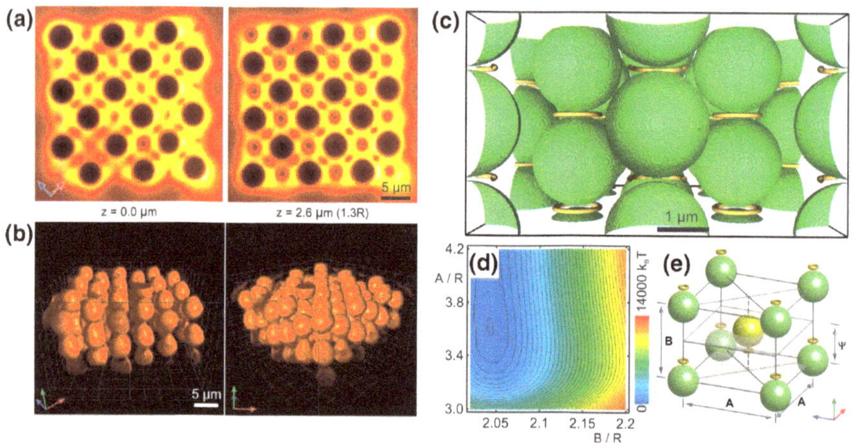

Fig. 4.24 Structure of a 3D dipolar nematic colloidal crystal. **a** Fluorescent confocal polarising microscopy image of two horizontal cross-sections of a 3D 6 × 6 × 3 dipolar colloidal crystal, assembled from 4 μm diameter colloidal particles in the homeotropic aligned nematic liquid crystal ZLI-2806. The images were acquired by refocusing along the z-axis direction by 2.6 μm. Scale bar, 5 μm. **b** The 3D representation of the fluorescent confocal polarizing microscopy image of a 6 × 6 × 3 3D dipolar colloidal crystal. Here, the fluorescence intensity was inverted to show the in-plane arrangement of the particles in the *XY*, *YZ* and *XZ* planes. Scale bar, 5 μm. **c** Numerical simulation of a 3D dipolar nematic colloidal crystal. Point topological defect opened into small loops, somewhat larger as observed in experiments, and are visualized as iso-surfaces of fixed nematic degree of order S. Scale bar, 1 μm. **d** Free energy of one colloidal crystal unit cell as a function of the lattice constants *A* and *B* in units of particle radius *R*. **e** Schematic drawing of the crystal structure showing the tetragonal Bravais lattice with basis

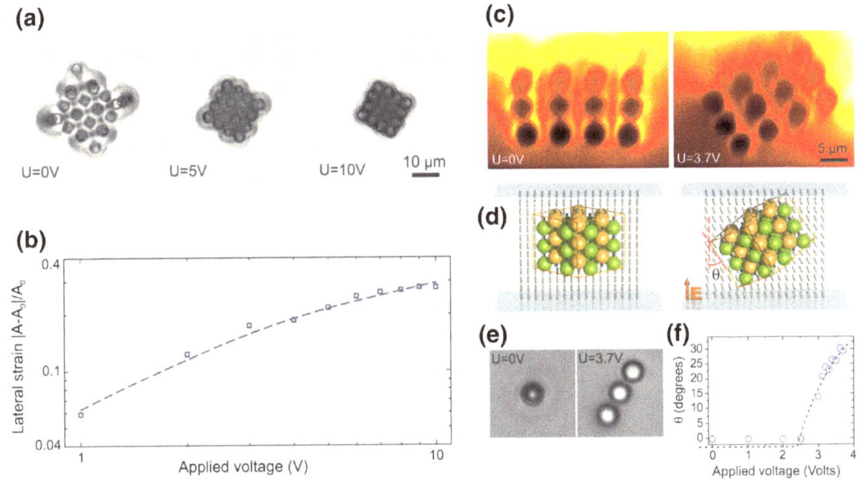

Fig. 4.25 Electric-field-induced shrinkage and rotation of a 3D dipolar nematic colloidal crystal. **a** $4 \times 4 \times 3$ dipolar nematic colloidal crystal made of 4.32 μm colloids in E7 shrinks under an applied electric field. The dielectric anisotropy of the host LC is positive, $\Delta\varepsilon > 0$. Scale bar, 10 μm. **b** Relative lateral shrinkage $|A - A_{\circ}|/A_{\circ}$, as a function of applied voltage for 3D crystal in E7 ($\Delta\varepsilon > 0$) nematic liquid crystal mixture. The electric field ($f = 1$ kHz) is applied along the B-axis of the elementary cell. Dashed line is guide to the eyes. **c** Fluorescent confocal polarizing microscopy images of electric-field-induced rotation of a 3D, $4 \times 4 \times 3$ dipolar nematic colloidal crystal in a ZLI-2806 liquid crystal with negative dielectric anisotropy, $\Delta\varepsilon < 0$. Scale bar, 5 μm. **d** Schematic views of 3D dipolar crystal and liquid crystal configuration without field (*left*) and after electric field is applied, inducing rotation of 3D nematic colloidal crystal as a whole (*right*). The crystal follows the liquid crystal molecules, which tend to align perpendicular to the field because of their negative dielectric anisotropy. **e** Microscope images of electric-field-induced rotation of a single colloidal chain in ZLI-2806 liquid crystal. The chain in the *left panel* points into the panel and is rotated from this direction by the field. **f** Angle of rotation of a colloidal crystal as a function of applied electric field

The 3D representation of colloidal crystals, as reconstructed from FCPM images is shown in Fig. 4.24b. Numerical analysis basically confirms the observed tetragonal structure of the 3D dipolar nematic colloidal crystal. It shows that such a crystal is composed of dipolar colloidal chains (see Fig. 4.24c, e), which are all pointing in the same direction. There is an additional chain in between four neighbouring chains, which has the dipole moment oriented in the opposite direction. These chains form the basis of the unit cell.

Similar to the electric field response of a 2D dipolar colloidal crystal, the 3D dipolar colloidal crystal shows a large response to an external electric field. If the field is applied along the optical axis, as shown in Fig. 4.25a, the crystal shrinks for nearly 30 % at a very low electric field of 0.37 V/μm, provided that the liquid crystal has positive dielectric anisotropy. An interesting phenomena is observed if the 3D dipolar colloidal crystal is assembled in a homeotropic cell of a nematic liquid crystal with negative dielectric anisotropy. In this case, the electric field induces a rotation of the crystal as a whole, as shown in Fig. 4.25c.

The experiments on 3D nematic colloidal crystals give clear evidence that it is possible to assemble photonic crystals out of colloidal particles in a nematic liquid crystal. Whereas the laser tweezers allow for manual assembly of such photonic crystals, other matters of manipulation of a very large number of colloidal particles could be used for the same purpose. For example, the microfluidic technique could be used to assemble 3D nematic colloidal crystals, and it is expected that advanced techniques such as the optoelectronic tweezers could be used as well.

4.7 Hierarchical Assembly of Nematic Colloids

In all experiments so far described in nematic colloids, colloidal particles of the same diameter were used. They might differ in surface properties, but in all cases their dimension was kept equal in order to reduce any possible mismatch in the colloidal sizes and positions. However, interesting phenomena are observed if one considers the interaction of two colloidal microspheres of very different diameters. For example, one could consider the interaction of 1 μm diameter homeotropically treated microsphere with a much larger, 19 μm colloidal microsphere, as presented in Fig. 4.26 [194].

In this image, one can see in panel (a) a non-polarised image of the interaction of a small colloidal particle, which is attracted into the Saturn ring of a much bigger colloidal particle. One can clearly observe that the small particle is attracted into the Saturn ring of the bigger colloidal particle over separation of the order of 20 μm. From the recorded sequence of images in Fig. 4.26a one can calculate the attractive potential between the smaller colloidal particle and the Saturn ring, which is shown in Fig. 4.26b. Similar to the colloidal pair interaction in nematic liquid crystal, the interacting potential of two very different particles is of long range and very strong, exceeding thousands of kT for 1.5 μm diameter dipolar colloidal particle.

This simple experiment of interaction of a small dipolar colloidal particle with a Saturn defect ring in fact reveals a simple physical mechanism, which is behind this phenomenon. If we consider well-separated particles, there are two contributions to the elastic energy in this system. The first contribution comes from the elastically distorted region around the small particle, and the second contribution is due to the elastic distortion of the Saturn ring. It is possible to lower the total elastic energy by bringing the smaller colloidal particle close or into the core of the Saturn ring defect. In this case, the two defect regions are shared between the two entities and this sharing lowers the total free energy of the system. Of course, this is a simple picture, which misses many details, such as the obviously curved trajectory of the particle and the mismatch between the dipolar symmetry of the smaller particle and the threefold symmetry of the cross-section of the Saturn ring. However, the experiments tell us this is indeed the essence of the mechanism of attraction.

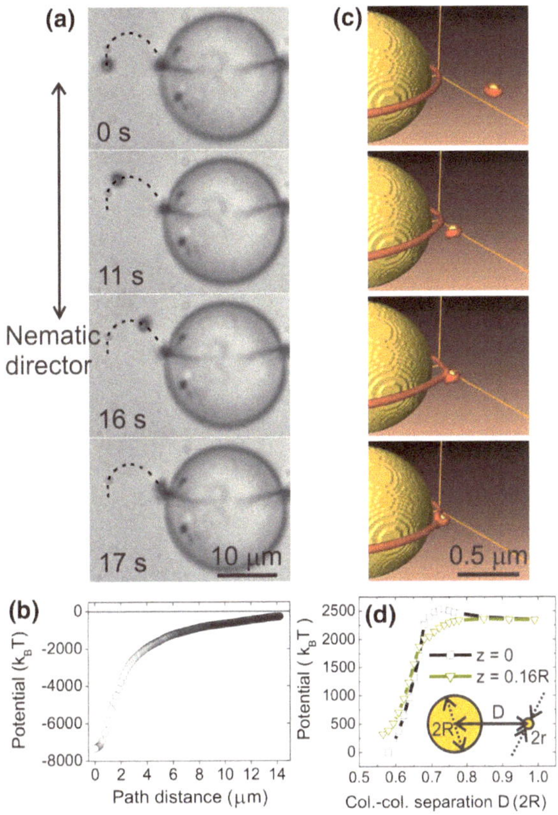

Fig. 4.26 **a** Small colloidal particle of 1.5 μm diameter is attracted into the Saturn ring of a 19 μm colloidal particle, following the black dashed path. **b** Binding potential of the small colloidal particle along the trapping path. **c** Numerical calculation: 100 nm nanoparticle is attracted into the Saturn ring encircling a 1 μm colloidal particle. **d** The calculated binding potential of the small particle as a function of the particle separation for trajectories at different of-centredness "z" from the central $x - y$ plane

Having this in mind, one could imagine that smaller particles should be attracted to any defect line in the nematic liquid crystals, created in quite different ways. Defect lines should therefore act as collectors of particles, which could be used for interesting purposes. The question arises, whether one could create artificially more sophisticated scaffolds or even a network of decorated defect lines, connecting or surrounding several large colloidal particles. Figure 4.27 shows a more complicated system of two big colloidal particles, which were brought together and forced to entangle their Saturn rings. The phenomenon of entanglement will be discussed in more details in Chap. 5. For the reason of understanding the present experiment, the reader can imagine that two colloidal particles in Fig. 4.27 are connected together by a single defect line, which forms a Figure-of-eight. In this figure a perpendicular alignment of the nematic liquid crystal was chosen for the measuring cell, so that the entangled rings appear as dark circles in the plain of the image. By bringing a smaller colloidal particle into the vicinity of such an entangled colloidal pair, one can clearly see it is attracted and bound to the figure-of-eight defect line (Fig. 4.27a).

Spatial organisation of smaller colloidal particles, which were manually assembled into the defect ring of the figure-of-eight, is shown in Fig. 4.28 and clearly

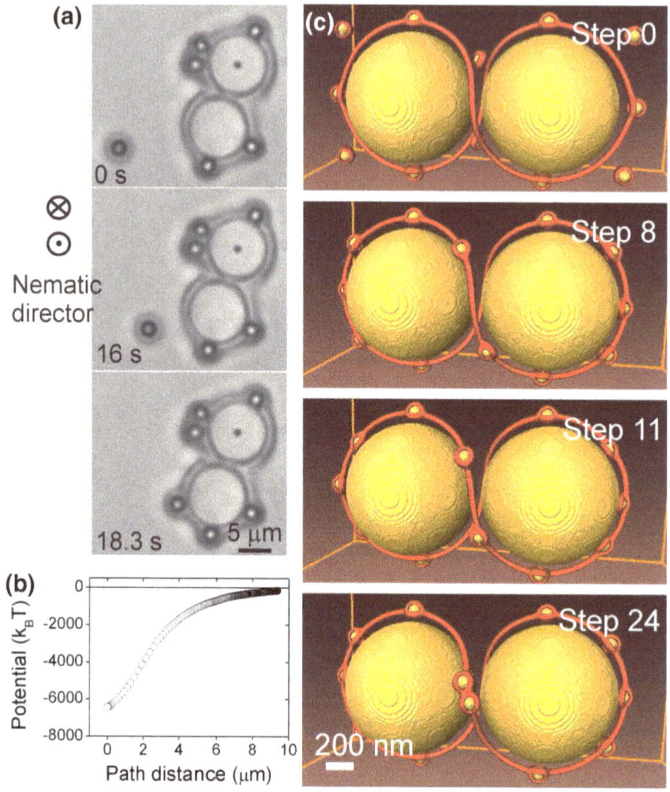

Fig. 4.27 **a** Sequence of images showing the attraction of 2.32 μm dipolar colloidal particles into the entangled defect loop around two colloidal particles. **b** The potential of the smaller colloidal particle in the vicinity of the figure-of-eight defect line. **c** Trapping sequence of smaller 100 nm particles into the figure-of-eight defect line entangling a pair of 1 μm particles. Step 0 corresponds to the final configuration and step 24 to the final configuration in the numerical simulations

shows the spiralling shape of the line. The images were taken in different focus of the microscope, which brought into focus particles of different heights. Whereas the binding energy of micrometre-size colloidal particles into the Saturn ring of a much bigger particle is nearly 10 000 kT, the question arises, what is the smallest diameter of the particle that could still be trapped into the Saturn ring defect. This was first numerically calculated [194] for rather small 100 nm particles interacting with a Saturn ring around 1 μm particle, and that is also shown in Fig. 4.26d. Binding energies depend on the surface anchoring strength and are of the order of 1000 k_BT. By reducing the dimension of the smaller particle, binding energy decreases monotonously to zero at approximately 50 nm small-colloid diameter. The theoretical prediction is that nematic nanocolloids of tens of nanometres diameter should still be effectively trapped into nematic defect lines. This prediction was in fact confirmed several years later in the experiment performed by Ryzhkova et al. [118, 119]. Using dark field

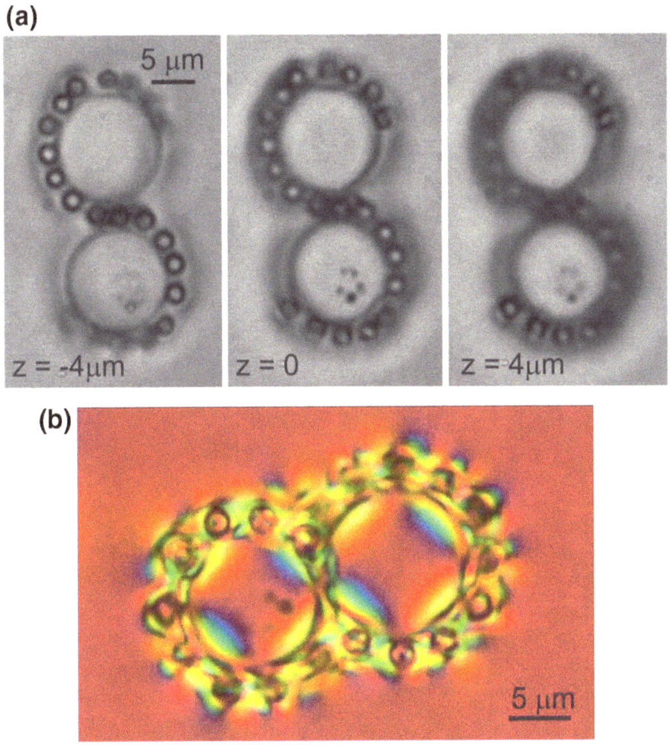

Fig. 4.28 **a** Smaller colloidal particles are trapped into the defect loop, twisting around a larger colloidal pair (diameter 10 μm). Images were taken at different heights of the microscope focus. **b** Red-plate image of an entangled colloidal pair with smaller colloids trapped into the entangled defect line

microscopy, it was possible to follow the motion of tens of nanometre diameters silica particles into the defect line of bigger colloidal particles. This is discussed in Sect. 2.6.

The experiments with colloidal particles of different sizes clearly demonstrate that a rich variety of colloidal superstructures could be formed or even self-assembled in nematic colloids. Some of these hierarchical structures are interesting for possible applications, such as the ring-like structure in Fig. 4.29. Here, the Saturn ring of the bigger colloidal particle is completely filled with smaller colloidal particles which form a kind of necklace encircling the particle. Providing that these smaller particles are made of conductive material, such a structure could function as a distributed split-ring-resonator (*SRR*) with the electrical scheme shown in the insert to Fig. 4.29b. Here, the capacitor *C* is formed of the two conductive surfaces of the

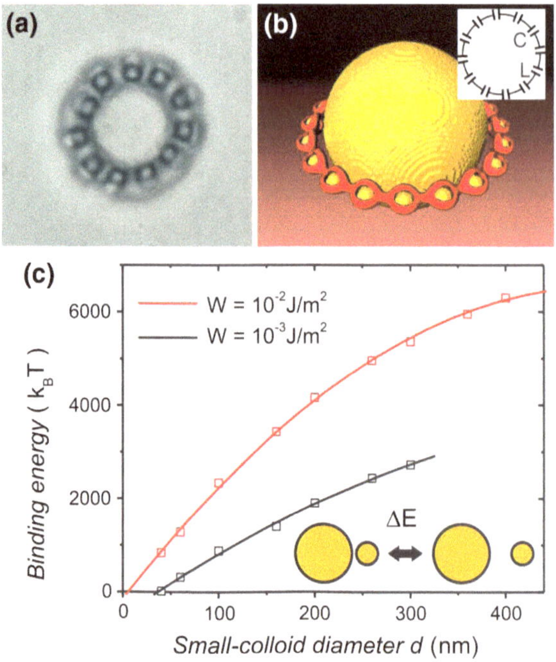

Fig. 4.29 **a** Possible split ring resonator superstructure, formed of small colloidal particles trapped into the Saturn ring of the large particle. **b** A calculated stable colloidal ring superstructure of small 100 nm and a large 1 μm particles. The *inset* shows schematic electrical circuit corresponding to a Saturn ring filled with conductive colloidal particles. **c** Calculated binding energy of a small particle with diameter *d* into the core of the Saturn ring, formed around a 1 μm colloidal particle for two different anchoring strengths *W* at the particle surfaces. Lines are parabolic fits

neighbouring smaller colloidal spheres in close proximity. The inductance L comes from the smaller conductive particles themselves. Such a colloidal ring therefore represents a distributed LS circuit which is in fact an electromagnetic resonator. The natural resonances of such micro-rings filled with nanometre particles should be in the teraHertz region, corresponding to the wave length of tens of micrometres.

Chapter 5
Entanglement of Nematic Colloids

Abstract This chapter describes topologically nontrivial nematic colloids, starting with a simple entanglement of two spherical particles in the nematic liquid crystal. We continue with the description of knotted and linked nematic colloids, where the full spectrum of experimental topology is demonstrated. The topological aspects are further discussed in the experiments of topological charge production and colloidal entanglement by optically controlled quench around a micro-fiber in the nematic liquid crystal.

5.1 Entanglement of Colloidal Particles in a Homogeneous Nematic

So far, we have been discussing the interaction of colloidal particles in nematic crystals, mediated by isolated topological defects, such as hedgehog point defects or Saturn rings. In all these cases each colloidal particle was accompanied by its own topological defect, which generated the necessary force of assembly. The first evidence of colloidal interaction, mediated by topological entities much different than isolated topological defects could be traced back to the numerical experiments of Guzman et al. in 2003 [195]. They were studying possible defect structures around two closely positioned colloidal particles with homeotropic surface anchoring in a nematic liquid crystal. By using molecular simulation and a dynamic field theory approach, Guzman et al. found evidence of new defect structures neither predicted nor observed before [195]. Instead of two separated Saturn rings, belonging to each particle, each pair of particles was connected by an unusual "3-ring" defect configuration, as shown in Fig. 5.1. This image shows the contour plot of the tensorial ordering field, corresponding to the order parameter of $S = 0.26$, thus tracing the regions of well-reduced order parameter. Because of the depressed order, these regions must correspond to the topological defects which are known to have reduced order in their cores.

This pioneering numerical study of two particle disclination rings was followed by a numerical study by Araki and Tanaka in 2006 [196]. They found a similar topological structure in a form of a single defect line encircling closely spaced

I. Muševič, *Liquid Crystal Colloids*, Soft and Biological Matter,
DOI 10.1007/978-3-319-54916-3_5

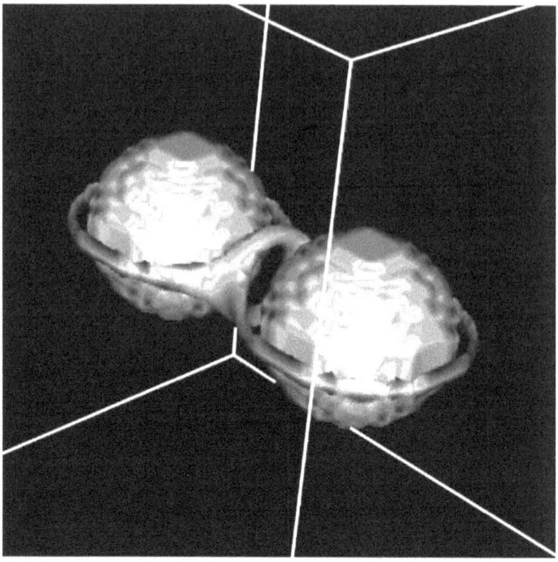

Fig. 5.1 A 3-ring defect loop structure around two spheres, showing the contour plot with $S = 0.26$. Reprinted figure with permission from O. Guzmán, E.B. Kim, S. Grollau, N.L. Abbott, J.J. de Pablo, Phys. Rev. Lett. **91**(23), 235507 (2003). Copyright (2003) by the American Physical Society

colloidal particles in a form of the figure-of-eight. At an early stage of their numerical separations, Araki and Tanaka also observed a 3-ring structure, similar to the one in Fig. 5.1, but the structure was unstable and only transient. In the course of time, this 3-ring structure developed into a new structure with a single-stroke declination line, surrounding and embracing both colloidal particles, as shown in Fig. 5.2.

Nearly simultaneously with the work of Araki and Tanaka, the Ljubljana group was independently performing numerical studies of a numerical quench of a pair of colloidal particles using LdG approach and reported at the Keystone ILC Conference [197]. The nematic was numerically rapidly cooled down from the isotopic to the nematic phase. These numerical experiments, which were performed prior to real experiments on real colloids, predicted an unusual phenomenon, where two colloidal particles were spontaneously entangled by the cooling surrounding liquid. This entanglement was soon observed in the experiments performed by Muševič and Škarabot in 2006 and published in 2007 together with numerical analysis [198].

A snapshot of images taken from the movie showing the numerical quenching experiment is shown in Fig. 5.3. At the beginning, a pair of colloidal particles, each of one micrometre diameter, are closely spaced and surrounded with the isotropic melt of the nematic liquid crystal. When the "numerical temperature" is suddenly decreased, the liquid crystal undergoes the isotropic-nematic phase transition, which nucleates a huge number of densely packed topological defects in the newborn nematic phase (first panel (a) in Fig. 5.3).

During the course of time, this dense tangle of topological defects annihilates and at a later stage of coarsening dynamics leaves behind a single topological loop. The loop is now encircling both colloidal particles and forms a figure-of-eight structure, which strongly binds the two colloidal particles together.

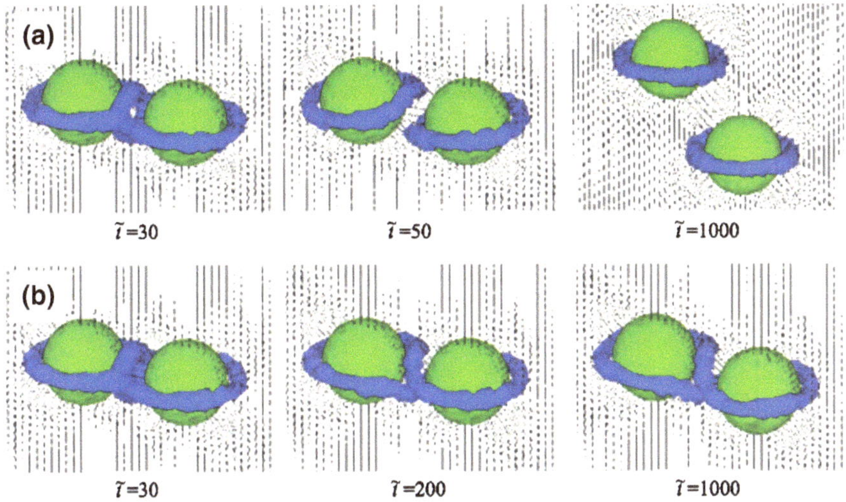

Fig. 5.2 The formation of defect loops around two colloidal particles in the nematic liquid crystal. At early stage, there is a transient defect structure, which is formed in both cases, shown as the first snapshot in **a** and **b** ($\tilde{t} = 30$). This structure is unstable and rapidly transforms into either: **a** two separated colloidal particles, each having its own Saturn ring, or **b** new type of topological configuration, where a single defect loop is entangling both colloidal particles. Reprinted figure with permission from T. Araki, H. Tanaka, Phys. Rev. Lett. **97**(12), 127801 (2006). Copyright (2006) by the American Physical Society

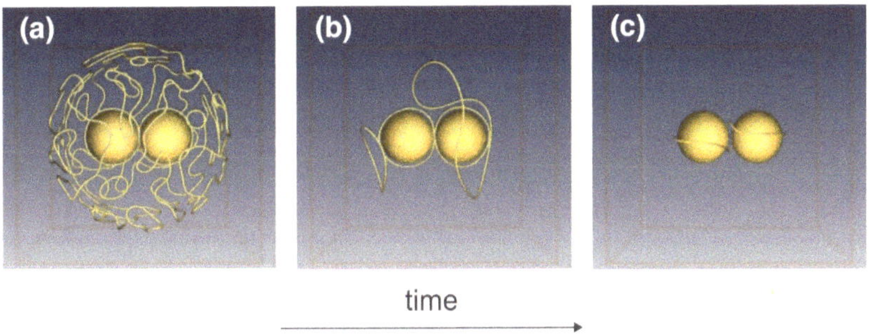

Fig. 5.3 Snapshots from the movie, showing numerical quenching of a pair of colloidal particles in the nematic. **a** Just after starting the quench, a dense mosaic of nematic patches comes out from the isotropic melt. **b** In the course of time, this mosaic of patches coarsens into a network of topological defects of densely packed topological defect lines, surrounding the two colloids. **c** At a later stage, just before reaching the equilibrium state, only few isolated defect loops still exist. Among them, the most prominent is the figure-of-eight defect loop, which is a truly stable topological state that exists indefinitely. Defects are represented by drawing the surfaces of constant order parameter S (coloured *yellow*, corresponding to $S = 0.5$). Courtesy of M. Ravnik

This was an exciting numerical prediction and it took little time to be realised in real experiments. These were performed by using the laser tweezers to position two independent colloidal particles of 19 μm diameter and homeotropic surface anchoring in a planar nematic cell with a homogeneous thickness of 21 μm. Because the thickness of the nematic layers is only slightly bigger than the microsphere's diameter, the topological defect now appears as a Saturn ring, encircling each of the two colloidal particles. The laser tweezers were used to position the two particles as close as possible.

By using a strong light of the laser tweezers, the nematic liquid crystal was locally molten, as shown in Fig. 5.4a. This melting is due to the absorption of the laser tweezers light in a thin layer of indium-tin-oxide, deposited on one or both glasses of the cell. Because of the first-order phase transition, the isotropic melt is clearly separated by a well visible interface with the rest of the nematic liquid crystal. Then, the strong light of the laser tweezers was shut down, and the molten isotropic island began to rapidly cool down because of the ceased absorption of light. In a fraction of a second the isotropic melt reaches the phase transition temperature, where the newborn nematic phase starts to appear in the course of time. In first stages, this newborn nematic phase appears as a very fine mosaic of disordered patches of the nematic liquid crystal, oriented randomly in space. This situation is shown in Fig. 5.4b. By further cooling down with time, this fine mosaic gets gradually more rough, which is due to the expansion and growth of each micro-

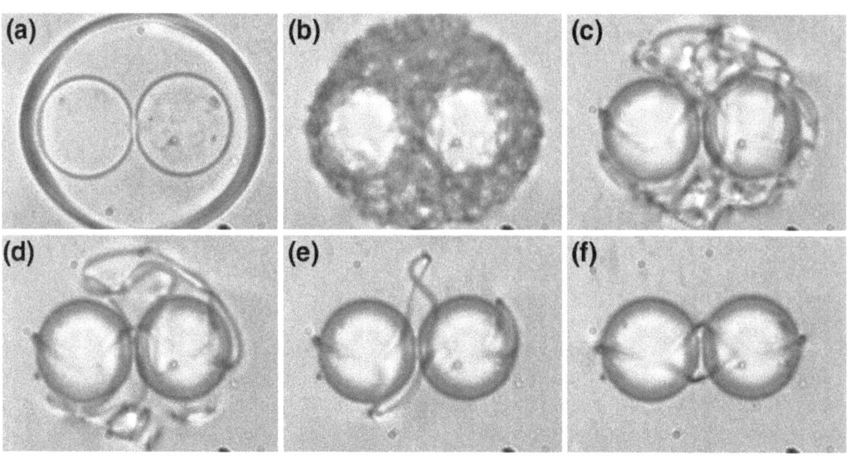

Fig. 5.4 Snapshots from the movie, showing real quenching of a pair of colloidal particles in the nematic. **a** In the isotropic phase, the particles are surrounded by the disordered liquid crystal. **b** Just below the isotropic nematic phase transition, a dense mosaic of nematic defect patches comes out from the isotropic melt. **c** In the course of time, this mosaic of patches coarsens into a network of topological defects of densely packed topological defect lines, surrounding the two colloids. **d** At a later stage, just before reaching the equilibrium state only few isolated defect loops still exist. Among them, the most prominent is the figure-of-eight defect loop, which is a truly stable topological state shown in **e**, which exists indefinitely **f**

domain (Fig. 5.4c, d). At this stage, a dense network of topological defects is formed where the newly formed nematic grains meet each other. At a certain time, the temperature of the nematic reaches its equilibrium value, but the nematic sample is still out of equilibrium because of the dense remnant tangle of topological defects. Then, the process of coarsening dynamics takes place, where this topological network partially and gradually annihilates, which is accompanied by the growth of a uniform nematic texture, as shown in Fig. 5.4d. Just before reaching the equilibrium state, only a small number of defect loops are present; most of them annihilate into the vacuum, but there is one loop which remains. This is shown in the experimental image in Fig. 5.4f, which is in excellent and nearly one-to-one correspondence with the numerical images shown in Fig. 5.3.

This mechanism of formation of dense network of topological defects is called the Kibble–Zurek mechanism [31] and applies at an early stage of defect texture formation. It was first invented by Kibble in 1970s to explain the emergence of cosmic strings in the Universe. Because of the mathematical correspondence between the cosmic fields, and fields relevant to condensed matter, the same mechanism was applied to a condensed matter system by Zurek [32]. In condensed matter, the Kibble–Zurek mechanism is used to describe the emergence of topological defects of different material fields, such as the magnetisation in magnetic systems or electric polarisation field in ferroelectrics [199].

A rather striking similarity between the predictions of the LdG theory and real experiments on real samples gives us confidence that theory is indeed able to direct experiments in many situations. In this case, it clearly led to the discovery of colloidal entanglement in nematic liquid crystals, which was the the foundation for many topological studies in the following decade. The reason for this is the fact that we could realise for the first time the manipulation of individual topological entities, such as the two Saturn rings in our case, into more complex topological objects. In this experiment, shown in Fig. 5.4, it was the entanglement of two colloidal particles by a single topological loop, which formed the geometrical shape, reminiscent of number 8. For this reason, this new topological entity was called figure-of-eight.

By performing a large number of experiments (124 in this case), other types of topological entanglement of a single colloidal pair were found. The first topological state, which was the most stable (36%), was the figure-of-eight. The second observed state, which was the second most stable, (13% of final states) was the figure-of-omega, which is presented in Fig. 5.5. In this case, we also observe a single topological loop, which is encircling and strongly binding both particles, but this loop now has the appearance of the Greek letter Omega (Ω). Instead of being simply twisted as in the case of the figure-of-eight, the figure-of-omega makes an additional small loop in between the two colloidal particles and in the plain perpendicular to the plain of the cell. Again, a striking agreement is found between the experimentally obtained and numerically calculated structure as shown in Fig. 5.5.

In most of the experiments the figure of-omega configuration was found unstable and transformed slowly into another more stable configuration, shown in Fig. 5.5d. This state is called the entangled hyperbolic defect and was obtained in 3% of direct quench experiments. The inspection of the structure shows that there are two

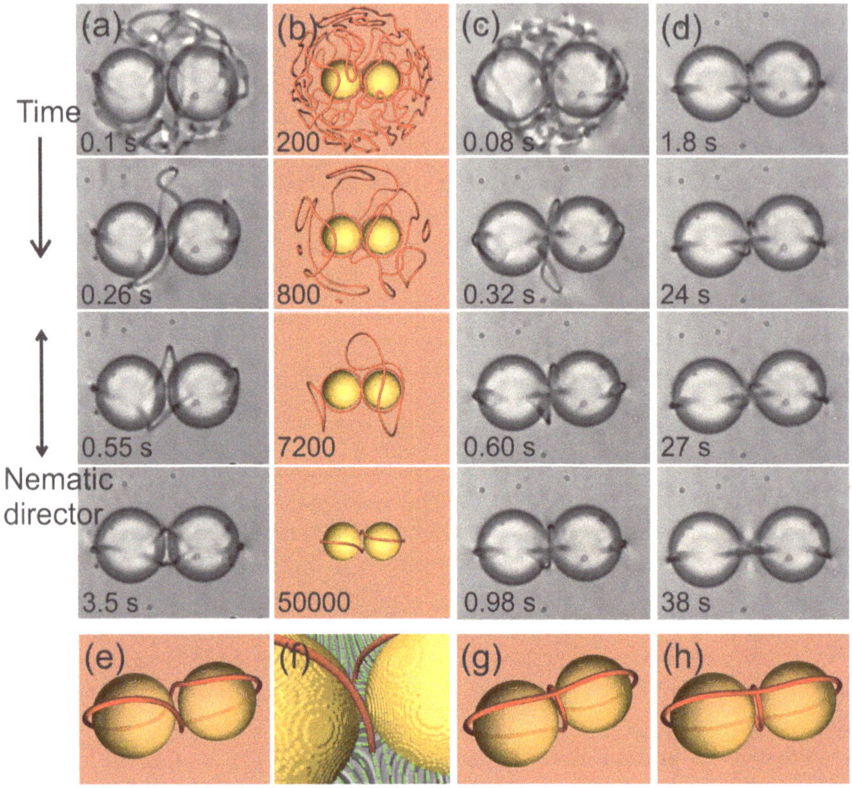

Fig. 5.5 Assembling entangled colloidal pairs by thermal quench using light. The diameter of the particles is 19 μm, the cell thickness is 21 μm. **a** "Figure of eight" entangled state. The disclination loops are visible under non-polarizing optical microscope due to the scattering of light. **b** Numerical simulation of the time evolution of entanglement, measured in the number of iteration steps. Defects are represented by drawing the surfaces of constant order parameter S (colored *red*, corresponding to $S = 0.5$). **c** Evolution of the "figure of omega" entangled state. **d** "Figure of omega" was unstable and transformed into the "entangled hyperbolic defect". **e** The calculated "figure of eight" structure. **f** Close look at the director field in the gap between the microspheres for the "figure of eight". **g** Calculated "figure of omega" structure. **h** Calculated "entangled hyperbolic defect" structure

topological defects in this case. The first defect, most readily visible from the numerical simulation in Fig. 5.5h, is a simple loop, entangling without twisting both colloidal particles. However, in order to satisfy the law of conservation of topological charge, there must be a second defect, which is observed as a small loop or even a point defect, located in-between the two colloidal particles. The formation of this defect is clearly visible from the snapshots of images in Fig. 5.5d, where the formation of the defect in between the particles pushes the two particles slightly apart. This way, there is enough room between the surfaces of the two colloidal particles to accommodate the point defect.

It is intuitively clear that the entanglement of two colloidal particles generates the force, which binds the two particles together. This is due to the fact that any forced separation of the two colloidal particles increases the overall length of the topological defect line entangling both defects. Now, this increase in length of the defect line leads to the increase of the energy of the system, because additional length of the defect has to be created by elongation. This means that external work is put into the system and the total elastic energy of the entangled colloidal pair is increased. The increase of the elastic energy upon colloidal displacement is the basic definition of a force. In other words, the entangling defect line exerts a force on both colloidal particles, which pushes them together simply to reach the smallest possible length of the defect loop to reach the lowest possible energy state.

This structural force due to colloidal entanglement was measured in the experiments as shown in Fig. 5.6. Instead of a single laser tweezers trap two independent light traps were formed, each grabbing one colloidal particle. After the optical traps were separated, the two colloidal particles followed that trap movement and separated from each other. This is shown in the first panel of Fig. 5.6, where one can see two light traps indicated by two small red crosses on each side. The colloidal particles are well separated from the natural equilibrium separation and one can clearly see a pair of defect lines, appearing as two dark strings, connecting both particles. These two dark lines are simply the two projections of the figure-of-eight loop, which is stretched by the laser tweezers. In the next step, the tweezers are shut down and one can clearly see from the sequence of panels in Fig. 5.6a that the two particles approach each other. The same experiment was performed on the entangled point defect colloidal pair, shown in Fig. 5.6b. From the recorded video images of the two colloidal particles approaching each other by the force of the entangling defect line, one is able to calculate the instantaneous force on each particle, and the integration of work of this force along the path of the colloidal particle results in the elastic potential as a function of the colloidal pair separation. This binding potential is shown in Fig. 5.6c for the figure-of-eight structure, and the insert to this figure shows the force of entanglement as a function of separation. The binding forces in the figure-of-eight structure are of the order of 10 pN and the force of the equilibrium state is zero because the force of the entanglement is balanced by the contact repulsion between the two colloids.

The binding potential of entangled colloidal pairs are of the order of 10 000 $k_B T$ and are much stronger than the binding potential, generated by isolated point defects. This is shown in Fig. 5.6c and d for the figure-of-eight and the entangled hyperbolic defect structure.

The entanglement is not limited to a single colloidal pair. The quenching experiments rapidly demonstrated that it is possible to entangle an arbitrary number of colloidal particles, which readily form entangled colloidal chains. Figure 5.7 shows two examples of colloidal wires assembled by entangling 4, 6 and 5 colloidal particles. Two different types of wires were observed, the figure-of-eight colloidal wire, shown in Fig. 5.7a, b and the entangled hyperbolic defect colloidal wire, stretched by the laser tweezers in Fig. 5.7c.

There are several important aspects related to the observation of entanglement of colloidal particles in the nematic liquid crystal by closed loops. First, there is an interesting signature in the formation of the entangled topological defect, as they

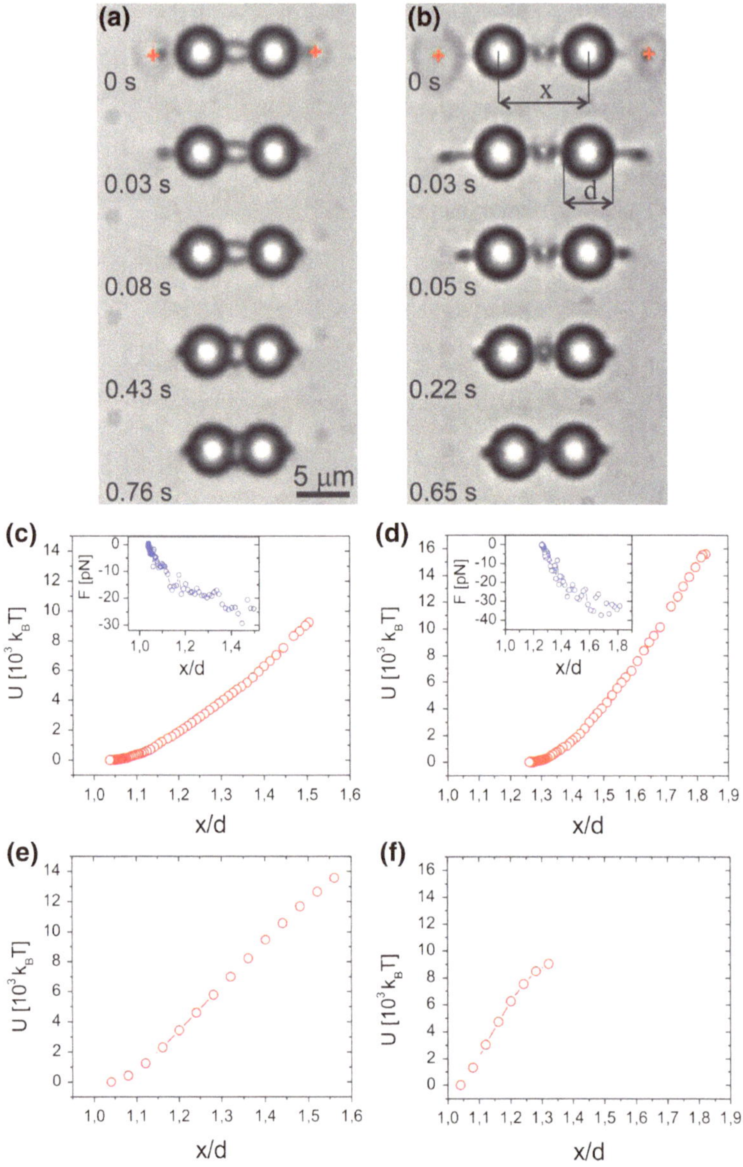

Fig. 5.6 Measuring the force of entanglement by stretching and releasing the colloidal pair using light. **a** Using focused light of the laser tweezers (two *red crosses* on *left* and *right*), the "figure of eight" colloidal pair is first stretched and then released by switching-off the light. The diameter of the particles is 4.7 µm, the thickness of the cell is 6 µm. The force and the binding energy are calculated from the video frames (**a**), and shown in **c**. **b** The same is done for "entangled hyperbolic defects", shown in **d**. **e** The pair binding energy, calculated for the "figure of eight", as a function of particle separation x, normalized to the particle diameter d. **f** The pair binding energy, calculated for the "entangled point defect" as a function of separation x, normalized to the particle diameter d

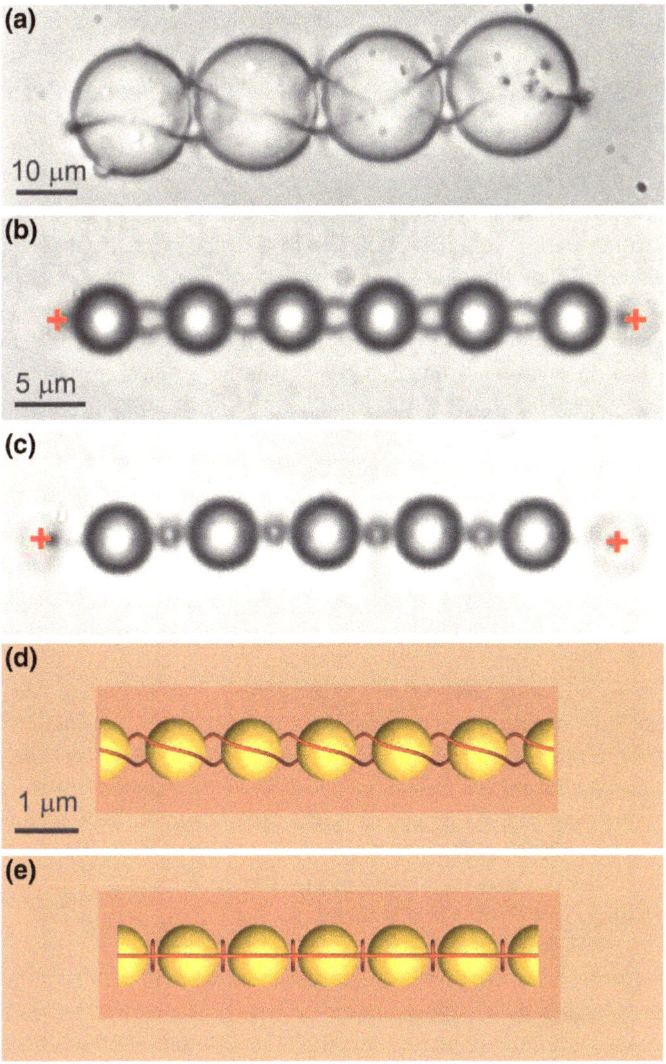

Fig. 5.7 Colloidal wires assembled by entanglement. **a** Right-handed "figure of eight" colloidal wire assembled from 19 μm glass spheres. **b** "Figure of eight" colloidal wire made of 4.7 μm microspheres, stretched by light of the laser tweezers (*red crosses*). In the absence of light, the average equilibrium surface-surface separation between neighboring spheres is 160 nm. **c** "Entangled hyperbolic defect" colloidal wire, stretched by laser tweezers. Without light, the average equilibrium surface-surface separation between neighboring spheres is 1100 nm. **d** Calculated structure of stretched "figure of eight" colloidal wire. **e** Calculated structure of stretched "entangled hyperbolic defect" colloidal wire

are created nearly exclusively by quenching from the isotropic phase. Whereas the conservation of the topological charge assures that in this system the topological charge of defect loops equals the number of colloidal particles, it does not prescribe the nature of the loops. The spontaneous formation of entangled defect loops is therefore ultimately connected to the symmetry breaking through the Kibble mechanism. In contrast to an empty space, where all relics of the disorder phase annihilate after coarsening, in our case, the relics condense at colloidal particles in a form of nematic braids of various complexities, as discussed by Ravnik et al. [200]. This topological condensation brings the physics of this experiment into close relation to other observations of entanglement, such as the entanglement of vortices in superconductors and superfluids, theory of strings and cosmological Kibble mechanism of string formation. The observation of entanglement raised new questions about the nature of topological charge in liquid crystals and the forms of the topological charge materialisation. Eventually it triggered the development of novel topological analyses of the charge in nematic braids by Čopar et al. [201, 202].

The second important aspect is the simple observation that originally isolated defect loops, such as Saturn rings could be fused and reshaped via thermal quench. This has led to the discovery of colloidal knots and links in chiral nematic colloids which will be described in the next section. It should be noted that in planar nematic cells only linear entangled colloidal wires were observed. Any attempt to entangle colloidal particles in 2D, i.e. in structures different from wires, absolutely failed. The underlying reason is quite simple and is due to the fact that these topological lines can propagate only in a direction perpendicular to the director. Eventually they could be deviated from that direction by a small angle to allow bending of lines along the surfaces of colloidal particles. However, it was absolutely impossible to direct such a line along the director of a planar cell. If a chiral structure is taken instead of a homogeneous one, the lines can propagate practically in any direction in the sample by choosing proper position with respect to the helical structure.

The third important observation is related to the photonic character of entangled colloidal wires. If the refractive index of microspheres is higher than the surrounding, each microsphere acts as an optical microresonator, which confines light. Because of the total reflection at the sphere-liquid interface, the optical waves inside the spherical cavity get totally reflected each time they hit the interface. This results in multiple and subsequent total reflections (or bouncing of light) in the interface and the light is circulating inside the microsphere. If the light reaches the point of origin with the same phase and amplitude, we have a condition for an optical resonance. These optical waves, which can be loosely described as circulating light waves inside the high reflective index object, are called Whispering Gallery Mode resonances. If two or many such resonators are positioned close to each other, such as the case of our entangled microspheres, the light could be transferred from one resonator to the other and back. We therefore have a series of optical resonators which are coupled together and the electromagnetic modes of series of resonators are also described as a photonic molecule. These photonic molecules are therefore easily realised by entangling colloidal microsphere in a nematic liquid crystal.

5.2 Knots and Links in Chiral Nematic Colloids

In this section we demonstrate that the richness of phenomena observable in chiral nematic colloids is much larger than in non-chiral, homogeneous nematic liquid crystals. By choosing chiral nematic crystals and colloidal particles with homeotropic surface conditions it was possible to realise knots and links of topological defects, which could be reconfigured by the light of the laser tweezers.

Knots are fascinating topological objects and historic symbols of complexity that have fascinated the human mind since the dawn of our history. Knots and links are treated within the mathematical discipline of topology, but have always played a prominent role in physical and life sciences. In supra-molecular chemistry, complex links could be demonstrated as interlinked molecular rings [203, 204], called catenanes, and interlocked molecules, called rotaxanes. Polymer molecules can be knotted and entangled and this has proven to be essential for the crystallization and rheological properties of polymers [205]. Knotted fields have also been predicted in classical field theory [206] and knotted structures were observed in interfering light beams [207]. Knots and links of light appear as lines of zero intensity [208]. In hydrodynamics, hydrodynamic vortices could be entangled [209], and in biological systems, molecular knots and links are very important. The entanglement of DNA molecules [210] plays a crucial role in vital processes or replication, transcription, and recombination.

In chiral nematic liquid crystals, knots and links and even Möbius structures were first observed by Yves Bouligand in his detailed studies of topological defects in cholesteric liquid crystals by using an optical microscope [211]. However, these features remained unexplored because of the difficulty of their precise control and the lack of means to manipulate them. With the invention of laser tweezers and fluorescent confocal microscopes it was possible to get a completely new insight into the 3D structure and obtain perfect control over the creation and manipulation of defect lines in nematic liquid crystals by a strong light of the laser tweezers. Liquid crystals proved to be a particulary suitable system for study of topological properties of knots and links of defect loops, because they appear at microscopic scales and could therefore be observed and analysed with an optical microscope.

In a work of Tkalec et al. [212] and later Jampani et al. [213] it was demonstrated for the first time that microscopic loops of topological defects could be knotted and linked into complex patterns of arbitrary complexity. The medium that supported knots and links is a chiral nematic liquid crystal with colloidal inclusions. An individual colloidal particle in a chiral nematic cell requires the well-known single-defect loop structure, known as a Saturn ring and shown in Fig. 5.8a. Because the chiral nematic liquid crystal is twisted for 90° in the measuring cell, the Saturn ring is tilted at 45° with respect to the molecular alignment at the cell walls, and is slightly twisted because of the twisting environment.

When the laser tweezers were used to bring together two or several identical colloidal particles, they spontaneously fused their rings, leading to the formation of longer loops that entangle two particles. By bringing together three or more of several

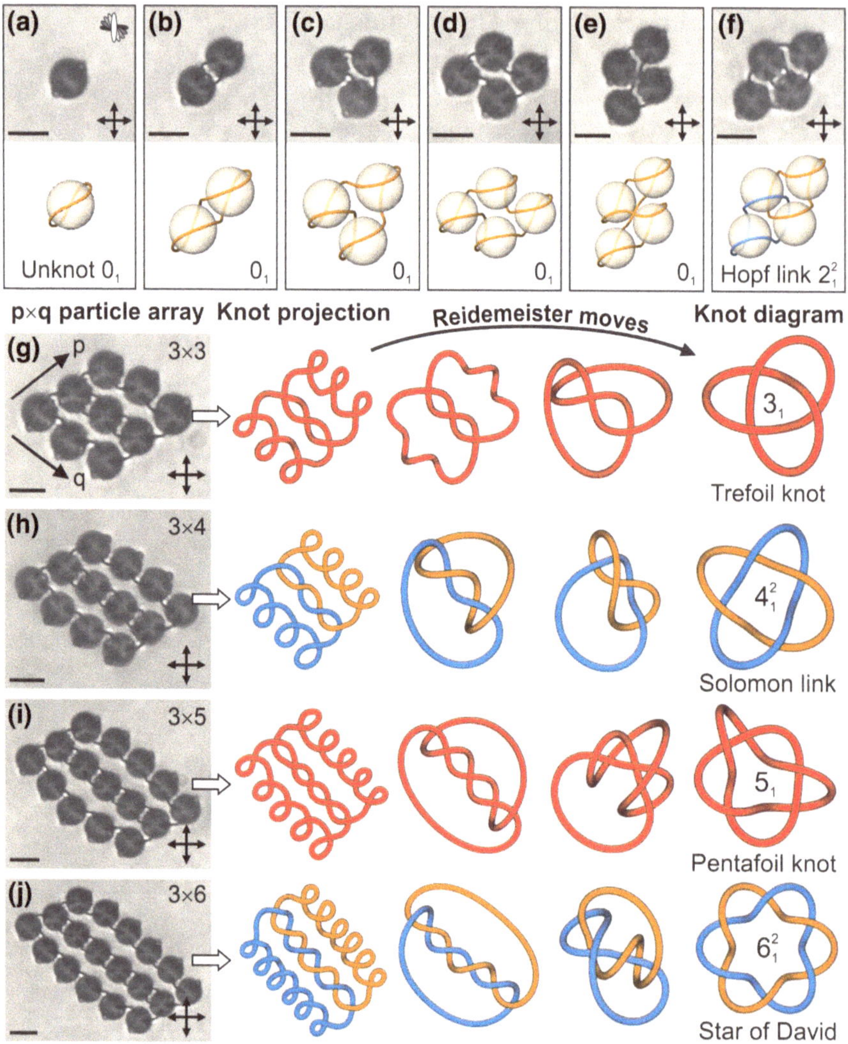

Fig. 5.8 Topological defect lines tie links and knots in chiral nematic colloids. **a** A twisted defect ring is topologically equivalent to the unknot and appears spontaneously around a single microsphere. The molecular orientation on the *top* and *bottom* of the cell coincides with the orientation of the crossed polarizers. **b–e** Defect loops of colloidal dimer, trimer and tetramers are equivalent to the unknot. **f** The Hopf link is the first nontrivial topological object, knitted from two interlinked defect loops. In **a–f**, the corresponding loop conformations were calculated numerically by using the Landau-de Gennes free-energy model. **g–j** A series of alternating torus knots and links on $3 \times q$ particle arrays are knitted by the laser-inducted defect fusion. The defect lines are schematically redrawn by using a programme for representing knots to show the relaxation mapping from the initial planar projection to the final knot diagram, which was performed by the sequence of Reidemeister moves. The designations of knots follow the standard notation C_i^N, where C indicates the minimal number of crossings, i distinguishes between different knot types, and N counts the number of loops in multicomponent links. These panels were drawn using KNOTPLOT 1.0. Scale bars, 5 μm

particles, it was necessary to use a rather strong light of the laser tweezers to locally heat the nematic liquid crystal, which resulted in heat-assisted fusion of defect loops into more complex colloidal structures shown in Fig. 5.8b–f. By choosing the focal plane of the microscope at different heights, it was possible to follow the defect lines through the liquid crystal and determine the resulting shape of the loop. Lower panels in Fig. 5.8a–f show the numerically calculated equilibrium of the colloidal superstructure together with defect loops. One immediately notices a one-to-one correspondence between the experimental panels in the upper row and the numerical panels in the lower row. A more detailed analysis of the shape and interrelation of topological loops reveals that the structures with up to 4 colloidal particles are topologically trivial. They are all formed of a single loop which embraces one, two, three or four colloidal particles without being knotted and linked.

Knottedness or linkedness can be analysed by performing moves called Reidemeister moves, also shown in panels (g–j). One is allowed to twist, stretch or perform other actions on the loops, apart from cutting and rejoining them. One can immediately see that the loop in Fig. 5.8e is in fact a simple loop, in topology also called the unknot. The first topologically non-trivial structure is observed in four entangled colloidal particles, shown in panel (f) of Fig. 5.8. There are clearly two loops which are not separated, and could never be separated without being cut. They form the famous topological object called the Hopf link.

The true richness of knots and links in chiral nematic liquid crystal colloids is revealed when the colloidal clusters are extended to the regular colloidal array of $p \times q$ particles, as shown in the first example of Fig. 5.8g. The laser tweezers were used on selected points on the array, which resulted into the reconnections of topological defect loops, which will be described in more detail in the Fig. 5.9. Series of nematic knots and links, realised on $3 \times q$ particle array is shown in panels (g–j) of Fig. 5.8. To identify the topology of the entangled defect loops, we performed a sequence of Reidemeister moves, which preserved the overall topology of the tangle. These Reidemeister moves virtually transform the real conformation of the loops into its planar projection with the minimum number of crossings by smoothing the twisted parts of the loops. It was found that the negative or left-handed crossings are favoured in a left twisted nematic liquid crystal because of the overall chirality of the measuring cell. The resulting mappings of the topological loops reveal a rather surprising result. There is a series of alternating torus knots and link, starting with the Trefoil knot, the Solomon link, the Pentafoil knot and the Star of David. This series of generic knots and links shows that the confining regular lattice of colloidal particles allows for the production of torus links and knots of arbitrary complexity. This can be achieved simply by adding and interweaving additional rows of colloidal particles, that is by increasing q.

The knots and links could also be retied by using the laser tweezers, which is illustrated in Fig. 5.9. The first panel in Fig. 5.9a shows an array of 3×4 particles and we concentrate on the encircled region. This region shows a local detail of the local crossing of the two defect loops which appears to be situated in-between the four colloidal particles. On this side, the two defect loops are simply crossing each

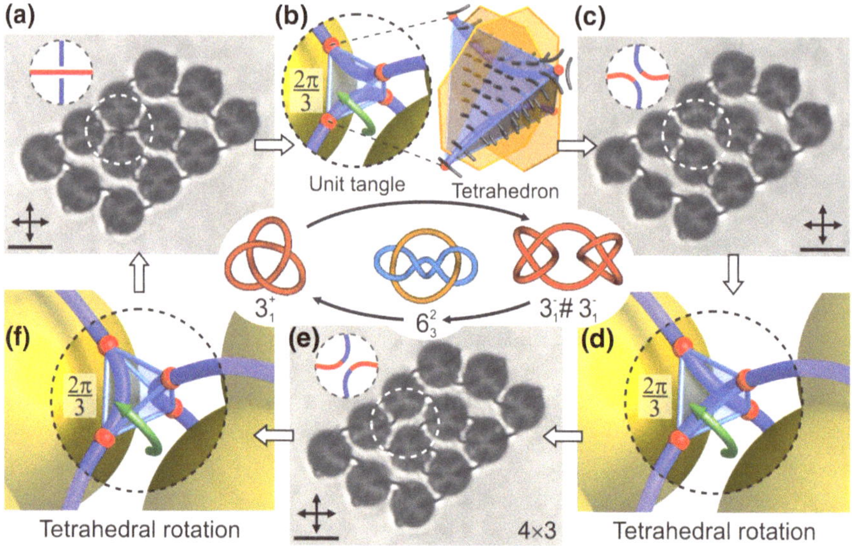

Fig. 5.9 Rewiring of knots and links by use of laser tweezers. **a** A right-handed trefoil knot is realised on a four-by-three colloidal array. The *dashed circles* indicate a unit tangle that can be rewired with the laser beam. The tangle consists of two perpendicular line segments and the surrounding molecular field. **b** By rewiring the unit tangle that corresponds to a $2\pi/3$ rotation of the encircled tetrahedron, a new composite knot, shown in **c**, is knitted. The sequence of tangle re-wirings in **b**, **d**, and **f** results in switching between knots and links, demonstrated in **a**, **c**, and **e**. Scale bars, 5 μm

other and one is positioned above the other, as shown in the red-black crossing in panel Fig. 5.9a.

If we look at this region from a perspective, we will see what is shown in the Fig. 5.9b. There are two colloidal particles painted yellow and there are two blue coloured defect lines crossing each other one on top of the other. If we further concentrate on the artificially created tetrahedron ending with red points, we can see that the director field is practically perpendicular at each of the surface of this tetrahedron. This is illustrated in a zoom-out drawing in Fig. 5.9b. An interesting observation here is that this tetrahedron could be rotated by 120° around each of the four of its symmetry axes. This transforms the director field on the surfaces of the tetrahedron into itself, but there is a huge change in the configuration of both sections of defect lines: they are moved to a different position by this rotation. This moving to a different position by rotating the tetrahedron around any of its symmetry axes means that the unit tangle (this is the name of the local crossing) is reconnected.

Although this is only a possible mathematical operation, it can indeed be realised in practice by the action of the laser tweezers. So by focusing the light on the encircled unit tangle in Fig. 5.9a, the local action of the light transforms this unit tangle into another configuration, shown in Fig. 5.9c. Note that the unit tangle configuration is now different, as shown by two bypassing and curved defect loops, which originate in the originally crossed red and blue defect loops. Under the action of the laser tweezers, this crossing was broken and reconnected in a way that each red section is now connected to blue sections. Looking at the panel in Fig. 5.9c, one can immediately notice that the topology of the defect tangle in this colloidal array is now changed. By applying the next transformation of this unit tangle, the third possible tangle state is reached, as shown in Fig. 5.9e. Again, mathematically this represents a rotation of the virtual tetrahedron around a symmetry axis for 120°, as shown in Fig. 5.9d. Local rotation results in rewiring of the unit tangle into the third possible configuration, shown in the insert to Fig. 5.9e. If this is continued, one can again reach the point-of-origin colloidal structure presented in Fig. 5.9a.

It is clear from the experiment described above that there are three possible and topologically different unit tangles at each side surrounded by four colloidal particles. Using the laser tweezers one can transform these unit tangles one into another and this results in a global change of the topology of the tangled colloidal cluster as a whole. For our 3×4 array shown in Fig. 5.9 this means that the topology was changed from the original trefoil knot Fig. 5.9a into the left-handed composite knot in Fig. 5.9c and the two-component link (Fig. 5.9e). Therefore a simple modification of a single crossing site results in three different topological structures. One could immediately guess there is a huge variety of all possible topological states, which could be reached by applying the rewiring technique at any other unit tangle. During this process, one might reach the same structure after performing different operations, and there is a given number of possible states that could be realised on a predetermined $p \times q$ array.

The optical retying of knots and links is directly related to the changes in the orientational field of the nematic host. The defect loops that we are knitting and tying into knots and links are not only structure-less things, but are surrounded by the nematic director field. They therefore posses a threefold rotational symmetry of the hyperbolic molecular orientational profile in the plain, which is locally perpendicular to the defect line. This is shown in the insert to Fig. 5.10a. This threefold symmetry pattern can twist as we move along the disclination loop and makes the defect loops three-sided strips which are a generalisation of the well-know Möbius strip. Because we always have closed loops, it is clear that only fractional values of the internal twist of the loop are allowed. This is because the director field should be continuous everywhere except in the core of the loops, and should match its original pattern as we move away from and return back to the same point on the loop. This allows rotation of the director field pattern by an integer multiple of 120°. In topology, this rotation by a discrete angle is called the self-linking number and counts the number of turns of the binom around the curb tangent for algebraic curves. In our case, it counts the number of discrete turns of the director field around the defect loop. When we have multiple loops in the system, the summation of the self-linking number of all the loops is used as a natural generalisation of this invariant. The quantisation of the self-linking number is directly related to the geometric, or Berry's phase.

The self-linking number SL and the number of loops N can be used for a unique classification of all possible loop conformations - that is all the available knot types on a given $p \times q$ particle array. Figure 5.10b shows the classification of the possible

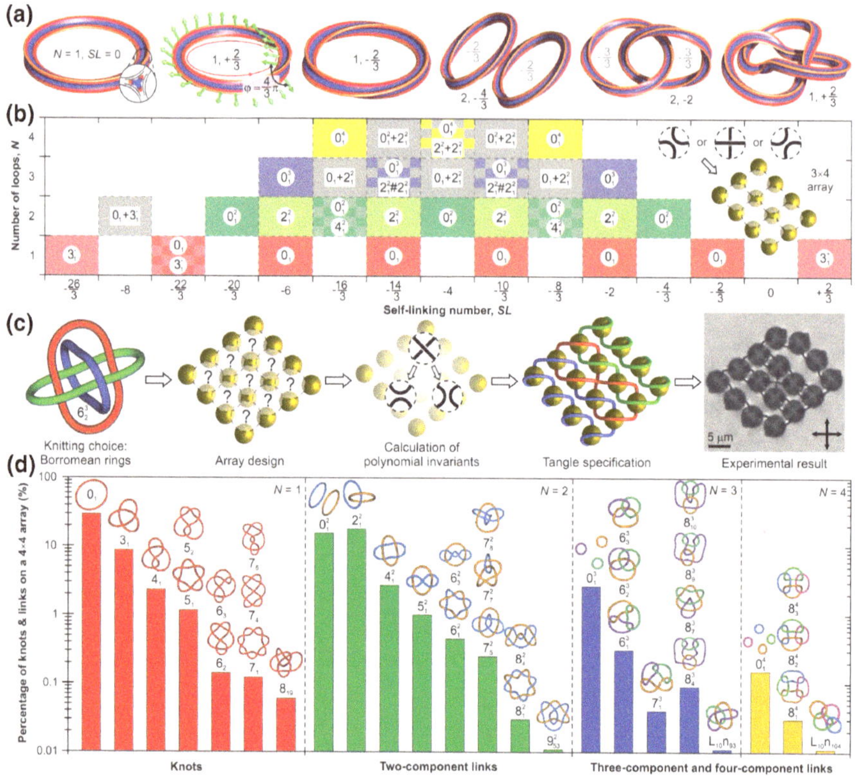

Fig. 5.10 Topological classification and made-to-order assembly of linked and knotted nematic braids. **a** The defect loops have a local threefold rotational symmetry of the hyperbolic cross section and correspond to three-sided strips, analogous to the Möebius strip. They can be distinctly characterised by the number of loops N and the self-linking number SL of fractional values. **b** The classification of all possible knots and links on a three-by-four particle array by SL and N. The hierarchical ordering of knots and links, depicted by distinct colours and standard knot symbols, are shown. **c** Made-to-order assembly of Borromean rings on a particular $p \times q$ particle array. The feasible tangle combinations were tested with the numerical algorithm based on the Jones polynomials and Kauffman bracket approach. The selected configuration was identified by direct comparison with polynomials in the enumerated Table of Knot Invariants [214] and then assembled by using laser tweezers. **d** The distribution of prime knots and links on a four-by-four particle array shows a large diversity of topological conformations with minimum crossing numbers up to 10. The probability of occurrence of a particular knot or link decreases with its complexity, as measured by the minimum crossing number

topological objects, which could be realised on a three-by-four particle array. This was determined by testing all the possible configurations of unit tangles. Using the self-linking number and the number of loops, the knots and links arrange hierarchically and regularly alternate between the knoted/unknoted and linked/unlinked structures. This observation is promising for predicting the complexity of the knots and links that can be realised on a specific $p \times q$ particle array.

Because there is perfect control over the tangle crossings by using the laser tweezers and because there is a theoretical tool for finding all possible conformations of nematic base, it is possible to assemble a desired topological structure by demand. This is illustrated in the assembly of the well-known Borromean rings as an example of a complex topological structure. The smallest possible colloidal array for the formation of Borromean rings is calculated to be four-by-four. By using a computer algorithm based on calculating the polynomial invariants from knot theory, the type of the tangles were identified for each crossing for each side, as shown in Fig. 5.10c. Following this theoretical recipe, it was possible to reconnect all tangles into the desired conformation, which resulted in a colloidal array, linked by Borromean rings.

Figure 5.10d shows the diversity of possible topological structures, which could be assembled on a four-by-four particle array. Almost fourty different knot and link types were identified among 39 possible tangle combinations with minimum crossing numbers up to ten. For this array 35% are prime knots or links, 29% are unknots, 18% are unlinks, and 18% are more complex composite links. Such a large diversity of topological object suggests that it is possible to design any knot or link on a sufficiently large colloidal array [215]. Chiral nematic colloids are a system, especially rich in knots and links. In other soft matter systems and polymers, the occurrence of knotted structures is fairly low and rarely reaches several percents of possible structures.

The analysis of knots of links was later spread to higher chirality of the surrounding liquid crystal and the natural parameter which describes the degree of chirality in chiral nematic colloids that is the ratio of the helical period p to the diameter of colloidal particles d. The experiments were performed in planar cells with their rubbing directions which determined that the surface alignments of the liquid crystal were set parallel on both surfaces. At a fixed thickness of the measuring cell h and fixed diameter of the colloidal particle d, the cells were constructed so that the thickness was just slightly bigger than the colloidal diameter d. The chirality was then varied by varying the helical period of the chiral nematic liquid crystal. This was done by adding the proper amount of the chiral dopant to the non-chiral nematic crystals. This way, the measuring cells were constructed, where the total twist of the chiral nematic structure was set to the multiples of π (π, 2π and 3π cells). The colloidal particles were chemically treated to induce perpendicular surface alignment of the liquid crystal. After introducing colloidal dispersion, laser tweezers and optical microscopy were used to analyse the structure of topological defects around the colloidal particles. At a later stage, the interaction and the entanglement of colloidal particles in cells with variable chirality were studied.

Figure 5.11 shows the structure of defect loops in π-, 2π- and 3π-twisted cells together with the results of the LdG numerical analysis displayed in the far right panels. These structures were determined by grabbing defect loops with laser tweezers and observing the defect loop structure under an optical microscope with a variable height of the focal plane, as shown in Fig. 5.12. In all cases, one could clearly determine the winding patterns of defect loops, most clearly visualised in panel (d) and (e) of Fig. 5.11. It is then easy to understand why this winding occurs. Let us begin from the simple, non-twisted cell and a colloidal particle, which is encircled by the

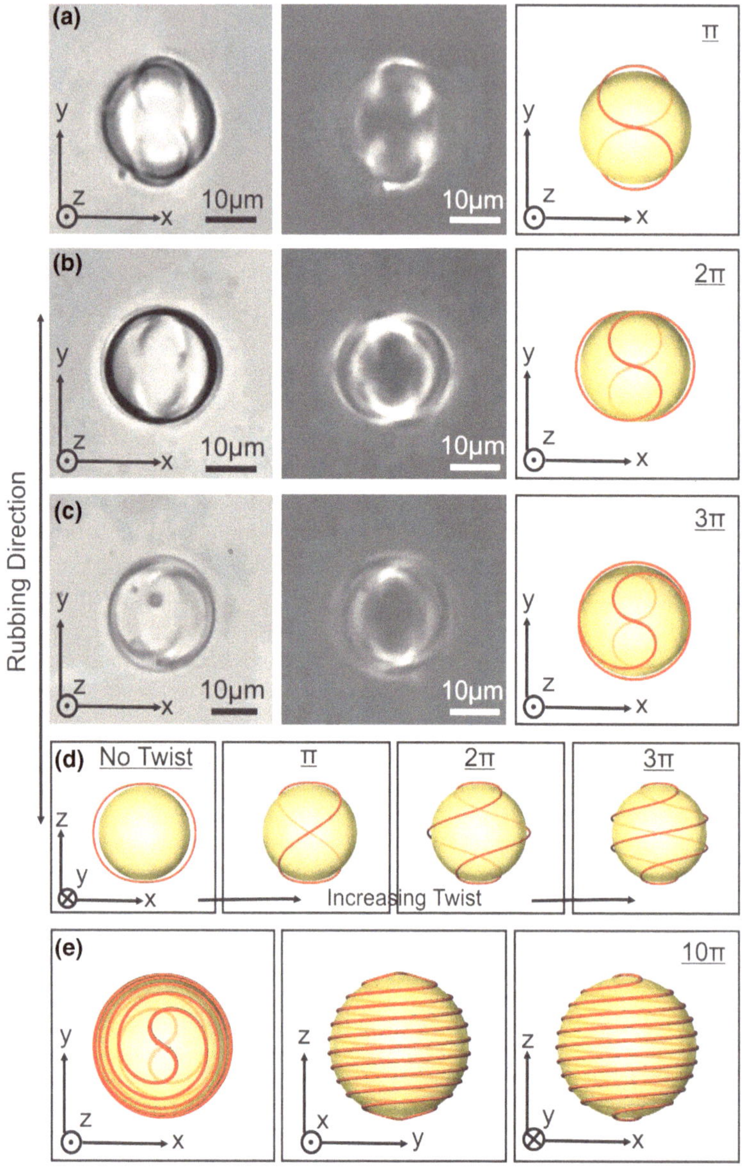

Fig. 5.11 **a** Unpolarised and polarised images of 20 μm silica colloidal particle with homeotropic surface anchoring in a $h = 22\,\mu$m thick and right-handed π-twisted cell of 5CB. The *right panel* shows the result of LdG numerical analysis. The *red line* presents the regions of the CLC with the order parameter $S = 0.51$. The size of the colloidal particle is 3 μm and the cell thickness is 3.2 μm. **b** The as in **a**, but the twist of the cell is now 2π. **c** The same as in **a**, but the twist of the cell is now 3π. **d** Numerical LdG calculation of winding of the defect loop in planar, π-, 2π-, and 3π-twisted CLC cells. Defects are shown in *red* as iso-surfaces of $S = 0.51$. **e** *Top* and *side view* of the winding of the defect line in a 10 π-twisted CLC cell

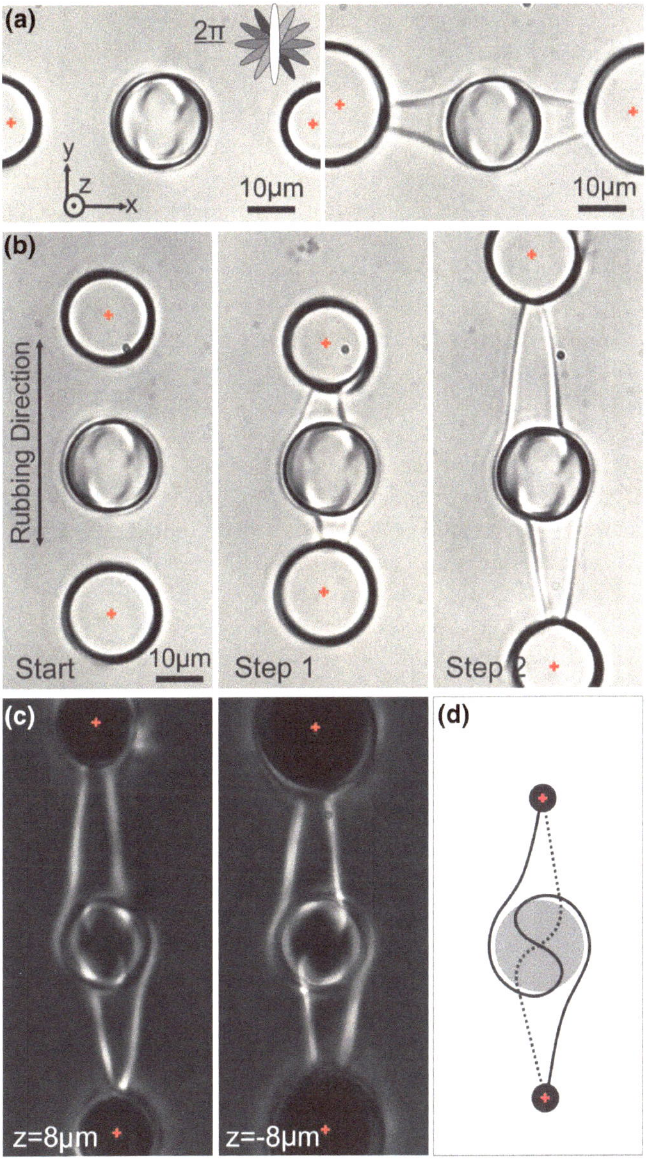

Fig. 5.12 **a** Silica microsphere of diameter $d = 20\,\mu\text{m}$ with homeotropic surface anchoring in a right-handed 2π-twisted nematic cell with thickness $h = 22\,\mu\text{m}$. Note the defect line wrapping around the colloidal particle. The circular object on the *right side* of the *left panel* is the isotropic island of 5CB, produced by local heating with high-power laser tweezers, indicated with a *small cross*. The laser power was 170 mW for each trap. The *right panel* shows how the molten island of 5CB attracts the defect line and stretches it in the x direction. **b** Two equal beams of the laser tweezers were used to stretch the defect line in the y direction. **c** The defect line is stretched by two beams in the y direction, and the images between crossed polarisers are shown for two different z positions of the focal plane of the microscope, separated by $16\,\mu\text{m}$. **d** Schematic view of the laser-tweezers-stretched defect line, wrapping the colloidal particle in the 2π-twisted nematic cell

Saturn ring shown in the side view on the first panel of Fig. 5.11d. Now we imagine that the upper glass is turned for π, thus forming a π-twisted chiral nematic structure. Because the sample was twisted, the originally flat Saturn ring is also twisted for π and forms the figure-of-eight side projection pattern, shown in the second panel of Fig. 5.11d. Further twisting of the cell results in a more chiral structure and a more twisted Saturn ring. This originally flat Saturn ring is therefore winding around the colloidal particle and can form highly twisted and densely packed winding patterns, as shown in panel (e) of Fig. 5.11.

The winding of the defect loop around colloidal particles in the chiral nematic liquid crystal has a pronounced effect on the pair interaction of colloidal particles, as discussed in Sect. 2.9 of this book. However, it also strongly affects the entanglement of a pair of colloidal particles, which was proved in quenching experiments. Whereas in nonchiral nematic cells, three different colloidal entangled states are stable, the richness of the entangled states is much bigger in chiral nematic cells. Because of the twisted environment, more entangled colloidal structures are stable and can be more easily formed due to more freedom of propagation of defect lines through the twisted structure of the chiral nematic liquid crystal. Similar defect structures were analyzed by Čopar et al. in densely packed 3D colloidal lattices [216].

The entangled pair colloidal structures were studied in detail in π-twisted cells, as shown in Fig. 5.13. In these experiments, the elements of a pair of colloidal particles were positioned close to each other in a π-twisted cell by using the laser tweezers. Then, the laser power was increased to melt the liquid crystal around the colloidal pair into the isotropic phase. After switching off the light, the colloidal pair was found to be spontaneously entangled by defect tangles of different complexity. The quenching experiments showed at least 17 different entanglements in a π-twisted cell. Some of them are shown in Fig. 5.13, together with results of numerical LdG calculations. A closer inspection of the entangled colloidal dimers reveals that they are all variations of the same structure. These variations can be obtained by applying the method of localised tetrahedral rotation to the orthogonal defect lines crossing (or tangles). The differences between the dimer structures can be considered as localised re-wirings at four tetrahedrally shaped rewiring sites, as schematically illustrated in Fig. 5.14.

The topological analysis using tetrahedral rotations shows that there are 81 possible but only 36 topologically different dimers. Due to different geometry and symmetry, these structures are energetically different. In real experiments, we observe six stable structures, all of them being predicted by the above mentioned analysis. The observed entangled colloidal dimers can be classified into two categories with different numbers of closed defect loops. Structures in Fig. 5.13a–d are entangled by a single closed defect loop. By performing Reidemeister moves on the defect lines, we observe the projection of the loop onto a plane, and find that no structure in Fig. 5.13a–d is a knotted one. In the structure, entangled by two defect loops in Fig. 5.13f, we find that this is the before discussed Hopf link. If the number of interacting colloidal particles in chiral nematic cell is increased to three or more particles, the topology becomes even richer, as shown in Fig. 5.15.

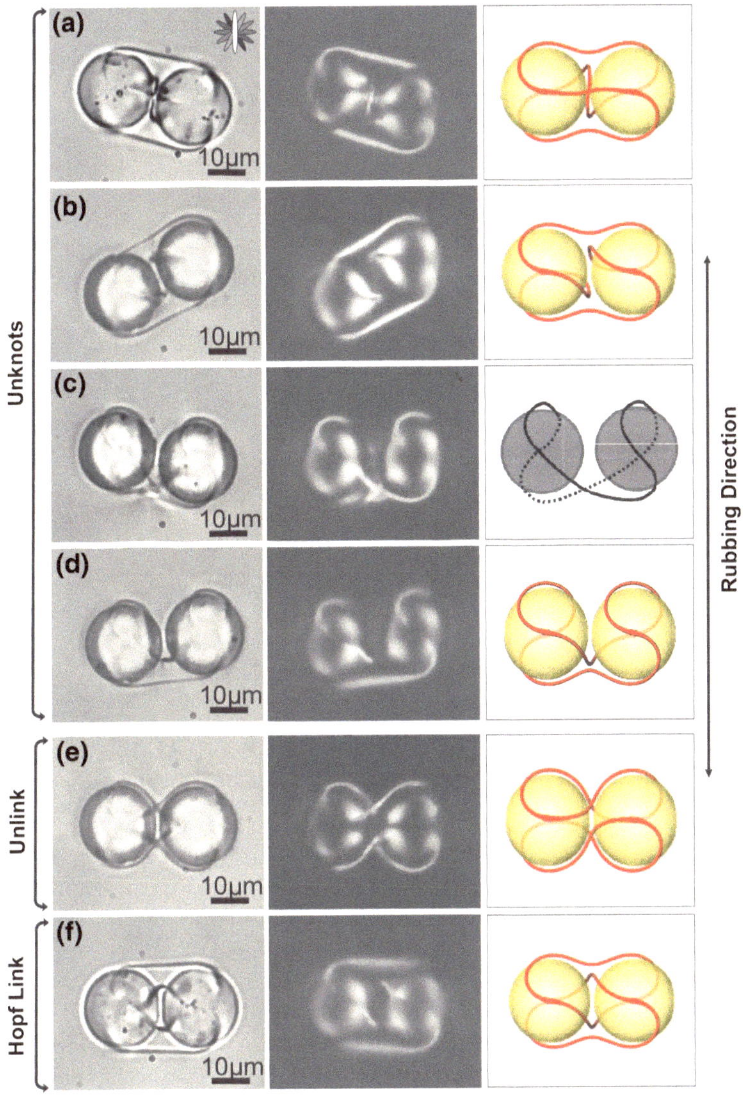

Fig. 5.13 Entangled colloidal dimers in the right-handed π-twisted CLC cell of $h = 22\,\mu$m thickness. All structures were obtained by quenching a colloidal pair from the isotropic phase, created by local heating with the laser tweezers. *Left panels* are taken in unpolarised light; panels in the middle are taken between crossed polarizers *right panels* present either numerically calculated colloidal dimers ($S = 0.51$) or just a schematic presentation of defect loop topology, as depicted from *left panels*. In a total of 176 experiments performed, the various structures **a**–**f** appear in **a** 6.8%, **b** 7.3%, **c** 6.2%, **d** 26.1%, **e** 5.1%, and **f** 10.8% of the experiments

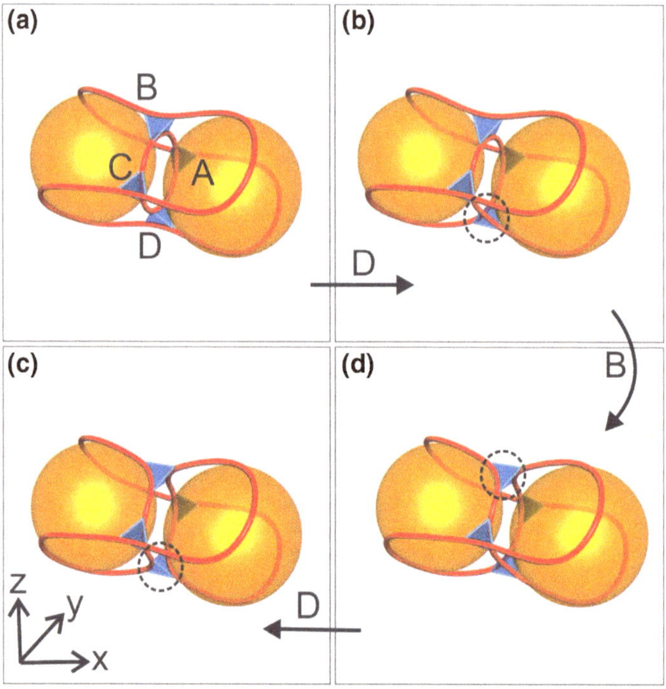

Fig. 5.14 Schematic depiction of entangled colloidal dimers with four rewiring sites labeled A–D. The rewiring sites are local tetrahedrons, where disclinations can take different paths, forming different entangled structures. The chosen structures (**a**–**d**) only differ in the *top* (B) and the *bottom* (D) rewiring sites, i.e., tetrahedrons. Rewiring of the tetrahedron D transforms the structure **a** into **b**. Subsequent transformations into **d** and **c** are performed in a similar manner, as shown in the figure. **a** is the most symmetric entangled dimer that is invariant to rotations for $2\pi/3$ around any of the principal axes

The colloidal trimer in Fig. 5.15a is entangled by a single defect loop, which is an unknot, similar to the tetramer in Fig. 5.15b. However, the second colloidal trimer in Fig. 5.15c is entangled by two interlinked defect loops, it is therefore a Hopf link. The latter is also shown for the colloidal tetramer in Fig. 5.15d. For this tetramer, another topological structure is found in Fig. 5.15e, where three defect loops are mutually interlinked, thus forming a short chain of topological loops. And finally, there is a knotted structure shown in Fig. 5.15f. It has three crossings with a single loop and is known as a trefoil knot.

The observation of knotted and linked defect loops in chiral nematic colloids, formed of a dispersion of spherical microparticles in a chiral nematic crystal, triggered large interest in the topology of liquid crystal colloids. Whereas all these experiments were performed by using a topologically simple microsphere of genus $g = 0$, further studies concentrated on topologically more complicated objects, such as tori, handlebodies, knotted and linked colloidal particles and colloidal particles with non-orientable surfaces.

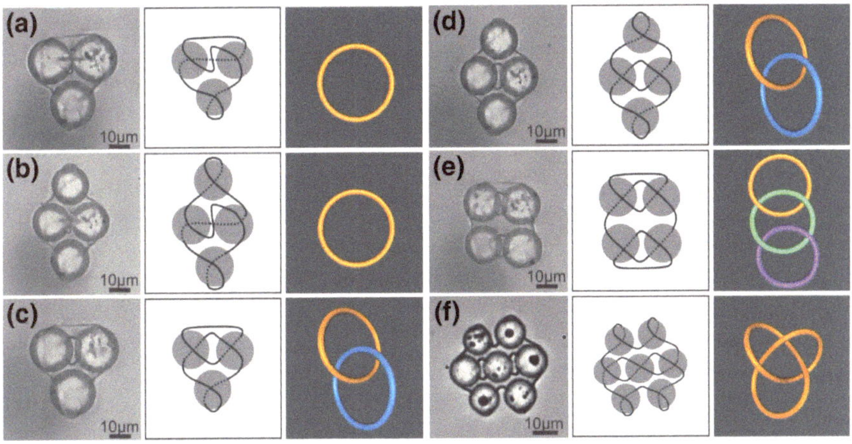

Fig. 5.15 Entangled colloidal clusters in the right-handed π-twisted CLC cell. The images on the *left* are taken under crossed polarisers. The panels in the *middle column* represent drawings of the defect loops, as deduced from images on the *left*. The *right panels* show the topologically minimised defect loop structure, after performing Reidemeister moves. These panels were drawn using KNOTPLOT 1.0. **a–e** Colloidal particles of diameter d = 20 μm in h = 22 μm thick cell. **f** Colloidal cluster of h = 10 μm silica microspheres in a h = 12 μm thick cell

5.3 Charge Production and Entanglement on a Fibre in a Nematic Liquid Crystal

Several years ago, topological charge was considered an exotic mathematical formulation with little importance to physical phenomena. However, in recent years, it started to play an important role in very different fields, materials, and systems, such as superconductors, a superfluids, systems of cold atoms, or soft ferromagnets. Topology, which is a mathematical discipline studying deformable surfaces, is now being considered as an important aspect in topological insulators, light transport, polymer chemistry, electromagnetism and acoustics. Strange enough, this field has witnessed a revival in liquid crystals, where it was intensively studied in the 1970s by Yves Bouligand and others [211, 217–221]. However, 40 years ago, there were no methods and instruments to video image in 2D, micro image in 3D, and grab micron-sized particles by laser tweezers. Today, we have methods of imaging 3D structures and fields on the sub-micron scale, visualise them in 3D and move and manipulate micrometre-sized objects in liquid crystals by using the laser tweezers. This is the reason why topological phenomena could be precisely studied in the experiments performed over the last five years.

An interesting question, connected to the conservation law for the topological charge, has emerged when considering possible applications of nematic colloids in photonics. The idea is to have microphotonic elements, such as lasers, fibres and optical microcavities dispersed in a nematic liquid crystal, where they are entangled together using the entanglement mechanism, described in this chapter. However, fibres are equivalents to rods an these are equivalent to spheres, because they can be smoothly transformed by shaping their surface continuously, which is allowed in topology. This means they are all objects of genus $g = 0$, which we know produces only one topological point defect. This is due to the conservation of the overall topological charge, and the newborn defect compensates for the topological charge of the inserted object with genus $g = 0$. However, in real application one needs to have more than one topological charge on the fibre, as we need to connect several photonic microelements to the fibre. The question arises of how to produce an arbitrary number of topological charges on a single fibre with genus $g = 0$?

The law of conservation of topological charge does not prohibit the existence of a multitude of charges on the fibre, as the only fundamental request is that the total topological charge be equal 0. This means that pairs of topological defects of opposite charges could exist on the fibre; the question is how to create them. This was explained in the experiments reported by Nikkhou et al. in 2015 [8, 101, 222].

It is well-known from particle physics that particle and anti-particle pairs could be created by smashing elemental particles together with high energies. A similar mechanism is considered to be responsible for the emergence of matter and anti-matter on the cosmological scale. This, the so-called mechanism of monopole creation on the cosmological scale was first considered by Kibble and was later on introduced to condensed matter by Zurek [32]. Monopoles, which are the singularities of a physical field, are produced by a rapid quench across the phase transition which is attributed to that particular field. Typically, there is a disordered phase of this field at high temperatures, whereas at lower temperatures, the field gets spontaneously ordered by the symmetry breaking, which occurs exactly at the phase transition. A quench is a rapid decrease of the temperature across the phase transition, which nucleates small regions of the newborn ordered phase, surrounded by disordered field around them. During the evolution, these regions grow and meet each other at phase boundaries, leaving behind topological defects in a form of points, strings, walls, etc. The very nature of these defects depends on the nature of the order in field. In liquid crystals, this is the field of orientational order it is a tensorial field and corresponding monopoles are topological point defects, loops of string-like objects and non-singular soliton walls.

Kibble–Zurek production of topological defects in liquid crystals was studied by many authors in the past [22, 223, 224], mostly concentrating on the time evolution of defects, their coarsening and annihilation. These studies were performed by using optical microscopy, and quenching, which could be a thermal quench, a pressure quench, or a sudden application of an electric field. In all cases, this quench was applied to rather large samples, where the boundary conditions of the quench were not really well controlled and only average properties of the sample topology were measured and studied.

Instead of applying a non-localised quench on an ill-defined sample, one can use the laser tweezers to apply a local temperature quench to a nematic liquid crystal, as shown in Fig. 5.16. We use the absorption of the focused beam of the laser tweezers to locally heat the nematic liquid crystal into the isotropic phase (Fig. 5.16a). This creates a 100 μm diameter island of a molten (isotropic) liquid crystal, which is

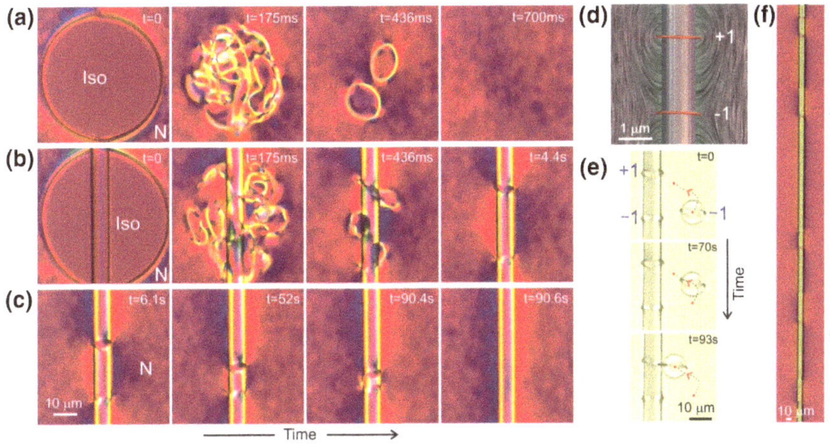

Fig. 5.16 Creation and annihilation of topological charges on a fibre. **a** The nematic liquid crystal is heated into the isotropic phase by the strong light of the laser tweezers, this creating an isotropic island (Iso). At $t = 0$ the light is switched off and the nematic liquid crystal is quenched into the nematic phase (N). The dense tangle of defects annihilates in less than a second. **b** The nematic liquid crystal is quenched from the isotropic island surrounding a fibre. A pair of defects is created, each carrying an opposite topological charge. **c** If let free, the pair annihilates into the vacuum. **d** LdG simulation of the Saturn ring and the Saturn anti-ring with opposite charges and windings. **e** The sign of the charge is tested using the repulsive force between like topological charges. **f** An arbitrary number of ring-anti-ring pairs can be created on a fibre. **a–c**, **f** were taken between crossed polarisers and the *red plate*, which shows the average molecular orientation in different colours

than rapidly quenched by shutting off the light. The island undergoes a rapid phase transition that leaves behind a dense tangle of defects. In less than a second, this tangle annihilates back to the uniformly ordered ground state of the nematic liquid crystal, which is the vacuum state with no stable defects.

However, there is marked change in the outcome of tangle coarsening, when the local melting experiment is performed in a sample containing a fibre. This is shown in Fig. 5.16b, where a rather long fibre was inserted during the fabrication of the sample, so that it is now embedded completely into the nematic liquid crystal. This liquid crystal was again locally molten by the laser tweezers, producing an island of the isotropic phase, which contains the fibre. When the light is switched off, in hundreds of milliseconds the dense tangle nearly annihilates completely and leaves behind a well-separated pair of topological defects, which are obviously stabilised by the perpendicular alignment of molecules on the fibre. Having in mind the law of conservation of topological charge, one immediately suspects that this should be a particle and its anti-particle or, in the case of liquid crystals, it should be a Saturn ring and the Saturn anti-ring. They should behave like a particle and anti-particle and should have topologically different structures and properties, such as the opposite winding number and the opposite topological charge. Only in this case, the overall topological charge is preserved, thus assuring the charge neutrality. If they are well-separated from each other, these two rings are indeed inherently individually stable and cannot be annihilated separately. Each of them could be grabbed by the laser tweezers and moved away from the other or closer to each other. If they are positioned close together using the laser tweezers and let free, they will be attracted to each other through the elastically distorted nematic liquid crystal. They start sliding and accelerating to each other, until they annihilate into the vacuum, as shown in Fig. 5.16c. The localised temperature quench could be repeated on other parts of the fibre, producing additional pairs, and practically an arbitrary number of ring-anti-ring pairs can be created (Fig. 5.16f).

Careful observation of each of the two rings (such as in the last panel in Fig. 5.16b) clearly resolves a quite different optical appearance of these two rings. Because this panel was imaged between crossed polarisers and with the red plate inserted, one can clearly see reddish and bluish regions close to the fibre, which alternate in side when one is moving along the fibre and across each ring defect. This alternation in colour means an alternation of liquid crystal orientation, clearly pointing to the different internal structure of the Saturn ring and the Saturn anti-ring. The molecular arrangement around such a pair of rings is illustrated in Fig. 5.16d by using the LdG theory. One can immediately recognise the characteristic three-fold cross-section of the familiar Saturn ring (lower ring in Fig. 5.16d). It has a winding number of $-1/2$ and has a topological charge of -1. The upper ring in Fig. 5.16d is the anti-particle of the Saturn ring; it has the opposite winding number of $+1/2$ and the opposite topological charge of $+1$. By observing Fig. 5.16d, one can see that the region between the two rings is actually in favour of both rings. This means that the rings prefer to approach each other as much as possible, to share as much of this common distortion and the use common elastic energy. This immediately leads to the conclusion that the attractive force, resulting in ring attraction and annihilation is actually due to the elastic distortion of the nematic liquid crystal between the two rings.

The topological charge of each ring could be tested using a small reference charge, such as the small colloidal particle with a Saturn ring in Fig. 5.16e. The Saturn ring of this small test particle has a -1 charge and is therefore repelled from the Saturn ring on the fibre and attracted to the Saturn anti-ring nearby. In reality, it takes tens of seconds for the particle to travel the separating path of several tens of micrometres and be attracted to the anti-ring on the fibre. This method of testing the topological charge is very simple and flexible and could be used for testing unknown topological charges in 2D systems, simply relying on the fact that equally charged parts of multipoles are repelled and the oppositely charged parts are attracted.

If the two rings are put far away from each other, they exhibit much suppressed Brownian motion due to thermal fluctuation of the surrounding liquid crystal. In this case, the rings are stable for a very long time; however, when they are closer than a critical separation of the order of several fibre's diameter, the resulting structural force starts pulling them together until they annihilate, which is shown in a sequence of frames in Fig. 5.17a. The dynamics of Saturn ring annihilation can be analysed by tracking their positions in an off-line analysis of recorded video frames. Using rather high frame rate, thousands of frames could be collected within the process of annihilation. This means that the trajectories of each of the rings can be very well

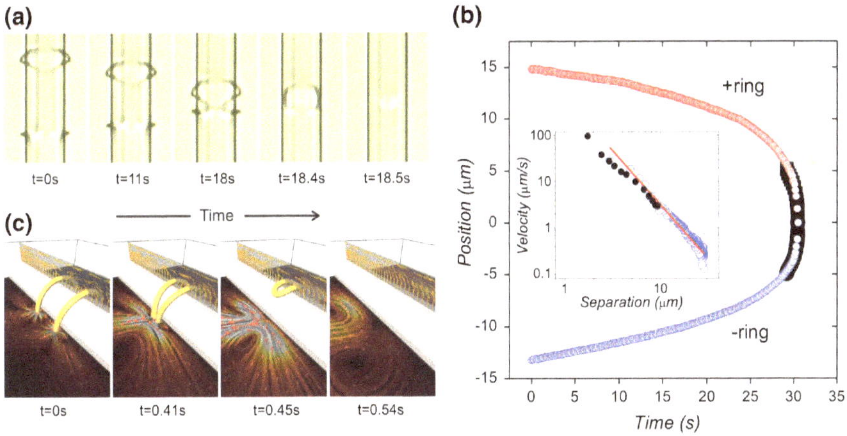

Fig. 5.17 Dynamics of charge annihilation on a fibre. **a** A pair consisting of a Saturn ring and an anti-ring is annihilated on a fibre. The length of the fibre is ~400 μm, the diameter is 8 μm and the cell thickness if 65 μm. **b** The positions of the + (*red*) and the − (*blue*) Saturn rings as a function of time during pair annihilation. The *inset* shows the relative velocity of the two rings. The *open symbols* are data from the experiment; the *closed symbols* (*black*) are data from the numerical simulation. The *red line* is a linear fit to the *blue data points*. **c** Numerical simulation of the ring and anti-ring attraction, showing the director (*yellow*, vertical plane) and velocity fields (*red*, horizontal plane)

reproduced as a function of time. An example of the annihilation event of the two rings is shown in Fig. 5.17b for the positively and negatively charged ring separately. There is clearly an asymmetry in the dynamics of the + and the − ring, which is commonly observed for opposite topological charges in liquid crystals. This asymmetry is most pronounced in the velocities of the + and the − ring, which is obtained by calculating numerical time derivative of positions in Fig. 5.17b. It turns out that the $+1/2$ ring is faster than the $-1/2$ ring and the ratio of their velocities is approximately $v_+/v_- \approx 1.5$. This result is similar to previously reported experiments on topological string attraction and annihilation in nematic cells, which also consistently reported faster dynamics of defects of positive winding number. The relative velocity of the two rings as a function of their separation shows a power-law dependance $v \approx 1/d^{\alpha}$ with the exponent $\alpha \approx 2.2 \pm 0.2$.

This experimentally observed dynamics of the annihilation of rings on the fibre can be compared against the rings' dynamics calculated from nematodynamics. In this case, the Beris–Edwards model of nematodynamics was used with the hybrid lattice Boltzmann method. The one-elastic-constant approximation was considered and the material parameters were those of 5CB. The rings were initially positioned at some distance and left to attract and annihilate, following numerical calculations. Some snapshots from the simulation of the attraction of the Saturn ring and the anti-ring pair are presented in Fig. 5.17c. These snapshots show the core of the defects and the string lines represented by arrows. The differences between the + and − rings are clearly shown in theoretical points added to the experimental data in Fig. 5.17b. The agreement is reasonably good.

Whereas in situation discussed above the fibre was set parallel to the overall orientation of the nematic liquid crystal in the cell, interesting topology of the monopoles on the fibre is observed when the fibre is rotated by 90° and set perpendicular to the bulk orientation of the nematic liquid crystal. In the experiments, this is done by using the laser tweezers to grab one end of the fibre and rotate it to the final direction by moving the laser focus. It turns out that the perpendicular position of the fibre with respect to the liquid crystal is unstable. If let free, this fibre slowly rotates by the torque of the surrounding liquid crystal, ending parallel to the liquid crystal after a long time.

In this perpendicular orientation one observes a gigantic Saturn ring, which encircles the fibre all along its long axis. One part of this really long defect loop is shown in the first panel of Fig. 5.18a. The existence of such an elongated loop defect is understood, as a fibre is topologically equivalent to a sphere. The genus of the fibre is therefore $g = 0$, and only one defect should spontaneously appear in the ground state, which is in this case a single $-1/2$ Saturn ring. Because it is long and very soft, this ring can be cut with laser tweezers and reshaped into an arbitrary number of isolated sections with different topologies and charges.

There are many ways of performing this cutting and sectioning, and one example is shown in Fig. 5.18a. Here, the gigantic Saturn ring with the winding number $-1/2$ is first cut into two separate loops by laser tweezers. This forms a smooth and narrow region in-between the two loops, which is called a topological soliton and is shown within the dashed ellipse in Fig. 5.18a, panel (*iii*). This soliton region connects the two loops which must have opposite winding numbers, in this case it is $+1/2$ for the left loop and $-1/2$ for the right loop. After the loop on the right is further cut into two separate loops (Fig. 5.18b), an isolated $-1/2$ loop is formed that rapidly shrinks into a point hedgehog carrying a -1 charge. Now, the right neighbouring loop has the winding number of $+1/2$, so that we have an alternation of the winding numbers sign, as we move along the fibre. By further performing the cut on the $+1/2$ right loop, another point hedgehog with a $+1$ charge is created, and so on. The sign of the charge of each point defect can be tested by a small dipolar particle, as shown in Fig. 5.18c, and finally an alternating sequence of $+$ and $-$ point hedgehogs with attached dipolar colloids can be formed, as shown in Fig. 5.18d.

Returning back to the first cut of the gigantic Saturn ring with a charge -1, one realises that an unusual loop must have been created during the first cut. Because of the conservation of the total topological charge, the total charge of all newly created point defects is always -1. Now, if one of the loops in Fig. 5.18a (iii) has a topological charge of -1, the other loop must be of zero topological charge. So the question arises whether zero charge loops could be created on a fibre. This is indeed observed in a very simple case of quenching the topological soliton, as shown in Fig. 5.18e. The topological soliton is a topologically smooth region which carries no topological charge but only propagates the topological flux from one end to the other end of the sample. By quenching the soliton, one could in some cases obtain a stable loop out of nothing, which is shown in panel (iii) in Fig. 5.18e. Because it was made from vacuum, this loop must have oppositely charged ends, which could be tested using a test dipolar colloidal particle, as shown in Fig. 5.18f. If the $+1$ end

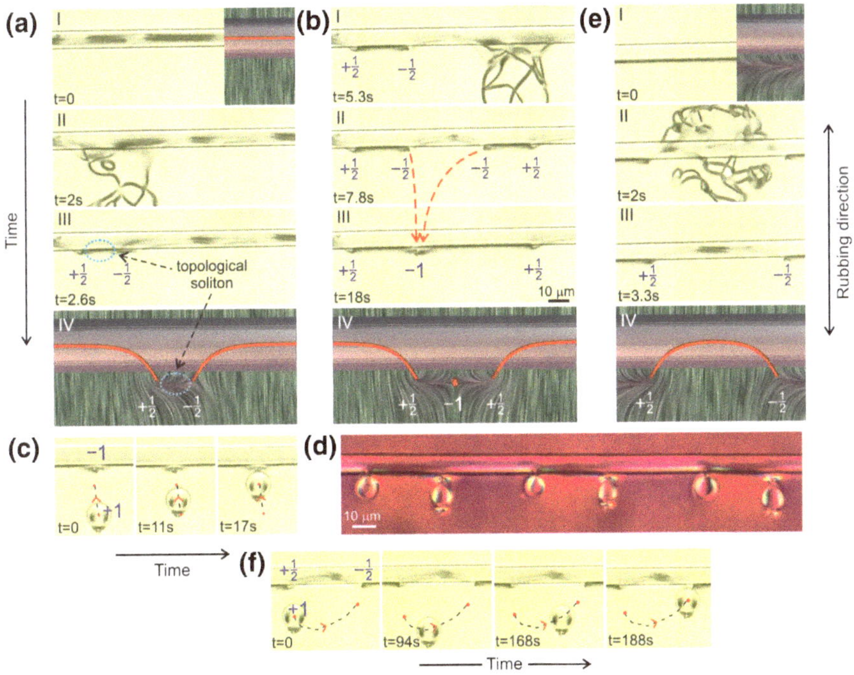

Fig. 5.18 Point charges and charge-neutral loops on a fibre. **a** After inserting a glass fibre into a thicker layer of the nematic liquid crystal, a gigantic Saturn ring with a −1 charge is created (*i*). The true image is on the *left*; the LdG simulation is on the *right*. This ring is cut by the tweezers (*ii*), creating a narrow region of a topological soliton in between the two loops with opposite winding numbers (*iii*). The LdG simulation is shown in (*iv*). **b** The second cut with the laser tweezers creates a second soliton on the *right*, isolating a closed loop in between. This loop shrinks into the −1 monopole (*iii*). The LdG analysis demonstrates the −1 monopole (hyperbolic hedgehog), with two closed loops on each side, with the winding numbers +1/2 (*iv*). **c** The topological charge of the −1 point defect is tested with an elastic dipole. The +1 part of the dipole is attracted towards the monopole on the fibre, identifying it as the −1 charge. **d** A sequence of alternating charges is created on a fibre, attracting a series of dipolar particles. **e** A topological soliton is seen as a *dark-shaded region* below the fibre, and the LdG simulation is shown on the *right* (*i*). A microquench produces a long-lived charge-neutral loop, surrounded by two solitons (*iii*). The LdG numerical simulation of a charge-neutral loop on a fibre is shown in (*iv*). **f** The charge of the charge-nautral loop is tested by the +1 end of the dipolar particle. This + end is repelled from the *left section* of the loop towards the *right section*, demonstrating oppositely charged sections

of this particle is exposed to the +1/2 end of this loop, the particle is weakly repelled from that part and attracted towards the oppositely charged end of the loop. So in total, this loop has two sections with opposite winding numbers which must meet and interconnect in two different points on the loop. A careful observation of the structure of this loop (Fig. 5.18e (*iii*)) indeed reveals some irregularities, but these are too small to be resolved optically.

Using the optical microscope, it is difficult to find and analyse the transition between the sections of the loop with the +1/2 and −1/2 winding numbers because

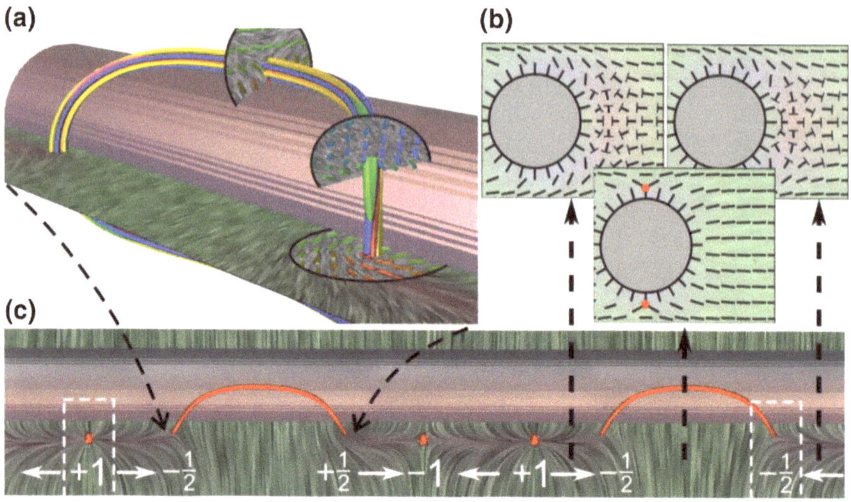

Fig. 5.19 Topological rules on a fibre. **a** A charge-neutral loop has a $-1/2$ winding number on one side and a $+1/2$ on the other, with two transitions through the twist profile. The colours indicate different local structures of the loop. The *yellow* and *blue* iso-surfaces indicate locations with a high bend and splay, whereas *green highlights* the twist deformation. Loops of this type are freely created and annihilated, as they do not contribute to the topological charge. **b** The fibre cross-section has three possible states: two states with an escaped disclination line of the winding number -1, with opposite escape directions, and the symmetric states with two $-1/2$ disclinations on the *top* and *bottom*. **c** The fibre can have any succession of the cross-sections from **b**. The transitions between cross-sections carry the topological charge, which can be assigned to entire point defects and loops, as well as to loop endings. The charge is closed in a *box* (two examples are shown), and the Gauss law measures the topological charge-that is, the number of escaped line exiting the box. The direction of the topological flux depends on the direction of the escape and is shown by *arrows*. The topological charges are marked

of complicated optical properties of deformed liquid crystal. This could be discerned by performing numerical calculations, which indeed find metastable charge-neutral loops which have a $+1/2$ winding number on one side and a $-1/2$ on the other. This is shown in Fig. 5.19a where two transition regions are presented in different colours showing how the $+1/2$ profile gradually transforms into the $-1/2$ profile.

The creation of alternating pairs of + and − charges and the formation of charge-neutral loops are governed by a simple set of topological rules. The only requirement is that any additional defect created on the fibre has to preserve the zero total winding number of the homogeneous direction field far away from the fibre. This can be achieved in three different ways: with two $-1/2$ disclination lines on the top and bottom of the fibre, or by having a soliton in the form of an escaped region with the winding number of -1 running on one side of the fibre with two possible directions of escape, as shown in Fig. 5.19b.

Similar to electrostatic charges, the topological charge of the loops and points can be determined by a Gauss law. In our case, the Gauss integral is reduced to counting the number of soliton lines carrying the topological flux away from or towards the defect. These solitons can originate from the charge and can terminate only at the topological charge. At each point of origin and end the only two possible charges are the $+1$ and -1 point defects and the $+1/2$ and $-1/2$ fractional charges, which can be assigned to the end sections of the loops. A whole loop can therefore either have the same-signed end and be a point monopole or have opposite-signed ends being a zero charged loop.

5.4 Elastic Interactions and Entaglement of Microspheres and Fibres in a Nematic Liquid Crystal

We have seen in the previous section that the multitude of topological charges could be created on a fibre using thermal quenching in different geometries of the fibre. During this process, loops or points of opposite charge or even zero charge loops are deliberately created and positioned to an arbitrary place on the fibre using the laser tweezers. The question then arises how do these topological defects on a fibre interact within an extra colloidal particle positioned close to the fiber. We have also observed in the previous section that such particles do elastically interact with charges on the fibre, and this serves us as a method to determine the unknown topological charge. However, there are other questions related to the entanglement of the external colloidal particles with the rings and loops on the fibre. Is it possible to entangle rings of opposite winding numbers, how stable is this, and what is the structure of the resulting entanglement? These and other questions were answered by Nikkhou et al. [8, 101, 222], who performed systematic experiments and analysis of entanglement and elastic interaction with differently charged rings and particles.

Let us first consider the geometry where the fibre is set on the rubbing direction, as shown in Fig. 5.20. After the quench, several pairs of topological charges are created, as shown in the last panel of Fig. 5.20.

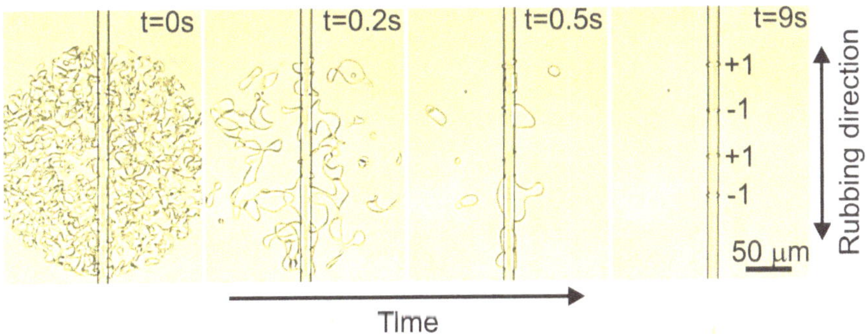

Fig. 5.20 Using a stronger light of the tweezers creates a larger isotropic island and two pairs of Saturn rings are created after the quench in this case

In the first set of experiments, the entanglement of a $-1/2$ Saturn ring on the microsphere is entangled with a $-1/2$ ring in the microfibre. The colloidal particle is brought close to the $-1/2$ ring on the fibre and the thermal quench is applied to the particle and to the $-1/2$ ring on the fibre. As a result, three entangled defect structures are found, as shown in Fig. 5.21. These entangled structures are identical

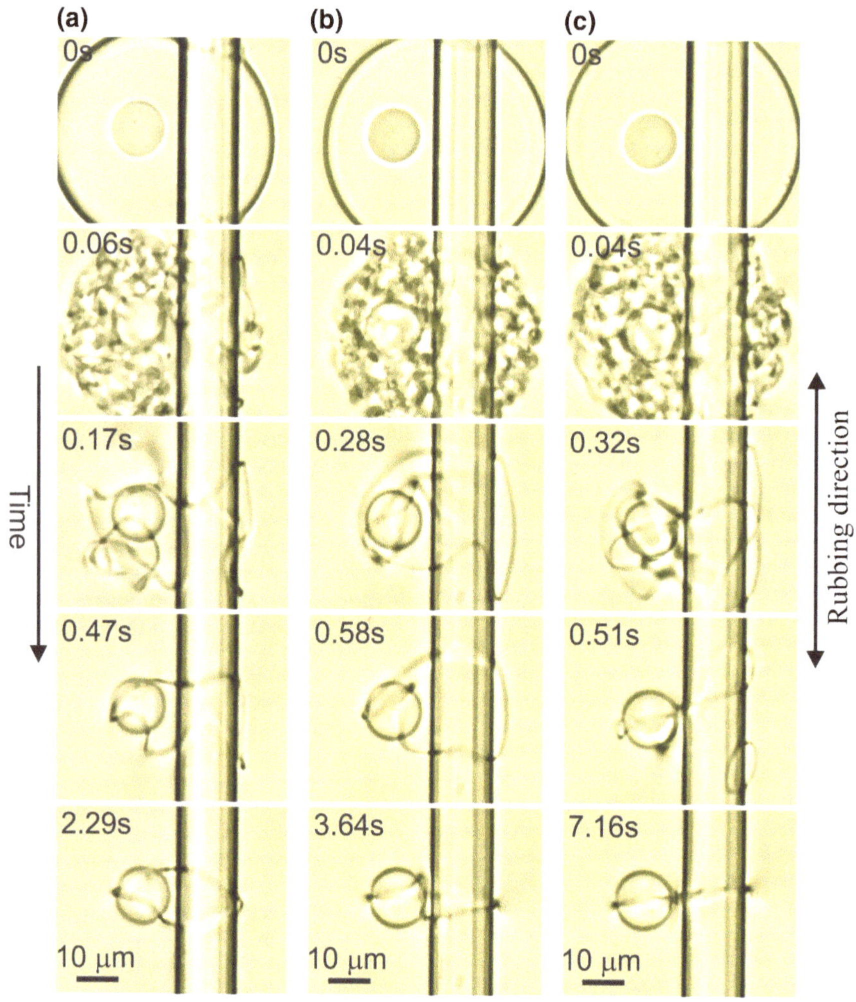

Fig. 5.21 The entanglement of the $-1/2$ Saturn ring of the fibre with the Saturn ring of a microsphere. Micro-fibre with the diameter of 12 μm and microsphere with the diameter of 10 μm are bound together by thermal quench using a focused laser light. **a** Figure-of-eight is created from one twisted loop. **b** Figure-of-omega is made from a single loop encircling the colloids with a twist segment between them, which is curved like an additional small loop. **c** An entangled hyperbolic defect is formed from two loops by a direct quench, or by transformation from an unstable figure-of-omega

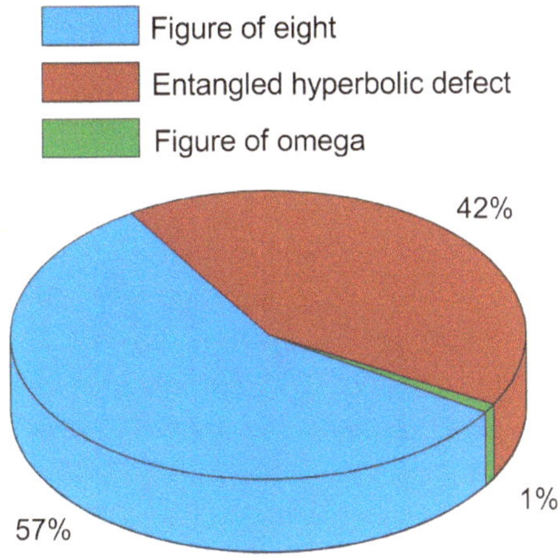

Fig. 5.22 Probability distribution for the formation of different fibre-sphere entanglement with −1/2 winding number for each starting ring. The analysis is done for one hundred experiments

to the entanglement of two microspheres which have been predicted by Araki and Tanaka [196] and then observed by the Ljubljana group [198].

Figure 5.21 clearly shows the evolution of the entanglement out of the disordered region after the quench. In Fig. 5.21a a single disclination loop is created out of a dense tangle of topological defects after the quench. The loop is encircling the microfibre and microsphere in the form of a twisted loop, known as a figure-of-eight. Another kind of binding with a single loop is shown in Fig. 5.21b, known as a figure-of-omega. This entanglement is unstable and usually transforms into a more stable configuration called the entangled hyperbolic defect, shown in Fig. 5.21c. In all these three cases, the winding number of both merging loops is the same, −1/2.

In the case of the figure-of-eight, the disclination line first starts from the front of the microsphere and then goes below the fibre and comes from behind over the fibre. Then, it goes again down below the microsphere and connects to the first end. From top view, this twisted loop looks like a number 8. For the figure-of-omega, the loop is more complicated, as it makes an additional small loop between the fibre and the microsphere (see last panel of Fig. 5.21b). This middle part of the loop is similar to the Greek letter omega when observed from the side.

The entangled hyperbolic defect is formed from the separated defect rings which are oriented perpendicular to each other. The bigger ring is encircling the fibre and the sphere and binds them together. The smaller ring is placed between the fibre and the sphere and is a −1 hyperbolic defect with an escaped core. By repeating the experiments, we have found that after the quench, the figure-of-eight is created most frequently. The second most frequent structure is the entangled hyperbolic defect, whereas the figure-of-omega is very rarely created. The probabilities for the formation of different types of entanglement are shown in Fig. 5.22.

The entanglement of the −1/2 ring on the fibre with the −1/2 ring on the colloidal particle is in fact topologically equivalent to the well known entanglement of two microspheres in the nematic liquid crystal. However, the question is whether one

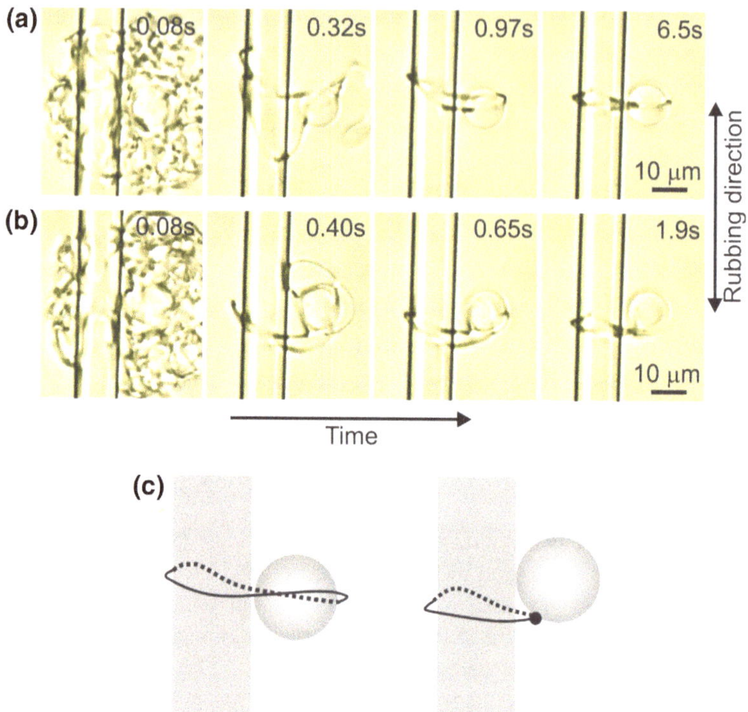

Fig. 5.23 The entanglement of the $+1/2$ Saturn anti-ring on the microfibre with the $-1/2$ Saturn ring on the microsphere. The nematic liquid crystal around the microfibre (diameter $12\,\mu$m) with a Saturn anti-ring and microsphere (diameter $10\,\mu$m), carrying the Saturn ring, is quenched. Two kinds of entanglement are created. **a** A twisted loop encircling both the fibre and the sphere. **b** A sphere looses its Saturn ring, which is attracted towards the fibre and transforms into a -1 hyperbolic point defect. This defect is connected to the Saturn anti-ring on the fibre. **c** Schematic representation of these two entanglements. The *left panel* shows the binding by a loop and the *right panel* shows the binding by a point defect

could entangle the $-1/2$ ring on the colloidal particle with the $+1/2$ ring on the microfibre?

In order to study the entanglement between the two rings of opposite winding numbers, the microsphere with $-1/2$ Saturn ring is brought closer to the $+1/2$ ring on the fibre and this region is then thermally quenched by the laser beam. Two different kinds of bindings are created, as shown in Fig. 5.23. In Fig. 5.23a the time sequence of images shows a creation of a single loop out of a dense tangle of defects. This loop is simply encircling the microfibre and the microsphere with a small twist in the middle sections. The first panel in Fig. 5.23c shows its schematic view. The disclination line is slightly twisted. It is interesting to note that this single ring, entangling the original $+1/2$ and $-1/2$ rings, has to be a charge-neutral ring. This means it has to be composed of a section having a $+1/2$ winding, which smoothly transforms into a section with a $-1/2$ winding. This change in the winding number in

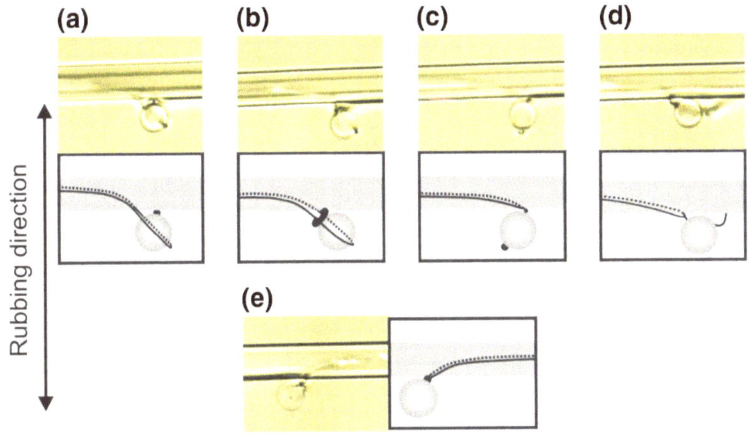

Fig. 5.24 Entanglement and binding of a microsphere with halves of the charged loop. The panels show the optical images in unpolarised light and the accompanying panels show the schematics. **a** The − end of the loop of the fibre embraces the microsphere. The hedgehog point defect of the microsphere is still there and is attached to the fibre. **b** The − end of the loop is encircling the microsphere and an additional smaller loop, originating from the point hedgehog, is encircling this part of the fibre's loop. **c** In a reversed orientation of the microsphere's hedgehog, the microsphere is now bound to the − end of the loop from its + end. **d** This binding between the − end of the loop and the hedgehog point defect is similar to the bubble-gum configuration. **e** The microsphere is bound to the + end of the loop with its − point defect

two parts of the loop somehow takes care of the conservation of the overall topological charge of two original loops, which is equal to 0.

The second type of binding of the $+1/2$ and $-1/2$ rings is shown in Fig. 5.23b. Interestingly, the $-1/2$ ring of the microsphere shrinks into a -1 point defect, which is now sitting in-between the fibre and the sphere, connecting the microsphere to the Saturn ring with a positive topological charge.

When the fibre is rotated from parallel to the orientation perpendicular to the overall nematic director orientation, the entanglement and binding of microspheres with defects on the fibre is even more complex. Figure 5.24 shows some selected examples, how a microsphere is connected to or entangled to the far segments of the loop, created on a fibre by cutting the original gigantic Saturn ring.

In the first set of experiments, shown in Fig. 5.24a–d, a microsphere is exposed towards the $-1/2$ end of the loop. By quenching the liquid crystal around this $-1/2$ end of the loop and the microsphere, four different kinds of binding or entanglement are created, as shown in Fig. 5.24a–d. The upper panels show the true microscope images of these bindings under unpolarised light, and the bottom panels show their schematic representation.

In Fig. 5.24a, the end of the loop encircles the dipolar microsphere, and the hedgehog point defect of the microsphere is located on the top end of the sphere and is attracted towards the fibre. Figure 5.24b presents another kind of binding, which is similar to the hyperbolic defect entanglement observed with two entangled microspheres. In this case, there is a smaller ring between the fibre and the sphere. The microsphere can also be attached to the end of this $-1/2$ loop segment from its $+$ end, as shown in Fig. 5.24c, or from the $-$ end as shown in Fig. 5.24d. In this case, the binding resembles the bubble-gum structure observed for two microspheres. Finally, we can observe the interaction of the microsphere with the $+1/2$ winding section of the loop. In this case, the sphere can be only bound in one way, as we can see in Fig. 5.24e. The microsphere is attached to the $+1/2$ end of the loop with the hedgehog point defect, carrying negative topological charge.

Chapter 6
Colloidal Particles of Complex Topology in Nematics

Abstract In this chapter, we discuss topological properties of the nematic director field around colloidal particles of complex shape with homeotropic and planar surface anchoring. We present experiments with colloidal handlebodies, knot- and link-shaped microparticles, Koch-star colloids, spiraling and ribbed rods in the nematics. Theoretical predictions for the topology of a nematic liquid crystal around non-orientable surfaces, such as a Möbius strip, are discussed as well.

6.1 Topology of Colloidal Particles Is Important

In previous chapters we discussed topological properties and interaction forces between colloidal particles of simple topology, such as microspheres and microfibers. These simple objects were studied from the beginning of research on nematic colloids, because there was no technology available for the production of colloidal particles with complex topology, such as tori, mutually connected tori forming handlebodies, mutually linked tori, colloidal spirals, Koch-star particles or knots made of polymers. With the advancement of technology, novel techniques of fabrication of complex colloidal particles were introduced into the field of liquid crystals, such as lithographical fabrication of silica handlebodies [43, 225] and spirals [226] or two-photon polymerisation of a polymer in 3D [227], which made it possible to fabricate microobjects of arbitrary topology, including knotted particles [228].

The topology of objects that are immersed in a nematic liquid crystal is certainly important for the topology of the director field of the nematic liquid crystal, which is forced to align on the closed surface of this object. We therefore have a situation where the topology of the objects is in a way imprinted into the topology of the surrounding nematic field. This interplay allows for testing of basic laws of topology in real experiments.

One of the first attempts to imprint the topology of nanoobjects into the surrounding nematic liquid crystal was described in an article entitled "Chiral Nematic Order in Liquid Crystals Imposed by an Engineered Inorganic Nanostructure" by Robbie, Broer and Brett [229]. They fabricated an array of densely packed helical columns of MgF_2 deposited on a glass substrate via glancing angle vacuum deposition. During

I. Muševič, *Liquid Crystal Colloids*, Soft and Biological Matter,
DOI 10.1007/978-3-319-54916-3_6

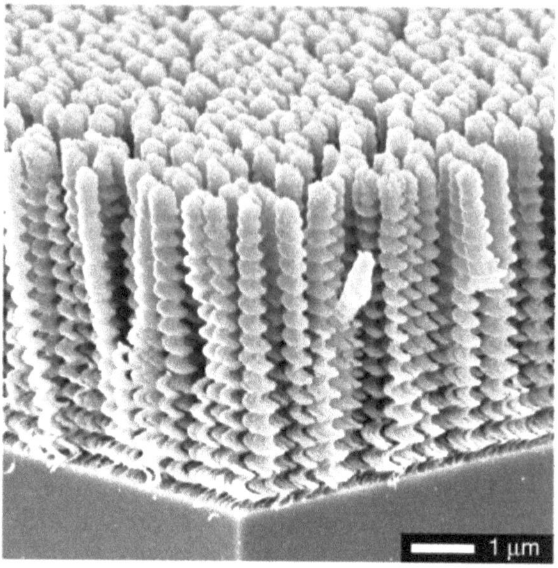

Fig. 6.1 SEM micrographs of MgF_2 structures, which are formed during the deposition and synchronised glass sample rotation. Helical, screw-like posts with as many as 15 helical turns are spontaneously formed, filling densely the glass surface. The helical period is 350 nm and the refractive index of MgF_2 is $n = 1.38$. Reprinted by permission from Macmillan Publishers Ltd: K. Robbie, D. Broer, M. Brett, Nature 399(6738), 764 (1999), copyright (1999)

the deposition, the substrate was slowly and steadily rotated, which resulted in a growth of well-isolated, well-separated helical columns with 15 turns and a helical pitch of 350 nm. SEM micrograph of a densely packed array of helical columns is shown in Fig. 6.1.

One can intuitively expect that the helical shape of each helical column of MgF_2 will induce a local helical structure of an achiral nematic liquid crystal which is infiltrated into this porous helical "forest". Because of the imprinted helicity, the nematic liquid crystal is expected to exhibit reflection of circularly polarised light passing through this film. For left-handed helices, as shown in Fig. 6.1, one expects the light with left-handed circular polarisation to be reflected, whereas right-handed circular polarisation will be transmitted through this film. This was indeed observed in the experiment.

These helical columns of inorganic material were therefore one of the first studies of the interaction of geometrically non-trivial colloidal particles (helices) immersed in the nematic liquid crystal. The induced chirality due to the surface anchoring of a liquid crystal is expected to be transmitted only over short distances. Nevertheless, this local chirality, induced by the helical shape of the colloidal particle, is important from two aspects: (i) Helical shape of the rod is expected to make the topology of the surrounding nematic much richer compared to a straight rod. Local turns could provide means of stabilisation of localised topological defects. (ii) Because of the more complex topological defects surrounding the helical rod, the pair interaction of these rods should be more complex and selective in chirality.

Whereas controlled evaporation of inorganic matter on a rotating sample results in a growth of macroscopic areas of densely packed chiral columns, this technique does not allow for the fabrication of micro-colloids of arbitrary shape and topology.

Recently, two quite different techniques have been used to fabricate complex micro-colloidal particles of microscopic dimensions: (i) photolithography of planar colloids, such as platelets, tori, and handlebodies, and (ii) 3D two-photon polymerisation of a light sensitive polymer. These two new techniques allow the exploration of complex nematic colloids, which are also named "topological colloids" by Senyuk et al. [43]. This chapter is devoted to the overview of studies of unusual colloidal particles in the nematics, such as handlebodies, polymer knots and links, Möbius rings, Koch stars, spirals and spiral rods.

6.2 Colloidal Handlebodies in Nematics

The motivation for studying topological properties of solid handlebodies made of silica and immersed in a nematic liquid crystal was to explore basic theorems of topology in real experiments. The central quantity that determines the topological properties of handlebodies is the genus of the handlebody. This can be illustrated by comparing a coffee mug and a doughnut [43]. Although they look quite different (and also serve different purposes), they are topologically equal from the point of view of topology. A coffee mug and a doughnut are objects with genus $g = 1$, and as such they are fundamentally different from, say, a ball or a solid cylinder, which are attributed the genus $g = 0$. A coffee mug can be smoothly reshaped, without cutting, into a doughnut or a torus, but there is no way of transforming it into a sphere. Obviously, the difference is in the number of "holes" penetrating the object, which also defines the genus of that object.

To study the interplay of particle topology and accompanying defects in liquid crystals, Senyuk et al. [43] have fabricated planar handlebodies made of silica with a number of handlebody holes (equal to the genus of the particle g) varying from 1 to 5. The surfaces of these particles were characterised by an Euler characteristic $\chi = -2g$, which ranges from $\chi = 0$ (tori) to $\chi = -8$. These silica particles were rather thin, with a square cross-section of $1\,\mu m \times 1\,\mu m$ and ring diameters ranging from 5 to $10\,\mu m$. The surfaces of these silica handlebodies were treated to induce perpendicular surface anchoring of the nematic liquid crystal. They were dispersed in the nematic liquid crystal and then introduced into cells with either perpendicular or planar surface anchoring.

Figure 6.2 shows several examples of silica handlebodies in homeotropic nematic cells of 5CB with the rings of the handlebodies aligned perpendicularly to the director and therefore parallel to the cell surfaces, as illustrated in Fig. 6.2i–l. Figure 6.2a shows $g = 5$ silica handlebody in homeotropic cell and between crossed polarisers. Far away from the handlebody, the field of view is dark because we are observing the sample along the optical axis and the polarisers are crossed. One can clearly see the transmitted light at the outer perimeter of the handlebody due to the perpendicular surface anchoring of the nematic liquid crystal at the edge of the handlebody, which is in fact a closed $-1/2$ ring. This outer ring is similar to the Saturn ring around a spherical microparticle with perpendicular surface anchoring.

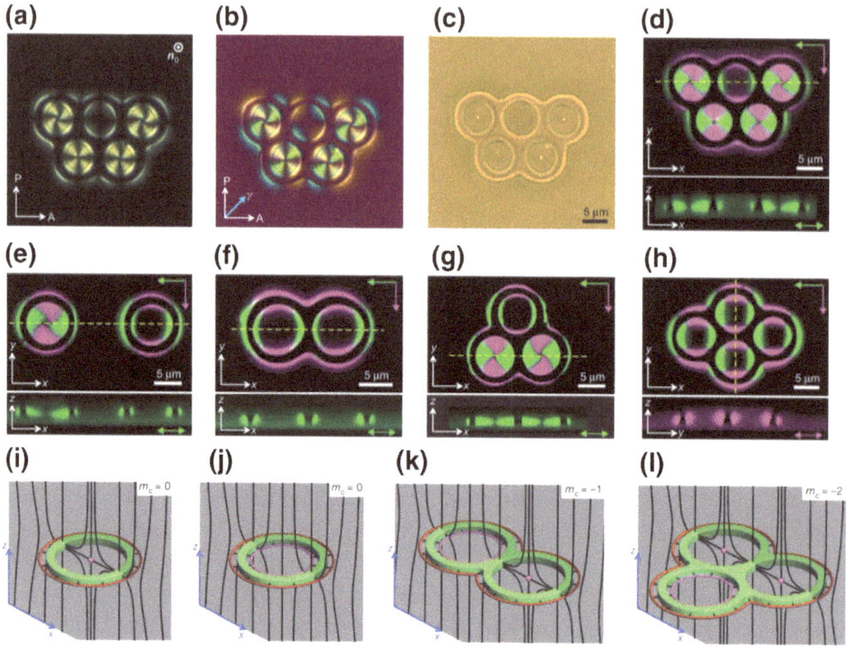

Fig. 6.2 Colloids in a form of handlebodies with perpendicular surface anchoring in a homeotropic nematic cell align perpendicularly to the far-field director. (**a–d**) Polarised images of handlebodies with $g = 5$ in a homeotropic nematic cell, imaged with different polarisation techniques. **a** Crossed-polarisers image. **b** A λ retardation plate was added between crossed polarisers at 45°. **c** Bright-field image. **d** 3 PEF-PM image. **e, h** The images of single (**e**), double (**f**), triple (**g**) and quadruple (**h**) colloidal handlebodies in a homeotropic nematic cell as observed by 3 PEF-PM. The images (**d, h**) are combinations of 3 PEF-PM fluorescence intensity images for two orthogonal polarisations of the excitation light. They are shown in *green* and *magenta* colours, corresponding to *green* and *magenta* arrows indicating the polarization direction. The cross section shown on the *bottom* of panels (**d, h**) are taken along dashed *yellow* lines. **i, l** The schematic drawing of the director field **n** (*black* lines) around the genus g colloidal handlebodies. The outer disclination loops are shown in *red* and they carry the $m = -1$ hedgehog charge. The inner disclination loops are shown as *magenta* lines and carry $m = +1$ hegehog charge. *Magenta* spheres are the hyperbolic point defects with the hedgehog charge $m = +1$. Reprinted by permission from Macmillan Publishers Ltd: B. Senyuk, Q. Liu, S. He, R.D. Kamien, R.B. Kusner, T.C. Lubensky, I.I. Smalyukh, Nature **493**(7431), 200 (2012), copyright (2012)

One can also see from Fig. 6.2a that there are two possible states in the interior of each ring forming the handlebody. One of the rings is dark in the centre, which indicates that the nematic liquid crystal simply penetrates the ring and there is only a remnant defect ring, following closely the inner edge of the handlebody ring. The other four rings have a different structure and appear bright, with a dark cross following the orientation of the polariser and the analyser. This indicates that the liquid-crystal molecules are partially parallel to the polarisation of the rings, which is only possible if there is a point defect in the centre of each of the bright rings.

The schematic structures of a bright and a dark ring are illustrated in the drawing in Fig. 6.2i and j. In both cases, there are two topological defects for each ring of the handlebody. There is always an outer ring with a $-1/2$ winding number cross-section and a hyperbolic point defect in the centre (Fig. 6.2i) forming a bright interior of the ring, or a $-1/2$ winding number defect loop attached closely to the inner edge of the dark ring. In this latter case, the nematic crystal simply penetrates the majority of the ring interior, resulting in a dark ring between crossed polarisers.

The internal structure of the nematic around handlebodies with different genus was studied with fluorescent confocal polarized microscopy. The examples of the polarised fluorescence captured from the nematic surrounding the handlebodies are shown in Fig. 6.2d–h. The cross sections through the plane of the handlebodies are coloured green and magenta, indicating horizontal (x) or vertical (y) direction of local orientation of the nematic. If we look at Fig. 6.2e, we can clearly resolve the point defect located in the fluorescent centre of the left-sided ring, whereas the ring on the right side shows the fluorescent signal only form the inner and outer edges of the ring. By comparing this panel with panels in Fig. 6.2i and j, one can realise two possible states of a single toroidal ring in the nematic. There are therefore two defects with the handlebody with the genus $g = 1$, and by the increasing the genus of the object, the number of topological defects accompanying that particle also increases. There are three defects for the handlebody with two tori and genus $g = 2$, shown in Fig. 6.2k: a hyperbolic point defect in the ring on the right side, a hyperbolic ring in the interior of the left ring and a single $-1/2$ closed loop encircling the handlebody on the outer side. Accordingly, a handlebody with three tori and genus $g = 3$, shown in Fig. 6.2l, has four topological defects: two points and two rings. In general, a handlebody with genus g shows typically $g + 1$ individual singularities.

Figure 6.3 shows crossed polarisers, red-plate and non-polarised images of tori and simple handlebody with homeotropic surface anchoring in a planar nematic cell, where the plane of the handlebody is parallel to the far-field director. For the case of a simple tori with one hole and genus $g = 1$, one observes two point singularities; a hyperbolic defect with -1 topological charge, situated outside the tori (a red sphere in Fig. 6.3j) and $+1$ hyperbolic point defect located inside the tori. The outer hyperbolic point defect can sometimes open into a small ring, as illustrated in Fig. 6.3k and shown on micrographs in Fig. 6.1b, e, h. If another ring is attached and we have a handlebody made of two connected rings having genus $g = 2$, an additional hyperbolic point defect appears in the interior of the added ring, as shown schematically in Fig. 6.3l.

The experiments show that the sum of the hedgehog charges attributed to induced point defects and disclination loops around the handlebody compensates for the hedgehog charge m_C, which is attributed to the colloidal particle, in this case a handlebody. The charge of the handlebody is uniquely determined by particle topology.

The relation between the topological charge of induced defects and the topological charge of the particle can be understood using Gauss–Bonnet theorem. The topological charge, which is attributed to the particle, is determined by the topology of its surface and is equal to the degree of the director field along the surface. This degree is calculated by integrating the Jacobian of the director field over that surface. Because the director aligns perpendicularly to each point of the surface, this integer

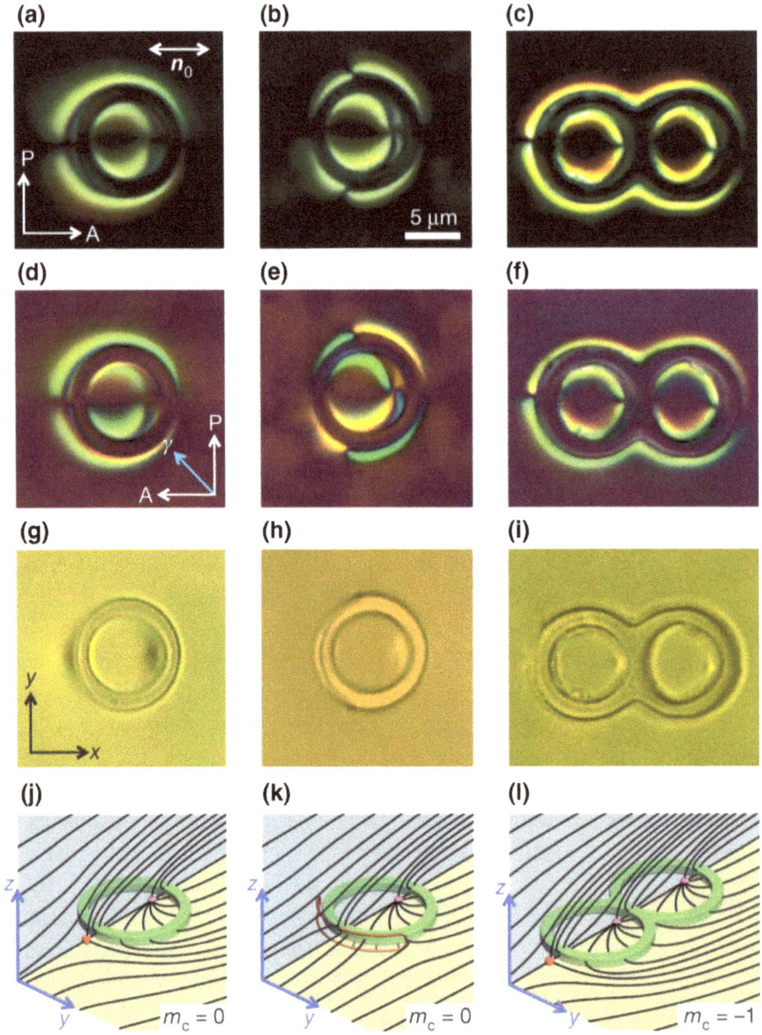

Fig. 6.3 Colloidal handlebodies in planar nematic cells are aligned along the far field director $\mathbf{n}_0$. **a**, **c** Crossed-polarisers images. **d**, **f** Crossed polarisers images with added *red* (λ) plate. **g**, **i** Images are taken in bright field. In polarised images (**a**, **c**), topological defects are clearly visible as dark discontinuities or intersections of dark brushes. Using the *red* plate (**d**, **f**), one can determine local orientation of the director and identify the structure of the director around the topological defects. In bright field images (**g**, **i**) topological defects are visible as bright or dark regions. The brightness of such region is due to scattering of light on topological defect and depends on the position of the defect with respect to the focal plane. **j**, **l** Schematic drawing of the nematic director (*black* lines) around colloidal handlebodies with perpendicular surface anchoring of the nematic. *Magenta* spheres indicate the $m = +1$ topological charge hyperbolic defects. The *red* spheres indicate the $m = -1$ hyperbolic point defects. The *red* loop in panel (**k**) shows a small disclination ring with hedgehog charge $m = -1$, which is obtained by opening the *red* dot in panel (**j**) into a ring. Reprinted by permission from Macmillan Publishers Ltd: B. Senyuk, Q. Liu, S. He, R.D. Kamien, R.B. Kusner, T.C. Lubensky, I.I. Smalyukh, Nature **493**(7431), 200 (2012), copyright (2012)

of the Jacobian reduces the total Gauss curvature of the surface divided by 4π. Here, one can use the Gauss–Bonnet theorem, which states that the total Gauss curvature of a closed surface is quantised in units of 4π and is equal to $4\,\pi(1-\mathrm{g})$. This is a conserved quantity, which remains unchanged against all continuous deformations of the surface. It therefore follows that the topological charge of the surface of the handlebody is $+/-(1-g)$. For a microsphere in a nematic, the hedgehog charge of the sphere is therefore $\pm\,1$, and has to be compensated by an external, topology induced hedgehog charge of opposite sign. It follows that the sum of all defect charges must exactly cancel out the degree on the colloidal surface S. This total hedgehog charge of point defects and disclination loops has to add up to $+/-(1-g)$, regardless of the orientation of the particle in the nematic liquid crystal.

Interesting considerations of the topological charge of hedgehog point defects induced around and inside a simple ring or two connected rings parallel to the far-field nematic director field are shown in Fig. 6.4a–d. If we first concentrate on a simple tori, presented in Fig. 6.4a and c, we see an unusual situation, where the opposite topological charge has to be attributed to the same structure of the nematic director field, which is schematically shown in details in Fig. 6.4e and f. Recall that the simple ring (tori), shown in Fig. 6.4a, has a genus $g=1$ and that the topological charge of the total Gauss curvature of such a ring is $(1-g)$, therefore equal to 0. This means that the total topological charge of the two hedgehog defects, which are observed in this case and shown in Fig. 6.3j, also has to be zero. Because of that, the two hedgehog point defects, shown by red and magenta spheres in panels (a) and (c), must have opposite topological charges. Their 3D projection, which is shown in Fig. 6.4e and f is from the point of view of director field (which is a headless vector) exactly equal. However, by attributing the lines of force to the director field, we clearly see that there is a distinction in the topological charge of the red and the magenta labeled defects in Fig. 6.4a and c.

Looking at the Fig. 6.4b and d, which show a handlebody made of two connected rings, we see three hedgehog defects. The topological charge of the handlebody is $(1-g)$, which equals to -1 in this case. The -1 charge of this handlebody has to be compensated with a total $+1$ charge of all defects which appear in this situation. There are therefore three hedgehog defects, one carrying the topological charge -1 (red sphere in Fig. 6.4d) and two identical hedgehogs, each with charge $+1$, residing in the interior of each hole. This makes the total topological charge of the ring and the defects equal to zero.

We should mention here that the topology sets the requirement for the minimum number of topological defects, which is $g+1$ in all cases. Of these, $g-1$ defects are of the same charge and are dictated by the particle topology and the two additional defects with opposite signs (carrying total zero charge) appear to relax the elastic distortion of the director and minimise the total free energy. This relaxation depends in detail on the elastic constants of the nematic liquid crystal. An interesting situation of the merging of point singularities into non-singular (escape) rings is shown for a simple tori perpendicular to the far-field nematic director. The topology requires the formation of two point defects of opposite charge (red and magenta spheres in Fig. 6.4g). However, these two point defects can transform into a non-singular

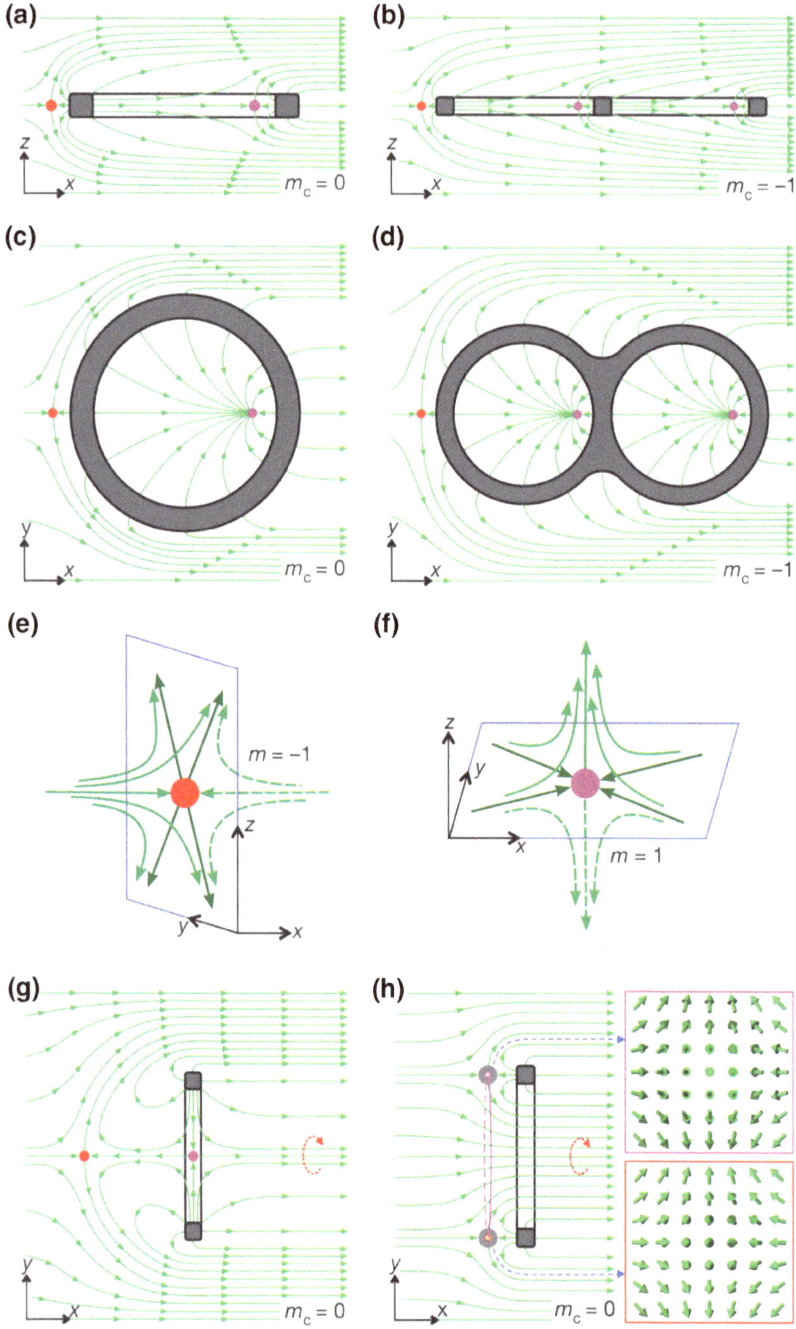

Fig. 6.4 (Continued)

◂ **Fig. 6.4** Topology of director field around genus g colloidal handlebodies with perpendicular surface anchoring of nematic liquid crystal in a planar nematic cell. Diagrams **a**, **f** show the vectorial presentation of the director field, where the arrows indicating direction have been added consistently. *Red* spheres are hyperbolic hedgehogs with negative unit charge. *Magenta* spheres are hyperbolic hedgehogs with positive unit charge, as shown in (**e**, **f**). **g**, **h** Director field in a vectorial representation showing an energetically unstable configuration with point defects of opposite hedgehog charges shown in (**g**). The two insets to (**h**) show details of the vectorial field of the escaped and axially symmetric configuration which in nonsingular and therefore has lower elastic energy. The nonsingularity is obtained by an "escape into the third dimension". Reprinted by permission from Macmillan Publishers Ltd: B. Senyuk, Q. Liu, S. He, R.D. Kamien, R.B. Kusner, T.C. Lubensky, I.I. Smalyukh, Nature **493**(7431), 200 (2012), copyright (2012)

entity, which was already described in Sect. 2.7: Vortices in nematic colloids. Briefly speaking, one first opens each point singularity into an open loop with corresponding winding number. We then have two loops of opposite winding numbers, which can merge together and become non-singular by escaping into the third dimension. The corresponding cross-sections of this merged non-singular loop are shown by the magenta and red boxes in Fig. 6.4h.

These optical studies of the number and structure of topological defects that are accompanying handle bodies, which induce perpendicular surface anchoring of the nematic liquid crystal, demonstrate the full strength of topological conservation laws and also clearly illustrate the ambiguity in defining the sign of the topological charge. Whereas for vectorial fields (such as the electric field) this can be uniquely prescribed, it is clearly shown that this is not possible for non-vectorial fields such as the director field of nematic liquid crystals. Because of the tensorial nature of this field, we see from the experiments on simple rings in the nematic liquid crystal that opposite topological charge has to be attributed to the same liquid crystal defect.

6.3 Knot- and Link-Shaped Microparticles in Nematics

Knots and links were first discovered in 2D arrays of microspheres with perpendicular surface anchoring of liquid crystals by Tkalec et al. [212]. Knots and links were formed from $-1/2$ disclination lines, which were originally present as Saturn rings encircling isolated microspheres. When brought together, these rings fused together either spontaneously or under an action of the laser tweezers into complex defect loops encircling several particles and forming either knots or links of practically arbitrary complexity.

Along this idea of having knotted and linked nematic director around colloidal inclusions, Martinez et al. [228] produced knot shaped and link shaped microparticles of solid polymer, which were then immersed in the nematic liquid crystal. The surfaces of these particles were previously treated chemically to induce either parallel or perpendicular orientation of liquid-crystal molecules. Because by inserting the particle into the nematic field a certain volume of a liquid crystal is excluded

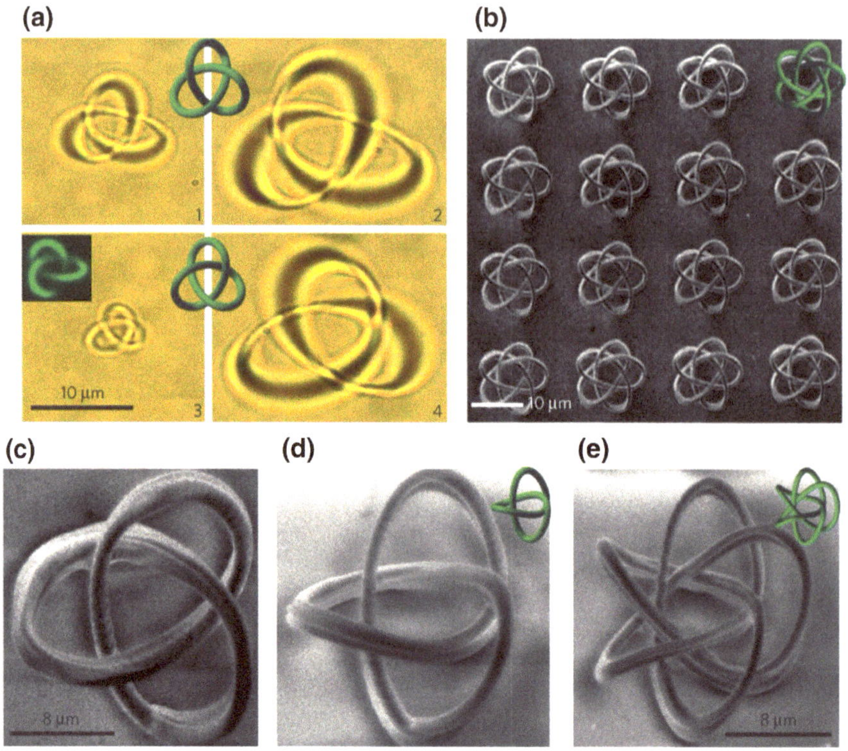

Fig. 6.5 Knotted particles made of polymer. **a** Optical micrographs of a photopolymerised trefoil colloidal torus knots $T(3, 2)$, which are either left-handed (panel 1 and 2) or right-handed (panels 3 and 4). Their models are shown in *green*. The inset to panel 3 shows 3 PEF-PM image of the particle. **b** SEM micrograph of an array of torus knots made of polymer. **c**, **e** Enlarged SEM images of the torus knots made of polymer. Reprinted by permission from Macmillan Publishers Ltd: A. Martinez, M. Ravnik, B. Lucero, R. Visvanathan, S. Žumer, I.I. Smalyukh, Nat. Mater. **13**(3), 258 (2014), copyright (2014)

and replaced by the polymeric object of complex topology. This particle induces, via surface anchoring of liquid crystal molecules, non-trivial configurations of the surrounding nematic director field. The topology of the surrounding field was studied for polymeric particles in a form of torus knots and torus links. In topology, a torus knot is a special kind of knot that lies on the surface of a torus, which is itself unknotted in R^3. Similarly, a torus link is a link which lies on the surface of a torus in the same way. Each torus knot and link is specified by a pair of integers p and q and is denoted by $T(p, q)$. These particles are formed of polymeric tubes that are looped p times through the hole of the imaginary torus and make q revolutions about the torus rotational symmetry axis. Several examples of torus knots are shown in Fig. 6.5, where most "popular" knots are the trefoil knot, shown in Fig. 6.5c, d and the pentafoil torus knot shown in Fig. 6.5e. These are SEM images of polymer torus knots, prepared by 3D two-photon polymerisation technique, explained earlier [227].

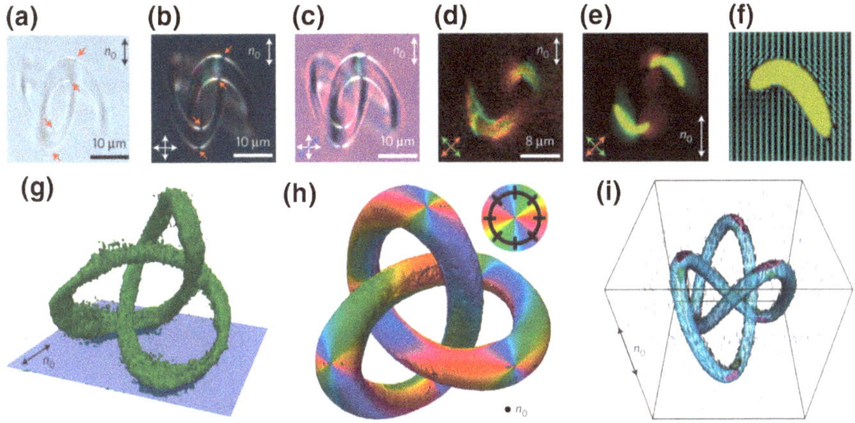

Fig. 6.6 A colloidal particle in a form of a trefoil knot with tangential surface anchoring of nematic liquid crystal. **a** Bright-field optical micrograph of the particle without any polariser. **b** The same particle in the nematic liquid crystal viewed between crossed-polarisers. **c** Crossed-polarised image with 530 nm retardation plate added. The slow axis of the plate is along the blue arrow. The *red* dots in (**a**, **b**) indicate the positions of surface boojums. **d**, **e** 3 PEF-PM cross-sections obtained for different positions of the imaging plane. These images are actually made by superposing two different images (coloured in *green* and *red*) taken at two orthogonal excitation polarisations indicated by crossed *green* and *red* arrows. **f** Numerically calculated director of the *top right* part of the cross section in panels (**d**, **e**). **g** Fluorescence intensity from the knotted colloidal particles represented in 3D. The *blue* plane indicates the imaging plane of panels (**d**, **e**). **h** 3D representation of the director field due to incorporated knotted particle. Surface boojums are located at the points where different colours meet. **i** Numerically calculated director field at the interface between the knotted particle and the liquid crystal. The areas coloured in *green* and *magenta* indicate reduced order parameter ($S = 0.42$) corresponding to the surface boojums -1 and $+2$ 2D defects. Reprinted by permission from Macmillan Publishers Ltd: A. Martinez, M. Ravnik, B. Lucero, R. Visvanathan, S. Žumer, I.I. Smalyukh, Nat. Mater. **13**(3), 258 (2014), copyright (2014)

After surface treatment, these particles are dispersed in a liquid crystal, which is in some cases fluorescently labeled and the resulting nematic liquid crystal deformation is observed by polarising or fluorescent confocal microscopy. In this case, the three-photon excitation fluorescence polarising microscopy (3PEF-PM) was used with the control of polarisation of the excitation beam which allowed for the full reconstruction of the 3D nematic orientation field.

When a trefoil polymer particle with parallel surface anchoring is inserted into the nematic liquid crystal, this knotted colloid induces strong elastic deformation of the nematic liquid crystal. It was found that in a stable configuration with a minimum elastic distortion energy, the trefoil colloids align with their torus plane perpendicular to the far-field nematic director. This is shown in optical micrographs in Fig. 6.6a–c, which show unpolarised (a), cross polarised (b) and red-plate (c) images of a trefoil polymeric knot in a homeotropic nematic liquid crystal cell. By inspecting unpolarised images in Fig. 6.6a, one can observe small dark spots, which alludes to the presence of point defects, located on the surface of the polymer particle inducing parallel surface anchoring. These boojums appear on that part of the surface, which is directly facing the view of observation and therefore the far-field nematic director. These surface defects are presumably surface boojums, which appear dark in bright-field micrographs because they scatter light.

The nature of these surface defects was revealed by probing the local azimuthal orientation of the director field around the particles by using 3 PEF-PM imaging. This method is highly sensitive to any rotation of the light-radiating dipole moment of the molecule and allows for reconstruction of the local orientation of nematic liquid crystal in 3D. The orientation of the nematic director in a given plane of imaging is colour-encoded, as indicated in Fig. 6.6d, e and more clearly presented in Fig. 6.6h. This colour coding immediately indicates the positions of surface boojums, which is very clearly presented in the numerically-generated image in Fig. 6.6h. Each surface boojum appears as a point, around which a rainbow-like pattern is observed. This allows for easy counting of the number of surface boojums and it appears that each trefoil polymeric knot has 12 boojums on its surface. The positions of these boojums are located to the regions of the particle, where the particle's surface is perpendicular to the far-field director. The colour encoding also allows for easy recognition of the winding number of each boojum defect, which is either -1 or $+1$. The sign of the winding number can be recognised by monitoring the sequence of colours as we move along a closed loop around the boojum. It turns out that half of these boojums have a positive winding number $+1$ and are accompanied by equal number of -1 winding number counterparts. These counterpart boojums reside on the directly opposite side of the knotted tube.

Whereas parallel surface anchoring results in the formation of localised point defects (surface boojums), it is expected that perpendicular surface anchoring of liquid crystal molecules on the surface of knotted polymer particles will display more complex topology of the nematic director field. This was indeed observed in the experiments with torus particle knots with perpendicular surface anchoring, dispersed in the nematic liquid crystal. These particle knots align with their torus plane perpendicular to the far-field director, which is their ground state. However, several metastable orientations were also observed and they could be reached by local melting of liquid crystal followed by laser tweezers manipulation. By using fluorescent confocal microscopy, "slices" of the director field could be obtained for different polarisations of the fluorescence-excitation beam. Some examples of trefoil polymeric knots with perpendicular surface anchoring in the nematic liquid crystal are shown in Fig. 6.7a, b. Because of 3D nature of these objects and high elastic distortion of the nematic liquid crystal, the optical images are not really clearly resolving the nature of the topological defects, associated with these trefoil knotted particles. This could be resolved by using 3PEF-PM, and two examples of the fluorescent images for two different directions of excitation polarisation are shown in Fig. 6.7d, e. The analysis of the images shows the presence of two defect lines, which are marked by red arrows in Fig. 6.7d, e. This means that the trefoil polymer colloid is accompanied by two singular defect lines with $-1/2$ winding number. It is clear that these lines must be closed in a loop and the question is what is the resulting topology of these singular loops attached to and following the

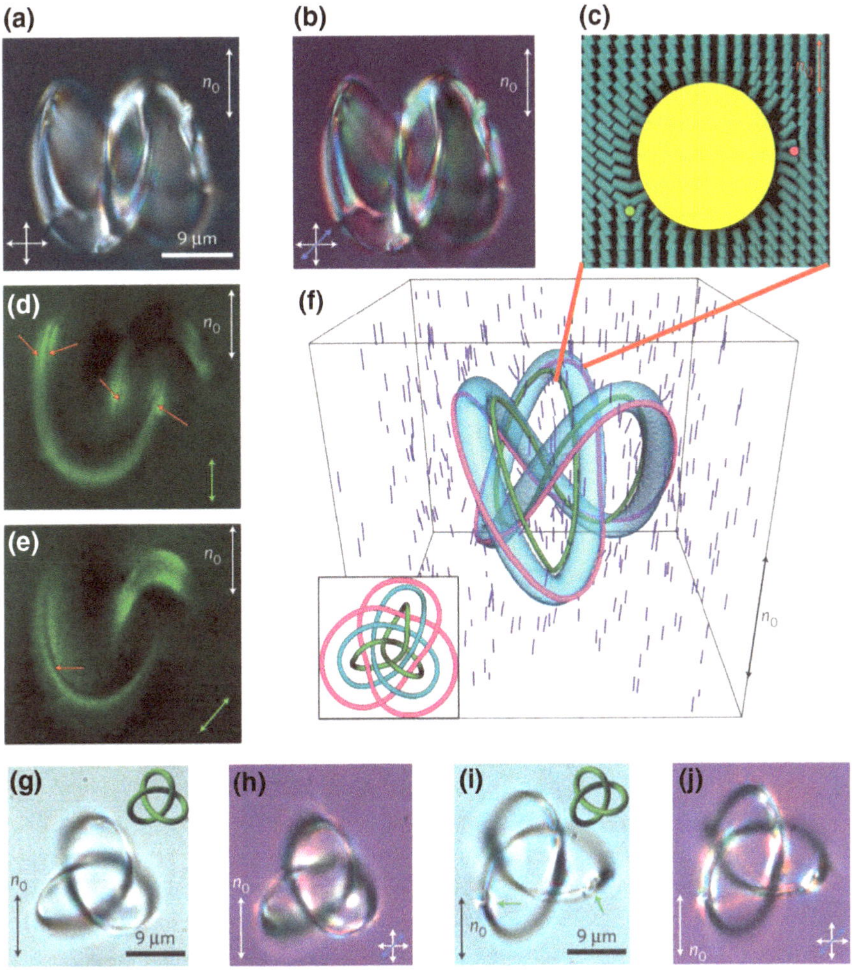

Fig. 6.7 Polymer trefoil knotted particle with perpendicular surface anchoring in a planar nematic cell. **a** Photomicrograph of the particle between crossed polarisers. **b** The same particle between crossed polarisers with a red plate added with its axis at 45°. **c** Numerically calculated director field in a cross-section perpendicular to the knotted tube, marked in panel (**f**). **d**, **e** 3 PEF-PM images of a knotted particle in the nematic. The excitation polarisation is different for each image and is indicated by the *green* double arrow. *Red* arrows indicate defect lines. **f** Numerically calculated director field around a polymer trefoil knotted particle. The torus plane is orthogonal to the far-field director $\mathbf{n}_0$. The *green* and *magenta* lines show regions with depressed order parameter and therefore visualise the defect lines. **g**, **i** Bright phase micrographs of trefoil colloidal knots with their torus plane parallel to the far-field director. **h**, **j** The same particle as in panels (**g**, **i**) as seen between crossed-polarisers and inserted full wave retardation plate. Reprinted by permission from Macmillan Publishers Ltd: A. Martinez, M. Ravnik, B. Lucero, R. Visvanathan, S. Žumer, I.I. Smalyukh, Nat. Mater. **13**(3), 258 (2014), copyright (2014)

surface of the trefoil polymer microparticle. A very convincing answer is given by numerical simulation of the nematic director field around a trefoil polymer knot with perpendicular surface anchoring, which are presented in Fig. 6.7f. One can see that two defect loops are closely following the surface of the trefoil-knotted microparticles and they themselves form two distinct trefoil knots. We therefore have three distinct objects, closely interlocked by the nematic director: (i) the trefoil polymeric knot, and (ii) two trefoil-knotted $-1/2$ singular loops. It is interesting to find from the experiments that these defect loops are actually both torus knots of the same type as the polymeric particle. Moreover, they are mutually linked into a two-component link. This link of line defects is additionally linked with the particle knot as shown in the inset of the numerically generated Fig. 6.7f.

We therefore have a complex topological situation with three distinct objects, each having its own topological charge, which are mutually linked knots. The question arises whether this is consistent with the requirement of a zero total topological charge in this system. A trefoil knot particle has Euler characteristic zero, which implies that the topological charge of the trefoil knot particle is also zero. This requires that the net topological charge of the two knotted $-1/2$ defect loops must equal to zero (under modulo 2, [40]).

6.4 Möbius Strips and Non-orientable Surfaces in Chiral Nematics

Knotted particles which we were discussing in previous Sect. 6.3 are examples of 3D bodies with orientable surfaces. In mathematics, orientability of the surface implies a consistent choice of normal vector which is attributed to the local part of the surface. If we choose a certain point on a trefoil knotted polymeric particle, shown in Fig. 6.7, and we move along the polymeric tube, we can return after one circulation to the same point of departure with the same orientation of the surface normal. However, there are also non-orientable surfaces, such as the one-sided Möbius strip, which is a classic non-orientable surface. By choosing a point of departure on a 180°-twisted Möbius strip and following the strip, we come to the same point of departure, but now the surface normal is flipped by 180°. This implies inconsistency in defining the surface normal and such a surface is said to be non-orientable.

In the context of topology, the classification theorem of surfaces states [230] that any compact surface can be classified by its genus, orientability, and number of boundary components, as discussed by Machon and Alexander [231]. We have already seen that the genus of the surface is equal to the number of holes penetrating the body or the number of handlebodies which are possessed by the surface. A sphere has genus $g = 0$ and a torus has genus $g = 1$. The orientability of the surface implies whether one can make a consistent choice of the normal vector to the surface. As we already mentioned, the one-sided Möbius strip is a well known non-orientable surface. The last descriptor of surfaces is the number of boundary components for a

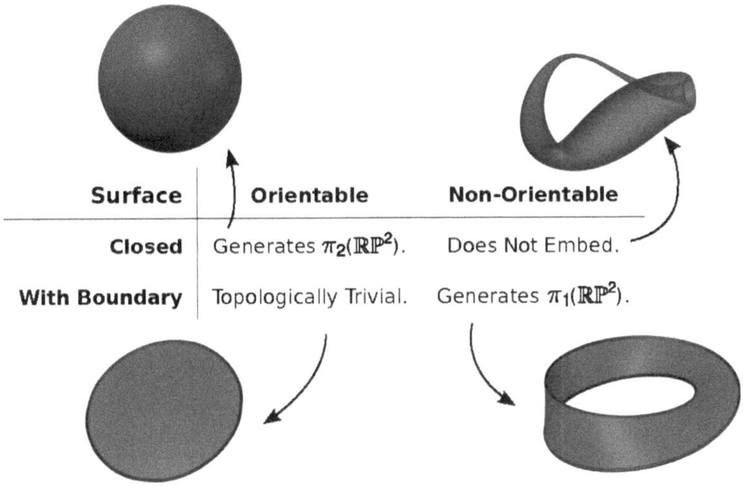

Fig. 6.8 Topological characteristics of different compact surfaces, which are embedded into a nematic liquid crystal in 3D. The surface prefers perpendicular anchoring of liquid crystal molecules. If the surface is closed and orientable (such as a sphere), it generates point defects or closed loops, which are elements of the second homotopy group. If the surface has boundaries and is non-orientable (such as the Möbius strip), it generates disclination lines in the surrounding nematic liquid crystal. Reprinted with permission of Proceedings of the Academy of Sciences of the United States, T. Machon, G.P. Alexander, Proc. Natl. Acad. Sci. U. S. A. **110**(35), 14174 (2013)

given surface, which is the number of distinct connected components in the surface boundary. For example, a 2D disc has one boundary component (closed line) and a torus has no boundary components.

All these elements that classify the surfaces should have important consequences for the type and number of topological defects generated when such a surface is embedded into a real 3D space. In our case, we are interested in topological defects, which should arise when different types of surfaces are embedded into a nematic liquid crystal and the tensorial field describing the orientation of a liquid crystal is forced to align locally on the surface. Clearly there will be a topological incompatibility between the surface and the ordering field, and the topological implications of each type of surface will generate different types of accompanying defects in the surrounding liquid crystal. The classification of defects is then separated into 4 different classes of surfaces [231]: orientable or non-orientable surface and closed surface or surface with boundaries, as illustrated in Fig. 6.8.

Closed and orientable surfaces such as spheres and tori are known to induce defects corresponding to the element $1 - g$ of the second homotopic group, where g is the genus of the surface. For example, spherical colloids can nucleate either a point or a disclination loop (Saturn ring) but this ring can always be shrunk continuously into a point, so it is properly classified by the second homotopic group. In general, orientable surfaces are never able to generate elements of the fundamental homotopic group, disclinations. Their surface orientability ensures that any disclination loop can always be removed in pairs or shrunk into points.

A very special example is a closed and non-orientable surface, such as the Klein bottle [231]. It turns out that such an object cannot be embedded in real 3D space without self-intersecting. This means that any representation of such a closed and non-orientable surfaces in a nematic liquid crystal is not possible and was not considered any further [231]. Orientable surfaces with boundaries, such as an infinitesimally thin (2D) disk, have trivial topological consequences for the surrounding liquid crystal. Because they are orientable, it is topologically necessary that no lines are generated in this case. Furthermore, they cannot generate any elements of the second homotopic group (points and loops).

Non-orientable surfaces with boundary necessarily generate a non-trivial element of the fundamental homotopic group, which is a disclination line. There are different kinds of disclination loops which can be generated in this way and there are four distinct homotopic classes of disclination loops (G.P. Alexander et al. [232]). In all cases, the disclination must have a zero hedgehog charge.

Whereas non-orientability of the surface requires the existence of a disclination loop, the precise form of these defects and their equilibrium configuration are dictated by the free energy of the distorted liquid crystal. The embedding of non-orientable surfaces in a nematic liquid crystal were first discussed by Machon and Alexander in 2013 [231]. When considering the realisation of a 2D non-orientable surface, they selected Möbius strips with different degrees of twisting, which were embedded in a nematic liquid crystal and enforced perpendicular orientation of liquid crystal molecules on the surface of the strips. However, practical realisation of such an object requires finite thickness of the strip, because it has to be fabricated from some material. The problem which arises is how to ensure a 2D behaviour in terms of the 3D nature of such an object. This can be resolved in theory by considering Möbius strips of finite thickness and taking different boundary conditions for the surface of the strip and its edge. In this case, the surface of the strip enforces perpendicular surface anchoring, whereas along the edge the surface anchoring of the liquid crystal is planar degenerate. This is illustrated in Fig. 6.9.

A Möbius strip with homeotropic boundary condition therefore necessarily generates a closed disclination loop of zero hedgehog charge, which entangles the surface. This can be explored by numerical simulations using the continuum Landau-de Gennes modeling, and for a non-chiral nematic liquid crystal the result is a single disclination loop entangling the Möbius strip. Because it has zero topological charge, the cross-section of this disclination loop shows a twisted $-1/2$ profile on the outside of the strip and changes into a $+1/2$ twisted profile on the inside.

Whereas the "canonical form" of the Möbius strip has one half-twist ($p = 1$), this can be generalised to a larger number of half twists the surface makes upon reconnecting. If the parameter p, measuring the number of half-twists, is even, the Möbius strip will be orientable surface with boundary and will have the topology of an annulus. However, if the number of half-twists is odd, then the surface is non-orientable and has the same topology as the canonical Möbius twist. The embeddings are in all cases distinct. For a single half-twist, the boundary of a Möbius strip is a

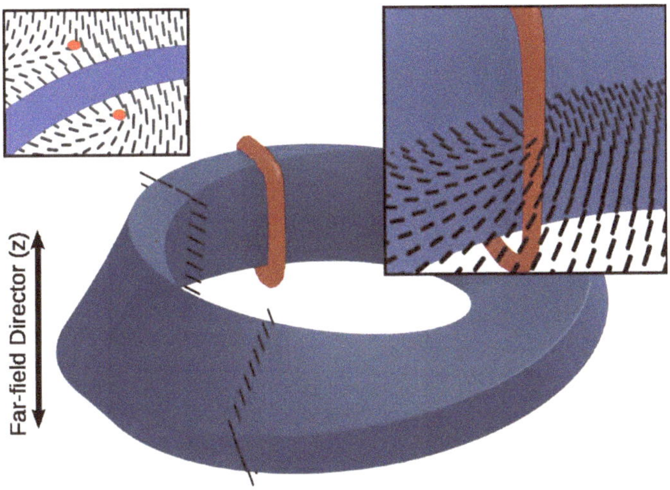

Fig. 6.9 Landau-de Gennes simulation of topological defects on Möbius strip, which induces perpendicular surface anchoring of the nematic liquid crystal on flat faces. Planar anchoring is considered on the short edge, so this object represents a 2D surface embedded into a liquid crystal. The defect line is encircling the strip, and it carries zero topological charge. This means that the defect line cross-section should change from a $+1/2$ winding number profile on the inside of the strip into a $-1/2$ profile on the outside. Reprinted with permission of Proceedings of the Academy of Sciences of the United States, T. Machon, G.P. Alexander, Proc. Natl. Acad. Sci. U. S. A. **110**(35), 14174 (2013)

simple knot. For p half-twists, this boundary is a torus knot $(p, 2)$ if p is an odd number and link if p is even. This can be realised by simply drawing lines along the edges of the Möbius strip with a different number of half-twists. Disclination lines that follow the surface of the multiply-twisted Möbius strip will have the same shape and properties as the boundary of the colloid. This means that these disclination line will form knots and links.

The results of numerical analysis of disclination lines following the surface of multiply-twisted Möbius strips showed that knotted and linked disclinations can indeed be realised in chiral nematic liquid crystal. These knots and links obey the fundamental topological requirements, but their stabilisation is not just a question of topology. The free energy of the elastically distorted liquid crystal is important as well, and it turns out that the chirality of the nematic liquid crystal has an important role in the energetic stabilisation of these configurations. This is very similar to the stability of knotted and linked nematic colloids, which are stable and observable in experiments only in chiral nematic liquid crystals. In the achiral nematic liquid crystals, knotted defects are unstable and topological defects transform into isolated loops and points. The results are presented in Fig. 6.10, where knotted and linked disclinations are clearly visible. For example, the structure in Fig. 6.10a is a Hopf link, where two closed loops are linked together. The next structure in Fig. 6.10b is a trefoil knot. The structure in Fig. 6.10c is a Solomon's knot and Fig. 6.10d shows

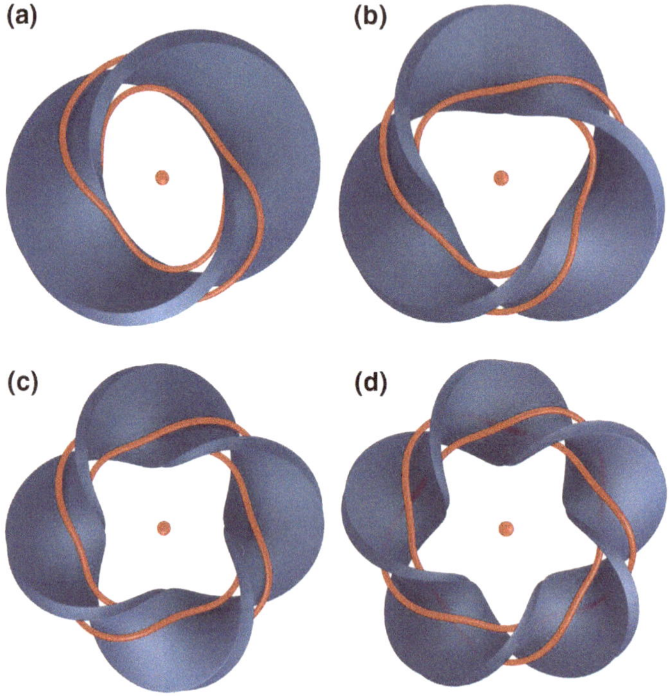

Fig. 6.10 When Möbius strips with p half-twists are embedded into a chiral nematic liquid crystal, the disclinations are knotted and linked. They belong to torus knots and links of the type $(p, 2)$. **a** Hopf link corresponds to $p = 2$, **b** Trefoil knot corresponds to $p = 3$, **c** Solomon's knot corresponds to $p = 4$ and **d** cinquefoil knot corresponds to $p = 5$. The *red* spheres in the centre indicate hedgehogs, which are positioned symmetrically above and below the strip. Reprinted with permission of Proceedings of the Academy of Sciences of the United States, T. Machon, G.P. Alexander, Proc. Natl. Acad. Sci. U. S. A. **110**(35), 14174 (2013)

a cinquefoil knot. In addition to these defect loops, isolated defects are observed in the centre of the structure, which turn out to be pairs of hedgehogs, positioned above and below the strip.

The disclination loops forming torus knots and links following the structure of the Möbius strips are not the only singularities generated by embedding a Möbius strip into a cell with fixed perpendicular boundary condition. In order to match the surface alignment on the cell walls, two hyperbolic hedgehogs are generated, which are positioned above and below the strip. The cross-section through the colloid has a profile reminiscent of a double-twist cylinder, which is also named "torus" texture, and is observable in different representations. All have in common a double-twist torus, which is a building block of blue phases.

6.5 Koch Stars Colloids in Nematics

We have discussed in Sect. 5.3 the production and stabilisation of a multitude of topological charges on a fibre in a nematic liquid crystal. Although a fibre itself is topologically equivalent to a sphere, both having genus $g = 0$, in contrast to a microsphere, a long fibre can carry a larger number of pairs of topological charges. Depending on the orientation of the fibre in the nematic liquid crystal, these topological charges appear in a form of one or several pairs of rings (i.e. Saturn ring and Saturn anti-ring). In another orientation, defects will appear in a form of one or several pairs of point defects with opposite topological charge. Altogether, these topological charges accompanying the fibre add-up to a single topological charge of a radial hedgehog. On a fibre, the charges can slide towards each other, and if let free, they will mutually annihilate, leaving only a single hedgehog as required by the conservation of the total topological charge. There are several ways of stabilising the positions of topological charges on an object in a liquid crystal, and they all rely on a creation of an artificial friction, which will prevent the sliding of charges along the object and their pairwise annihilation.

If we consider a single torus with homeotropic surface anchoring of a nematic liquid crystal, the conservation of topological charge requires the appearance of two topological singularities with compensating topological charge, which were observed as a defect ring encircling the torus and a hedgehog positioned inside the opening of the torus. However, this torus can be smoothly transformed in any other topologically equivalent object with the same genus but different geometrical shape. One of the interesting transformations is shaping fractal-like colloidal particle out of a smooth torus.

Fractals are ubiquitous objects based on a very useful concept of self-similarity on different scales. The main feature that characterises a fractal is an exact or a quasi self-similarity to a part of itself. In the case of a perfect fractal, this geometric shape is self-repeating over all length scales. As a result of this scale-invariance, no characteristic length can be assigned to an ideal fractal object. Fractal properties are observed in nature in different contexts such as polymer networks, growth phenomena, porous media, classes, brain networks and structure details of genomes. Among the elementary fractal shapes is the Koch fractal, which is introduced by a simple iterative method, often visualised with the shape of the Koch snowflake, as shown in Fig. 6.11.

By generating the Koch fractal, the number of geometrical irregularities grows very fast. If we consider an ideal Koch fractal particle immersed in a nematic liquid crystal, we see that each geometric irregularity might host a topological defect. A fractal particle with strong surface anchoring of liquid crystal molecules will therefore host a stable constellation of a large number of topological defects. The question then arises, how does this develop with increasing number of fractal iterations. In theory, this can be explored by constructing ideal Koch fractal particles with well-defined surface anchoring and the corresponding director field is obtained by the free energy minimisation taking in the account tensorial nature of the nematic order

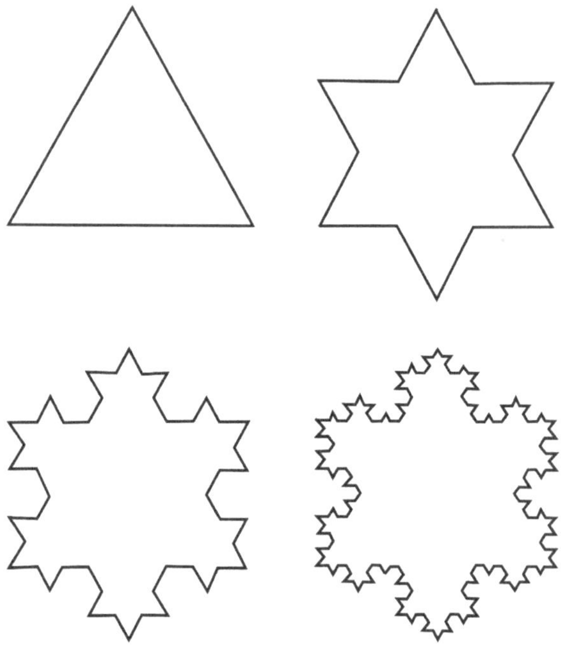

Fig. 6.11 Construction of a Koch fractal of different iterations. Note the rapidly increasing number of shape irregularities after each step of iteration

parameter. On the other hand, Koch fractal particles can be fabricated with two-photon polymerisation and their defect structure can be studied using the optical techniques.

To explore the topological properties of a nematic field induced by fractal geometry, the iteration of Koch fractals was used, which links together chosen parts of iteratively down-scaled equilateral triangles. The zero Koch iteration corresponds to an equilateral triangle and therefore has a threefold rotational symmetry axis, whereas the higher Koch iterations have a sixfold rotational symmetry axis.

Real Koch star particles were produced [233] by using 3D two-photon direct laser writing technique, which produces Koch star colloidal particles with very thin walls, as shown in Fig. 6.12. Scanning electron microscopy images of the four iterations of the particle, shown in Fig. 6.12, demonstrate perfect shape and surface smoothness of the polymer particles. This technique itself is able to reproduce four iterating families of Koch star particles, limited by the optical resolution of the system, which is around 120 nm cross-section of the voxel in the imaging plane. After the polymer Koch star particles were surface treated to induce homeotropic anchoring of nematic liquid crystal, they were introduced into 30 μm glass cells with strong planar surface alignment. Because of this different surface alignment of liquid crystal on the particle and the surfaces of the cell, the polymer particles are levitated inside the cell by the force of elastic distortion of the liquid crystal.

Optical images of Koch star colloidal particles in the nematic liquid crystal of low optical birefringence are shown in Fig. 6.13a–f. In all panels, the far-field nematic

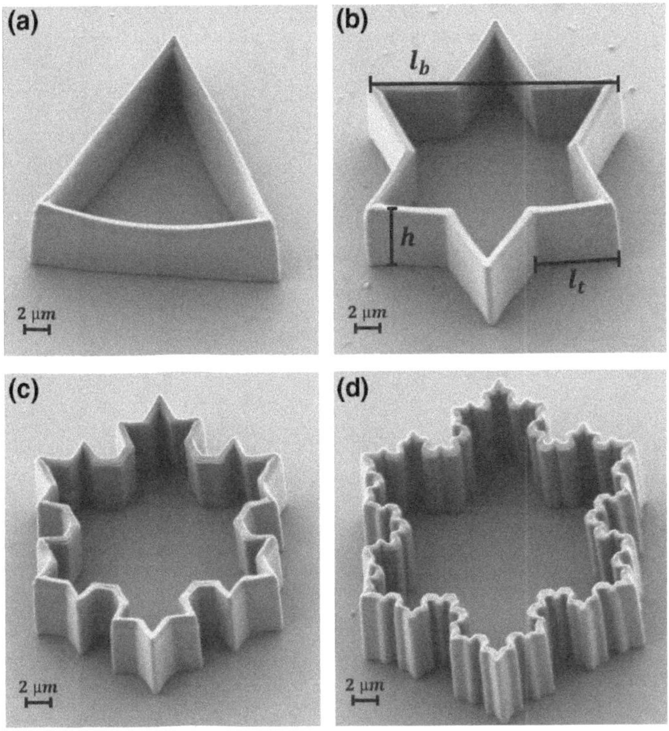

Fig. 6.12 Koch star fractal particles imaged by scanning electron microscopy. The particles are produced by direct laser writing into a photosensitive polymer using 3D 2-photon laser polymerization system Photonic Professional by Nanoscribe GmbH

director is vertical, as illustrated in Fig. 6.13a (II). Because the polymeric particles have very thin walls and are produced by moving the laser beams to the photo resist, there might be hidden optical artifacts, either because of direct laser writing or optical diffraction. For this purpose, the liquid crystal and the particles are first heated into the isotropic phase to observe any optical irregularities. The images of particles in the isotropic phase are shown in the first column (panels a I, f I). Although the liquid crystal and the polymer are nearly index matched, there are still visible optical artifacts due to optical diffraction and optical inhomogeneities caused by direct laser writing. However, these are minor optical distractions when compared to the strongly distorted and birefringent nematic liquid crystal pattern around and inside these particles when the system is cooled into the nematic phase, as is shown in panels (a II, f II) in Fig. 6.13.

The zero-iteration Koch star particles are shown in Fig. 6.13a, b. In the nematic phase, there are two stable orientations of this triangle-like particle in a planar nematic cell. In one of the stable positions, shown in Fig. 6.13b, columns II and III, one of the sides is perpendicular to the rubbing direction. In this case, this is the lower side of the triangular particles shown in Fig. 6.13b (II), (III). The nematic director

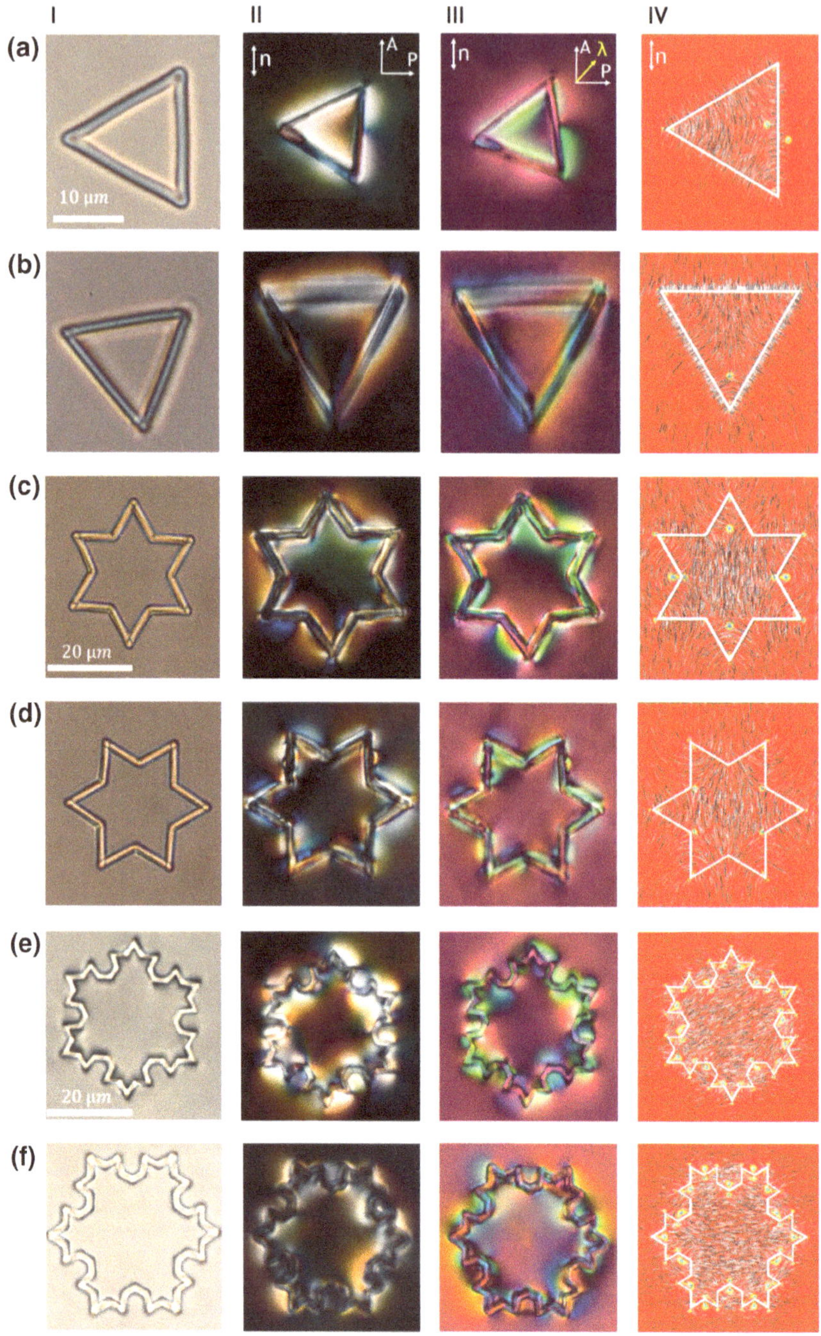

Fig. 6.13 (Continued)

◀ **Fig. 6.13** Nematic topological states stabilized by fractal Koch-star colloidal particles. **aI–eI**, Optical unpolarized images of Koch-star particles in the isotropic phase of the CCN mixture at 70 °C. **aI–eI** The same particles as in panels I, now observed between crossed polarizers. The rubbing direction, indicating the far-field planar orientation of the nematic is shown by arrows in panels (**aII–aIV**). Defects are recognized as point regions in the optical image, surrounded by rapidly varying color and intensity of the transmitted light, indicating strong director distortion. **aIII–eIII** The same particles as in panels **2aII–eII**, now viewed between crossed polarizers and *red* plate added at 45°. Different colors are due to different in-plane orientations of the nematic molecules. **aIV–eIV**, LdG numerical modeling illustrating contour plots of the scalar order parameter in the midplane of the particles with $l_b/\xi = 100$ and ξ is the correlation length of the liquid crystal molecules in the $x - y$ coordinate plane containing the coordinate center. The calculated director field in the $x - y$ plane of the contour plots is also superposed

is practically undistorted along this lower side and strongly distorted on the other two sides. There is also strong distortion of the director field inside the triangle, which is specially localised in the corners of the triangle. By rotating the polariser and ioniser, one can determine the location of topological defects in the corners of the triangle in Fig. 6.13b (II), (III). This orientation is also found to be stable using nematic calculations, which are shown in Fig. 6.13b (IV). These defects are actually pairs of defects with opposite topological winding numbers and charges. Each of the three pairs of defects in each corner of the triangle therefore compensates the winding, giving total winding zero, as expected for the total charge of the torus. Remember that any iteration of the Koch star particle is topologically equivalent to the torus. A torus has genus $g = 1$ and it is known that colloidal handlebodies with genus g are accompanied by defects with a total topological charge of $+/-(1g)$, which is 0 in our case. All Koch star particles should therefore have an even number of topological defects which mutually compensate their winding and charge, to keep the total charge of any Koch star particle 0 at all times. In addition to these defects, observable in the plane of the Koch star particles, an additional closed defect loop is observed, which is running all around the upper edge of the particle.

The first iteration Koch star particles (star of David) is shown in Fig. 6.13c. In the experiments, there are several stable orientations of these particles in the planar cell and some of them are more abundant than others. Elastic distortions are clearly visible between crossed polarisers and in red-plate images. Most of the distortion is concentrated in the inner and outer corners of the Koch particles, which are therefore hosting topological defects. By using the laser tweezers, it was not possible to detach any defect line from the particle, although it is predicted from theory to be running all along the edge of the particle. However, one is able to pull the two $-1/2$ defects, labeled by the two arrows in Fig. 6.13c (II). By counting the number of defects, one can see 8 pairs of defects in the corners, four inner corners do not show any defects, only splayed nematic profile. By using the laser tweezers, one is able to grab and move only one pair of defects labeled with arrows in Fig. 6.13c (II).

Unlike the zero and the first iteration, the second iteration Koch star particles show two stable orientations in the planar nematic cells, which are presented in Fig. 6.13c, d for different polarisations of the illuminating light. The configuration in Fig. 6.13c

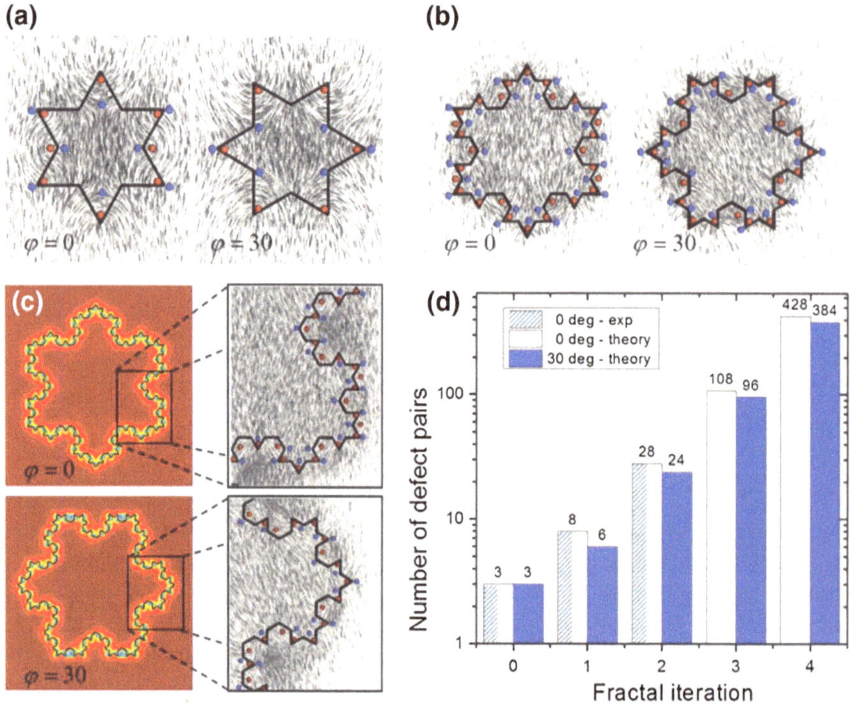

Fig. 6.14 Fractal geometry as generator of defect pairs. **a**–**c**, Pairs of +1/2 (in *red*) and −1/2 defects and surrounding director field (in *gray*) as generated by Koch particles of iterations $N = 1 - 3$ at angles $\varphi = 0$ and $\varphi = 30°$ of the particles relative to the far-field undistorted nematic director. The director and defects are plotted in the later mid-plane cross-section of the particle; these indicated 2D defect points are formally just two-dimensional cross-sections of actual 3D defect loops that entangles the whole particle. **d** Number of the defect pairs for different iterations as obtained from experiments and numerical modelling. Particles of size $l_b/\xi = 100$ are used in the numerical analysis

occurs with 70% probability and the symmetry axis of the Koch particle is parallel to the far-field nematic director. Defects for both second iteration Koch particles are different for each of the two stable orientations. They can be identified by observing the particles at different orientations of the analyzer. This shows the presence of 28 pairs of defects with mutually compensating winding and charge. This number of self-compensated defect pairs is also observed for the second stable orientation. The exact 3D morphology of the defects is strongly affected by the sharpness of the real particle edges, as they can pin or even locally suppress sections of defect loops. The generation of topological defects and corresponding topological states in the fractal Koch cavities is therefore a result of the interconnection between geometry of the object and its topology. Local fractal surface modulations induce local formation of defect pairs to minimise the elastic energy of distortion.

As already mentioned, Koch particles are topologically equivalent to tori, thus having zero total topological charge. This charge is also conserved when immersing these particles into a uniformly aligned nematic field. This requires that the net topological charge of all the surrounding defect structures must be equal to zero. Using numerical calculations, it was found that the number of defect pairs grows exponentially with iterations. Figure 6.14 shows comparison with theoretically predicted number of defect pairs and experimentally observed number for different iterations. Present state of the art of nano-manufacturing of Koch particles allows for production of three iterations of Koch star particles, and the number of observed defects perfectly matches the predictions of the LdG theory [233].

6.6 Spiraling Rods, Flat Spirals and Ribbed Rods in Nematics

The case of Koch star colloidal particles in the nematic liquid crystals clearly demonstrates how the geometry of the particles influences the number of topological defects accompanying the particle. Here we discuss similar examples of the generation and stabilisation of oppositely charged topological defects on genus $g = 0$ colloidal particles, which are shaped into 3D spiraling rods, straight but ribbed rods and flat, 2D-like spirals with perpendicular surface anchoring in the nematic liquid crystal. These particles are produced either by 3D direct laser writing (3D spirals and ribbed rods) or by photolithographic techniques. SEM images of 3D spiraling rods pro-

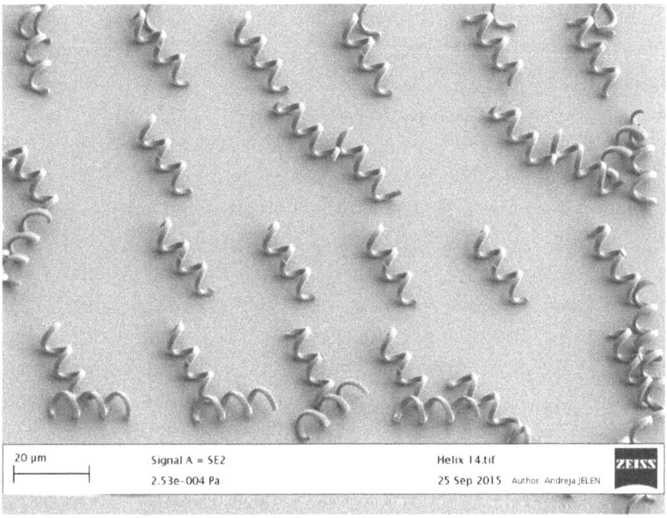

Fig. 6.15 SEM images of spiral colloidal particles produced by 3D direct laser writing technique. Image courtesy of M. Nikkhou [8]

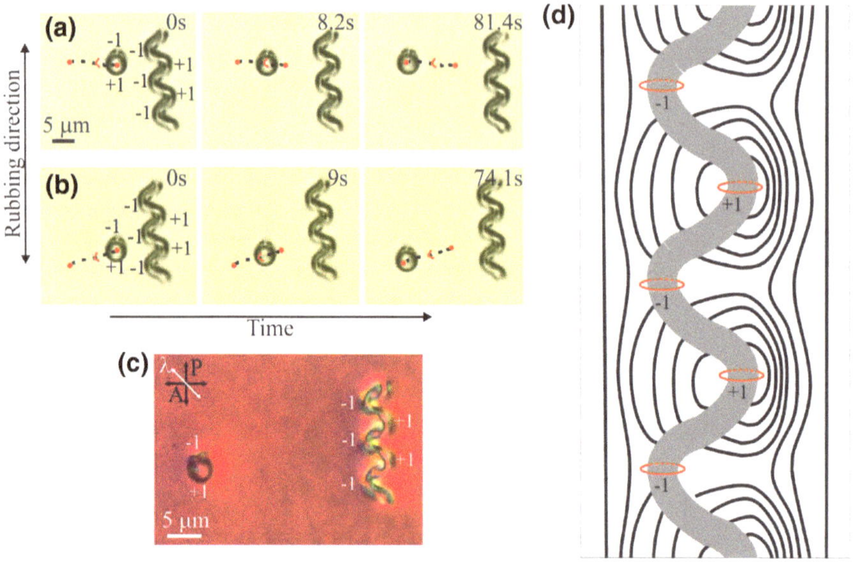

Fig. 6.16 Existence of several topological rings on a small microhelix. **a** Microhelix under crossed polarisers with inserted λ-wave-plate. By comparing the colours in the vicinity of the microhelix and a dipolar microsphere, one can recognise the sign of the defect rings around the helix in different regions. **b** Schematic representation of the director around the Saturn rings and Saturn anti-rings on the microhelix. **c**, **d** Dipolar microsphere with a point defect on *top* is repelled from the microhelix, which represents the opposite director field in the vicinity of the dipole and microhelix. The dashed lines show the repulsion trajectories. Image courtesy of M. Nikkhou [8]

duced by direct laser writing are shown in Fig. 6.15. The length of the spiraling rods is several tens of μm, typical diameter of the rod is 2 μm and there are several turns along the length of the spiral.

After the surfaces of spiral colloids are chemically treated to induce strong perpendicular surface anchoring of the nematic liquid crystal, they are dispersed in the nematic liquid crystal and this dispersion is introduced into a planar nematic liquid crystal cell. The particles are preferentially aligned with their spiraling axis along the far-field director, and there is a strong deformation of liquid crystal along the spiral. This deformation is clearly visible in the red-plate image shown in Fig. 6.16a. This red-plate image also shows a microsphere with perpendicular surface anchoring of the nematic liquid crystal in the vicinity. The presence of this microsphere helps us determine the director profile from the red and blue sections of the nematic liquid crystal around the spiraling colloids. Remember that the director field around the homeotropic microsphere with hyperbolic hedgehog defect is well-known (see Sect. 2.1).

Because of colour encoded information of the local orientations of the nematic liquid crystal in the red-plate image, one is able to reconstruct the nematic director around a spiraling colloid, which is shown in Fig. 6.16b. One can clearly recognise the presence of defect rings, which are encircling the spiraling microrod at the positions of its extreme. Because of the conservation of topological charge, this microrings must alternate in charge and winding number. In other words, we have the sequence of a Saturn ring (charge -1) followed by a Saturn anti-ring (charge $+1$) and so on. This situation is therefore identical to what we observed in the case of long fibers in a nematic liquid crystal, presented in Sect. 5.3. The difference with a long and straight fiber is that the spiraling rod stabilises the positions of the charges and anti-charges. It is easy to understand that moving, for example, a $+1$ ring towards a -1 ring requires an external force and energy. Spiraling rod therefore generates a kind of energy maxima and minima, where the topological charges and anti-charges are resting in their minimum free energy positions. Another way of considering this alternating charge stabilisation is that the spiraling rod increases the friction for sliding defect rings along the rod. This sliding, which is rather free in the case of a straight fiber, is here energy costly and accompanied with the dissipation of energy.

Besides recognising the sign of the topological charge from the red-plate images, these charges can also be tested by an external elastic dipole as shown in Fig. 6.16c, d. When the elastic dipole is oriented with its -1 hyperbolic hedgehog on top, the hedgehog will be repelled from the -1 charged region (Saturn ring) on the spiraling rod. If the elastic dipole is transferred to the other side of the spiraling rod, its -1 hegehog will be attracted to the $+1$ section of the rod, which is now exposed to the outside because it is residing in the neighbouring half-turn with respect to the -1 hedgehog. It should be noted here, that such a simple spiraling rod can actually host a variety of different combinations of topological charges, which can all be accessed by locally heating the liquid crystal with the laser tweezers. Different constellations of the topological charges can be obtained, such as a single ring spanning over two half-turns of the spiral and so on.

Similar stabilisation of topologically oppositely charged defect rings are obtained on micro-ribbed rods immersed in the nematic liquid crystal. These ribbed colloids are again produced by 3D direct laser writing, surface functionalized and introduced into the nematic liquid crystal. Because of their specific geometry, these micro-ribbed rods can host a sequence of defect rings with alternating topological charge. An example of such ribbed microrod is shown in Fig. 6.17a–c. The non-polarised image clearly shows light-intensity modulation along the rod, which indicates a presence

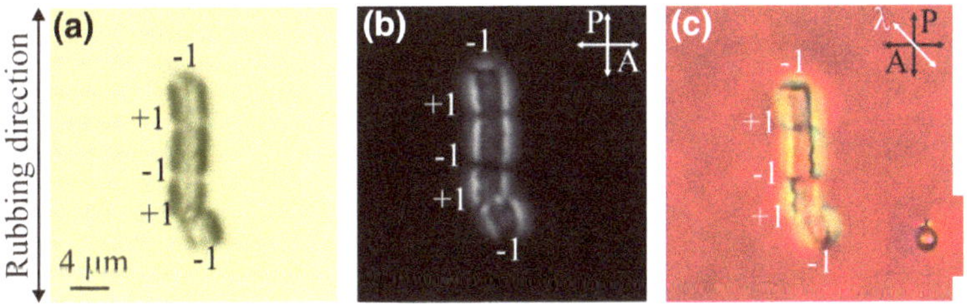

Fig. 6.17 Unpolarizednematic liquid crystal. **a** Unpolarizrd image. **b** Crossed polarizers. **c** *Red* plate image. Image courtesy of M. Nikkhou

of topological defect rings. These are further resolved in crossed polarised image in Fig. 6.17b. Finally, the red-plate image of the ribbed rod, shown in Fig. 6.17c, helps us reconstruct the director field.

Finally, similar stabilisation of a multitude of topological defects on a simple genus $g = 0$ colloidal particle in the nematic liquid crystal was demonstrated for flat colloidal spirals (Senyuk et al. [104]). These particles were produced using direct writing laser photolithography, which is a different technique from 3D, two-photon polymerisation. Because of lithography, these spirals are rather thin, made of 1 μm thick silica layer and shaped into one, two or three arms spiral using direct laser writing in the photoresist applied on top of silica layer. Similar to spiraling rods and ribbed cylinders, these 2D spirals host a multitude of different topological defects, which can be analysed by polarised and confocal fluorescent images and compared to theoretical predictions. Interestingly, surface point defects in a form of boojums with opposite topological charge were observed for tangential surface anchoring of the nematic liquid crystal. In contrast to simple microsphere, which is accompanied by two -1 boojums, spiraling geometry induces the formation of pairs of oppositely charged surface boojums. They appear because of the geometry and elastic deformation, but their topological charge is mutually compensated for each pair, so that in total, such a flat spiral with tangential surface anchoring is accompanied by topological defects with a total charge of -1. This charge is distributed between two surface point boojums.

Chapter 7
Nematic Microdroplets, Shells and Handlebodies

Abstract This chapter discusses the structure and topology of nematic microdroplets for different surface anchoring and chirality. The most interesting case are chiral nematic microdroplets with perpendicular surface anchoring, where knots and links were predicted. We show experimental results on these droplets, which show elastically assembled point defects, skyrmions, and torons in a liquid microsphere. Nematic shells and handlebodies are briefly discussed from the topological standpoint.

7.1 Structure and Topology of Nematic Microdroplets

Liquid crystal droplets, handlebodies and shells are made by dispersing liquid crystal in an immiscible fluid, such as water, glycerol or different sorts of oils. Because of the chemical incompatibility and immiscibility, the interface is formed between the liquid crystal and the surrounding fluid. Because any interface is associated with an energy cost of creating it, liquid crystal microdroplets tend to minimise the surface free energy, which results in perfect spherical shape for a typical nematic dispersion. However, the liquid crystal structure inside the droplet is also associated with elastic deformation and surface energy. While the energy of elastic deformations is usually much smaller than the surface energy, there are some examples when these two contributions are of comparable magnitude. This results either in a shape change and the spherical droplet transforms into a tactoid [132, 234, 235], or the surface anchoring breaks and the interior of the droplet releases the elastic distortion and becomes practically uniform.

In the equilibrium state of a liquid crystal droplet, one has to consider the minimum of the free energy functional, which can be separated into a volume and a surface contribution (O.D. Lavrentovich, [42]):

$$F = F_V + F_S = \int_V f dV + \int_S \sigma dS \tag{7.1}$$

I. Muševič, *Liquid Crystal Colloids*, Soft and Biological Matter,
DOI 10.1007/978-3-319-54916-3_7

The volume integral is taken throughout the volume of the droplet and f is the bulk free energy density of the liquid crystal. S is the area of the bounding surface, and σ is the surface free energy per unit area.

If the interior of the droplet is isotropic liquid, the surface free energy is constant and the equilibrium shape is found by minimising the surface energy at constant volume. The droplet is stable in a spherical shape of radius $R = (3V/4\pi)^{1/3}$. If such a droplet somehow starts growing, for example from a metastable isotropic phase that is surrounding the droplet, then the free energy of the bulk nematic is negative, $F_V = \frac{4}{\pi} fR^3 < 0$ and the surface term $4\pi\sigma R^2 > 0$ is positive. Negative bulk free energy means that it is more favorable to nucleate the nematic phase within the surrounding isotropic phase. In this case the volume and the surface energies will be equal at a critical radius of $R_c = 2\sigma/f$. If such an embryonic droplet has a radius R larger than the critical radius R_c, it will grow indefinitely, because the bulk energy is proportional to R^3 and is therefore always larger in magnitude (but negative in sign) compared to the surface energy, which is positive and increases as R^2.

In liquid crystal droplets, the energy balance between the bulk nematic and the surface energy is far more complex, because the surface energy is not a scalar quantity, but depends on the surface orientation of the molecules. The total surface energy, which is obtained by integrating the surface free energy density over the surface therefore depends on the local and global director orientation at the surface. Moreover, it appears that the surface and bulk energies are often comparable.

The surface energy of a nematic-liquid interface is considered in a form of two contributions. The first is the normal surface energy, which describes the energy needed to extend the area of the surface while preserving the equilibrium director orientation. This contribution to the surface energy is typically in the range of $W_a \approx 10^{-3} - 10^{-2}$ J/m^2 for the cyano biphenyl-glycerol interface. The anisotropic part of the surface free energy is several orders of magnitude smaller compared to the normal surface energy and is of the order $W_a \approx 10^{-6} - 10^{-5}$ J/m^2. It is also known that surfactants (detergents) are able to decrease the surface energy by an order of magnitude.

When considering various contributions of the surface and elastic terms, we consider that the isotropic part of the surface energy is of the order of σR^2, whereas the anisotropic surface energy is of the order of $W_a R^2$. The total energy of elastic deformation will be of the order of KR, where $K \approx 10^{-11}$ N is the bulk elastic constant. Because the elastic energy increases linearly with the radius R, the surface term will dominate and the realistic liquid crystal droplets are practically spherical. A typical length $l_\varrho = K/\sigma$ is of the order of a molecular length, which means that the shape could be different from spherical for very small liquid crystal nanodroplets. However, in some cases the isotropic surface tension could be quite small, like in the case ow water-based liquid crystals, such as in chromonic liquid crystals [234, 235]. In this case the elasticity is important and the shape of the droplets becomes non-spherical, and we have elongated "tactoids".

Nematic droplets were first theoretically studied by E. Dubois Violette and O. Parody in 1969 [236] and they predicted two possible configurations depending on whether the surface anchoring conditions are perpendicular or parallel to the nematic-

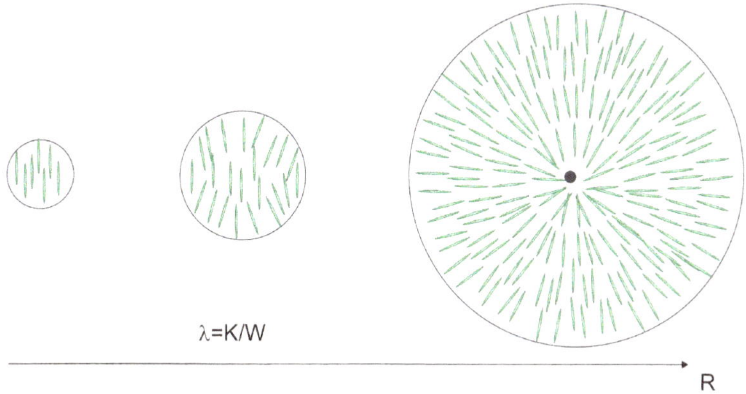

Fig. 7.1 Schematic structure of nematic droplets with perpendicular surface anchoring for different sizes. The characteristic length is the surface extrapolation length. This figure was inspired by the illustration of O.D. Lavrentovich in Ref. [42])

isotropic interface. Although liquid crystal droplets are nearly always of spherical shape, the internal structure of the director field inside the droplet is strongly influenced by their size. This is shown schematically in Fig. 7.1, where nematic droplets with perpendicular surface anchoring are shown for different sizes. The crossover length is the surface extrapolation length defined by the ratio of the elastic energy and the amplitude of the angular dependent anchoring energy, K/W, and is illustrated in Fig. 7.1. If the droplets are much larger than this surface extrapolation length, the interior of the droplet will adopt a radial configuration with a radial hedgehog defect in the centre. The elastic distortion is of the splay type.

If the radius of the droplet is decreasing, the anisotropic surface energy decreases as $\sim R^2$, whereas the energy of elastic deformation decreases as $\sim R$. This means that the surface energy costs decreases faster than the elastic contribution. As a result, it will be more favourable for smaller droplets to change the surface anchoring energy and relax the cost of the elastic deformation. This will result in a gradual approach of the director field inside the droplet to a practically uniform director for very small droplets. In these very small droplets it is simply energetically too costly to cause any elastic distortion inside such a small volume. The interior of a very small droplet will be practically uniform and the surface free energy will be slightly increased. The crossover diameter of a droplet is of the order of the surface extrapolation length $\lambda \approx K/W$ and is of the order of $1\,\mu$m. If the droplets are much larger than this crossover size, they are considered as being "large" and they will contain topological defects and will be elastically deformed. On the other side, there will be no defects and small elastic deformation for droplets which are "small" i.e. much smaller than the surface extrapolation length K/W.

This crossover in the internal structure is therefore a very useful criterion for the selection of the type of microdroplets. When considering very small droplets of the order of the wavelength of the visible light, they will strongly scatter light but their

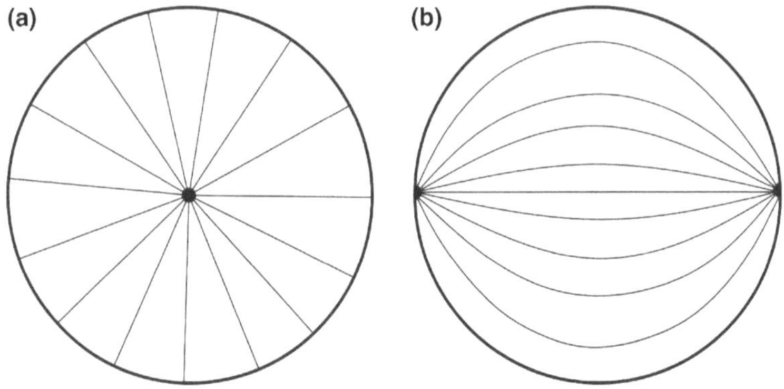

Fig. 7.2 **a** Schematic of a radial nematic droplet with homeotropic surface anchoring and a single radial hedgehog, located in the *centre* of the sphere. **b** Schematic of a bipolar nematic structure for a nematic droplet with planar nematic anchoring

interior will have little distortion and will therefore be topologically trivial. If one is interested in the topology of nematic or chiral nematic droplets, then very large droplets should be considered. They will be rich in topological phenomena and plenty of defects could be observed.

The number and type of topological defects in a spherically shaped liquid crystal droplets are determined by their net topological charge, which in term has to obey the Poincare and Gauss theorems (O.D. Lavrentovich [43]). The Poincare theorem states that the sum of all charges m of the vectorial field $\mathbf{T}$ on the closed surface is equal to the Euler characteristic of the surface, which is 2 in the case of a sphere. The Gauss theorem states that if the vector field is normal to the closed surface, then the sum of the topological charges N of all point defects inside the sphere is equal to one half of the Euler characteristic $E/2$, which is equal to one in the case of a sphere. For a nematic sphere with homeotropic surface anchoring, there should be one radial hedgehog located in the centre of the sphere, as illustrated in the Fig. 7.2a.

If the surface anchoring conditions are planar instead of perpendicular, the trivial possibility are two surface boojums, residing at the poles of the droplet, which is now called a bipolar nematic droplet, schematically shown in Fig. 7.2b. In principle, it is possible to have additional topological defects in both structures, which will obey the conservation of the total topological charge and nutually compensate their charges. These topological states will have higher energy and will not be stable, but will decay into the one of the two possible configurations shown in Fig. 7.2. The nematic droplets are topologically simple and the most interesting topological phenomenon is the transition of a droplet from a radial nematic into bipolar nematic and vice versa. This is achievable by varying the surface anchoring conditions by adding surfactant molecules.

One of the earliest studies of a transition from the bipolar to the radial structure and back was demonstrated by Volovik and Lavrentovich [237]. Starting from the bipolar nematic droplet of 30 μm radius, the change of the surface anchoring was

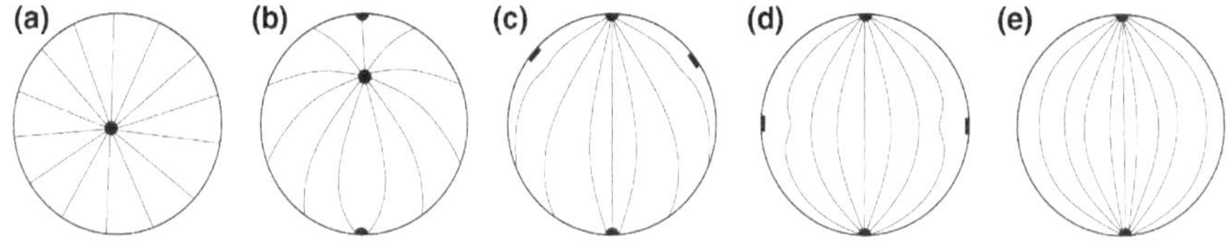

Fig. 7.3 **a**, **e** Schematic drawing of the transformation of the director field and the boojums from a radial nematic droplet shown in **a** to the bipolar nematic droplet shown in **e**. At the beginning the surface anchoring conditions are homeotropic and gradually change to planar in panel **e**. The cores of the hedgehogs are shown by *circle* the cores of the boojums are shown by *semi-circles*, and the cores of the defect rings by *rectangles*. The schematics was following the illustration in a paper by Volovik and Lavrentovich [237]

gradually transformed by adding the lecithin surfactant to the glycerol where the droplets were floating. At some point of added lecithin concentration, an additional ring was formed at the inner equator of the droplet, as shown in Fig. 7.3b, whereas the two boojums disappeared from the poles by "sinking" into the surrounding glycerol. The remaining ring then shrunk into a point hedgehog located at one of the poles, which was then expelled from the surface to the centre of the droplet, when the anchoring conditions changed to strong homeotropic anchoring (Fig. 7.3f).

Structural transformation of bipolar droplets into radial configuration could also be triggered by external agents and molecules. For example, Lin et al. have shown that the presence of endotoxin could trigger such a transition at already very small endotoxin concentrations of $\approx$1 pg/ml [238, 239]. This surface sensitivity could be used in an entirely new line of biosensors [240]. These microdroplet-based biosensors could be remotely read-out by using lasing emission from the Whispering Gallery Modes, as proposed by Humar and Musevic [9, 241–243]. They were analysing the changes of the spectrum of the WGMs, circulating in the interior of the nematic droplet. The droplets were floating in water, where increasing solution of the surfactant SDS was added, thus changing the surface anchoring from perpendicular to parallel. More details are presented in Sect. 9.2. Topological structure and dynamics of defects in nematic droplets was studied and discussed in a number of publications [36, 244–247].

Structure and optical properties of nematic microdroplets with various surface anchoring conditions were extensively studied within the context of Polymer Dispersed Liquid Crystals (PDLC), which was discovered by J.W. Doane et al. in 1986 [248]. A dense polymer dispersion of sub-micrometer nematic droplets appears in a form of a milky film which strongly scatters light because of the mismatch of the refractive index of polymer and that of the liquid crystal inside the droplet. When an external electric field is applied, it forces the nematic molecules in the droplet to align into the field direction. This is accompanied by a change of the refractive index of the interior of the droplet, which could be matched to the refractive index of the polymer by a proper selection of the materials. In this case, a strongly scattering PDLC film becomes transparent for particular direction of light propagation. The influence of external electric field on the director structure in nematic droplets was discussed in a number of articles [249–253]. Segregation of small particles in the topological defects of nematic droplets was studied experimentally and using Gay-Berne simulations [254].

7.2 Structure and Topology of Chiral Nematic Microdroplets with Parallel Surface Anchoring

It is clear that the topology and structure of chiral nematic droplets is expected to be much richer compared to nematic droplets, which exist in two different ground states, i.e. the radial and the dipolar droplet structure. The first studies of chiral nematic droplets were reported by Robinson and Ward in 1957 [255, 256], when studying concentrated solutions of polypeptides. They observed a spiral-shaped optical pattern with a radial disclination, which was latter explained by Pryce and Frank. According to that model, the internal structure of the chiral nematic droplets with parallel surface anchoring would look like a spiral or a series of concentric rings, winding from the surface to the centre of the sphere and back. An example of such a droplet is shown in Fig. 7.4.

The internal structure and topology [257, 258] of chiral nematic droplets is expected to be quite different for the two possible surface anchoring. Chiral nematic droplets with parallel surface anchoring at the interface with the exterior carries medium have been extensively studied and three different structures were found in the experiments: (i) radial spherical structure (RSS), also known as the spherulitic texture or the Frank-Pryce model, (ii) diametrical spherical structure (DSS), and

Fig. 7.4 A chiral nematic droplet with pitch $p \approx 1\,\mu\mathrm{m}$ and parallel surface anchoring, floating in glycerol. The series of concentric rings corresponds to spiraling cholesteric layers, which are topologically connected with a defect running from the surface to the centre of the droplet

(iii) planar bipolar structure (PBS). Schematic presentations of the three most commonly observed structures are shown in Fig. 7.5.

The director structure of chiral nematic droplets with degenerate planar surface anchoring were recently studied by Seč et al. [259] within the fully tensorial Landau-de Gennes module and six metastable orientational profiles were found. In addition to DSS, RSS and PBS, the new bipolar structure (BS), the Lyre and the Yeti structures were predicted. The latter two were never predicted or observed up to date.

The diametrical spherical structure shown in Fig. 7.5b is the most symmetric one. It shows cylindrical symmetry with the symmetry axis along the direction which connects the centres of the ring defects shown in red colour in Fig. 7.5. The director forms uniformly curved cholesteric layers with their normals along the radial direction. The ring defects form a series of concentrically positioned tori of the double twist cylinder, whereas the central torus resembles the triple twist torus structure similar to the one observed by Smalyuk et al. [185]. The series of ring defects appear in order to compensate for the director field imposed by the tori. The numerical calculation always finds an even number of defect ring regardless of the ration of the cholesteric pitch to the droplet diameter. This linear series of defect rings and surface boojums is found similar to the previously predicted +1 diametrical disclination line with one difference that in this study it actually appears in a form of alternating rings of tau and lambda lines. Such a structure is essentially different from the dissociated models of the +1 line in the cholesteric, where the line decomposes into non-singular lambda lines.

The radial spherical structure shown in Fig. 7.5a is most commonly observed in the experiments and consists of distorted double twist tori with a variable minor radii. There are no singular bulk defect in the nematic director in the RSS structure, but instead cholesteric λ disclinations appear running from the centre of the droplet to the surface. These escaped disclinations are found to be a double helix of two intertwined cholesteric lambda cholesteric lines as visualised in Fig. 7.5a. This double helix defect structure is quite different from the Frank-Pryce model where the non-singular disclination lines with the winding number +2 are predicted.

The third structure is the bipolar structure shown in Fig. 7.5d. It is cylindrically symmetric and characterised by only two surface defects which are positioned diametrically. This bipolar structure actually gradually evolves from the well-known bipolar structure in chiral nematic droplets and the cholesteric winding is switched on. It is essentially formed from a $\lambda + 1$ disclination line which is spanning diametrically along the symmetry axis of the cylindrically organised droplet. The central region is most uniformly aligned along the symmetry axis and is surrounded by distorted double twist tori.

The fourth structure is the planar bipolar structure. In this structure, the two surface boojums are attached to the interface and the chiral nematic structure winds through the interior of the droplet. Theoretical studies of chiral nematic droplets with parallel surface anchoring were performed by the group of Žumer et al. [249, 260]. There are quite some experimental studies of the structure and organisation of chiral nematic droplets with parallel surface anchoring. They were mainly driven by the interest in applying chiral nematic droplets in polymer dispersions, thus forming polymer

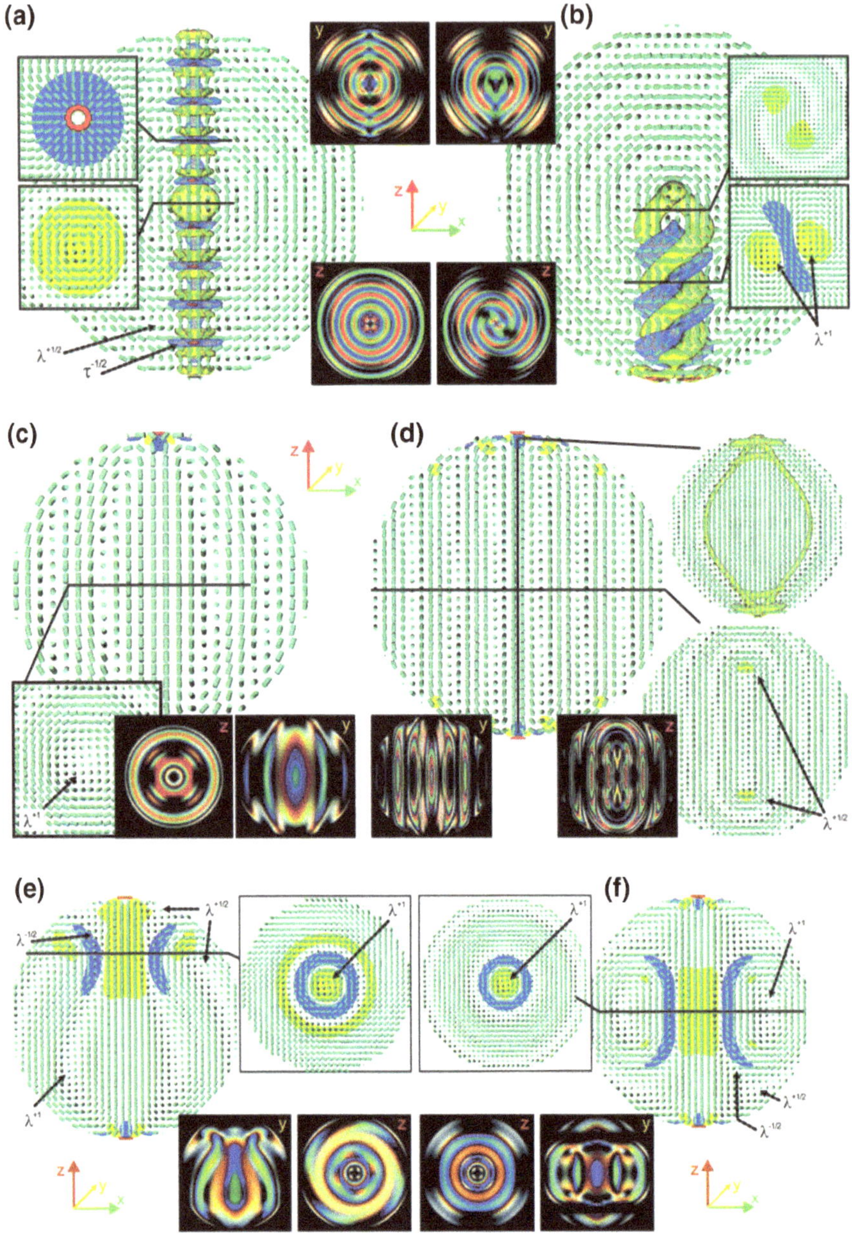

Fig. 7.5 Schematic view of **a** diametrical spherical structure (DSS), **b** radial spherical structure (RSS), **c** bipolar structure (PBS), **d** planar bipolar structure (PBS), **e** Lyre structure and **f** Yeti structure. Regions with reduced nematic degree of order are shown in *red* (iso-surface of $S = 0.48$), whereas *blue* and *yellow* regions show the splay-bend parameter. Reproduced from D. Seč et al. Soft Matter, 2012, **8**, 11982, with permission of The Royal Society of Chemistry

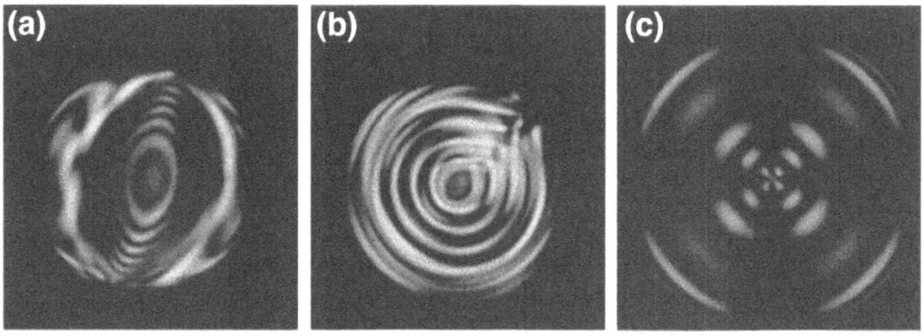

Fig. 7.6 Crossed polarizers images of chiral nematic droplets with different chiralities. **a**, **b** Droplet with low chirality, $P > R$. **c** Droplet with very high chirality $P << R$. Reprinted figure with permission from F. Xu, P. Crooker, Phys. Rev. E **56**(6), 6853 (1997). Copyright (1997) by the American Physical Society

dispersed cholesteric liquid crystals [250–252, 261, 262]. This kind of displays is both reflective and coloured and usually the direct anisotropy of the material is negative. The pitch of the liquid crystal is chosen to exhibit a visible selective reflection in a planar texture. When this material is placed in droplets in zero field, light is weakly scattered and the display appears dark. If a large voltage is applied, the droplets are reoriented into the planar texture with helical axes parallel to the field and the display appears coloured.

The structure of chiral nematic droplets with parallel surface anchoring with various radii R and an intrinsic pitch P were studied experimentally by Xu and Crooker [261]. They found different textures for different ratios of the pitch to the radii of the chiral nematic microdroplets. In this study, the diameters of droplets ranged from less than 1 to over 100 μm and two different mixtures with two different pitch values were used. Depending on the ratio of P/R, the droplets exhibited three different characteristic structures presented in Fig. 7.6.

For low chirality, the structures are similar but not quite identical to those predicted by Bezič and Žumer [263]. The droplets exhibit rotational symmetry and when their axis is perpendicular to the light, a series of concentric rings is observed with two dark regions near the edge. These features are similar to non-chiral nematic droplets with a twisted bipolar nematic configuration. Structural transitions in chiral nematic droplets with variable strength of a planar surface anchoring were studied numerically and experimentally by Zhou et al. [264].

For high chirality, Xu and Crooker [261] observed two different manifestations: either as a concentric ring pattern with a radial defect running from the centre to the surface, which is characteristic of the Frank-Pryce structure, or a singular defect line of the Frank-Pryce structure, which is converted to a nonsingular line by escaping the director in the radial direction, as shown in Fig. 7.7f.

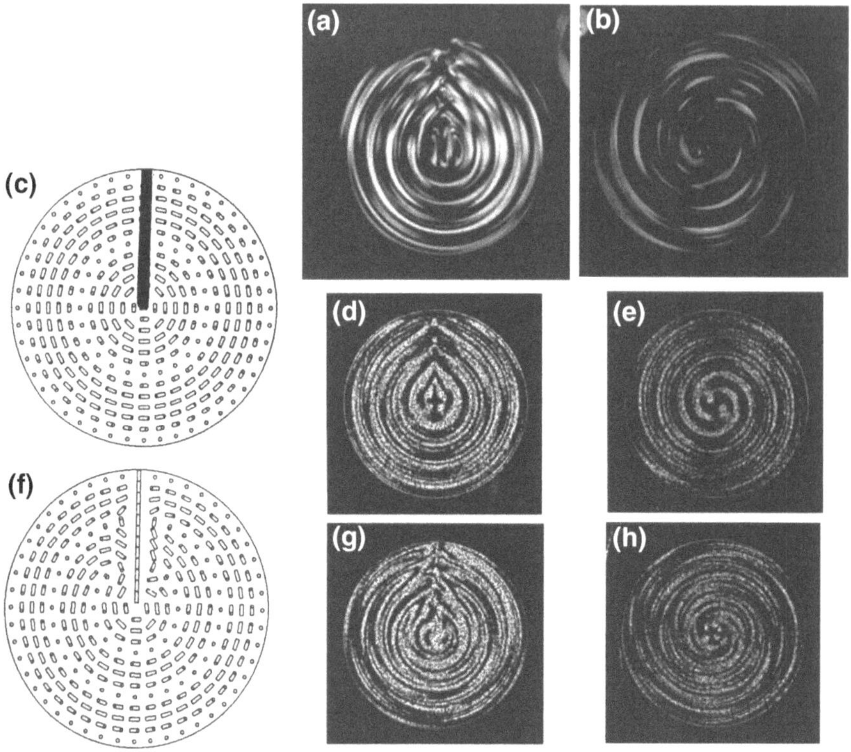

Fig. 7.7 Chiral nematic droplets with low chirality, $P < R$, between crossed polarisers. **a** View of the droplet, when the disclination line is parallel to polariser, **b** disclination line is here set parallel to light beam, i.e. running into the plane of this figure. **c** Schematic picture of the director field within the Frank-Pryce model with unescaped $\chi = 2$ line. **d**, **e** Numerical simulation of **a** and **b** from Frank-Pryce model in **c**. **f** Escaped Frank-Pryce model with escaped $\chi = 2$ line; **g** and **h** Numerical simulation of the structure in **a** and **b** from the model in **f**. Reprinted figure with permission from F. Xu, P. Crooker, Phys. Rev. E **56**(6), 6853 (1997). Copyright (1997) by the American Physical Society

7.3 Structure and Topology of Chiral Nematic Microdroplets with Perpendicular Surface Anchoring

Whereas chiral nematic droplets with parallel surface anchoring attracted a lot of interest because of their possible applications in polymer dispersed chiral nematic displays, there are few studies of the structures of chiral nematic droplets with perpendicular surface anchoring. It can be understood that this perpendicular surface anchoring causes frustration when the internal structure of the droplet is chiral. The most comprehensive theoretical analysis of the structure and topology of chiral nematic droplets with perpendicular surface anchoring was performed by Seč et al. [265]. They were using Landau free energy minimisation and topological theory

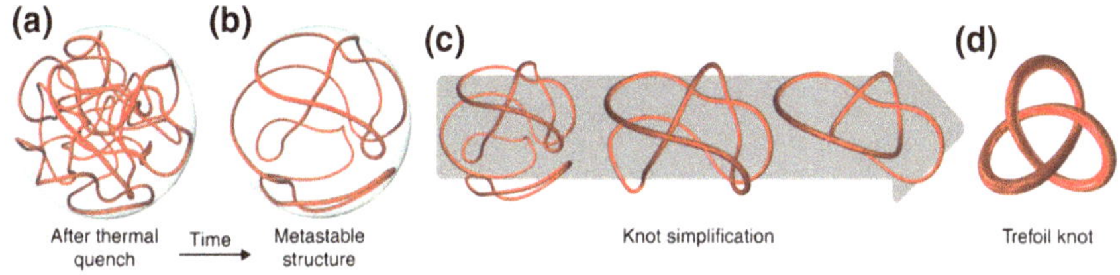

Fig. 7.8 Knots in homeotropic cholesteric droplets might be obtained by thermal quench. **a** Short after the numerical thermal quench a highly disordered state is obtained for $N = 5$. Disclination *lines* are visualized by showing the areas with depressed order parameter $S = 0.48$. A few millions of simulation steps later, the system relaxes to a metastable state, where it stays in a local free energy minimum. **c** The type of the knot could be deciphered by performing Reidemeister moves, therefore geometrically simplifying the disclination line. In this case, the disclination forms a trefoil knot shown in **d**. Reprinted by permission from Macmillan Publishers Ltd: D. Seč, S. Čopar, S. Žumer, Nat. Commun. **5**, 3057 (2014), copyright (2014)

to analyse the structure and topology of defect lines formed during thermal quench of the chiral nematic droplet. The tendency of the cholesteric to create layer-like helical structure competes with the spherical boundary conditions of the droplets, thus creating a frustrated environment with a multitude of metastable states. In such a droplet, the disclinations are not allowed to pass through the boundary, which results in confined closed disclination loops.

By performing a temperature quench and therefore rapidly cooling down the liquid crystal through the nematic-isotropic transition, metastable structures are observed, as illustrated in Fig. 7.8.

After a thermal quench, formation of liquid crystal domains results in a multitude of randomly positioned singular disclination lines with winding number $-1/2$ between the domain walls, as shown in Fig. 7.8a. After the structures are relaxed to the local energy minima, the resulting metastable structure often includes knotted and linked disclinations. After numerically extracting the curve of these disclination loops, the defect tangle is simplified using standard Reidemeister moves. The topology of the simplified tangle can be analysed. There is a large variety of states that can be obtained and it was predicted that the type of knots or links should depend on the ratio of the chiral pitch to the droplet radius. The chirality parameter N was introduced, which measures the number of π turns of the director along the droplet diameter. For chirality parameter larger than 3, the complexities of predicted topological states quickly increases, as shown in Fig. 7.9.

For $N = 4$, only multiple unlinked loops and Hopf links are predicted, while for $N = 5$, a set of links is predicted, as shown in Figs. 7.8 and 7.9: Trefoil knot (Fig. 7.8d), Solomon link (Fig. 7.9c), Hopf link (Fig. 7.9d), Whitehead link (Fig. 7.9e) and a three-component link (Fig. 7.9f). All these topological states were predicated for relatively small droplets with variable surface anchoring.

It is quite surprising that until recently there were only a few experimental studies of the 3D ordering of chiral nematic liquid crystal in spherical droplets with perpendicular surface anchoring. Candau et al. [266] studied the cholesteric droplets with

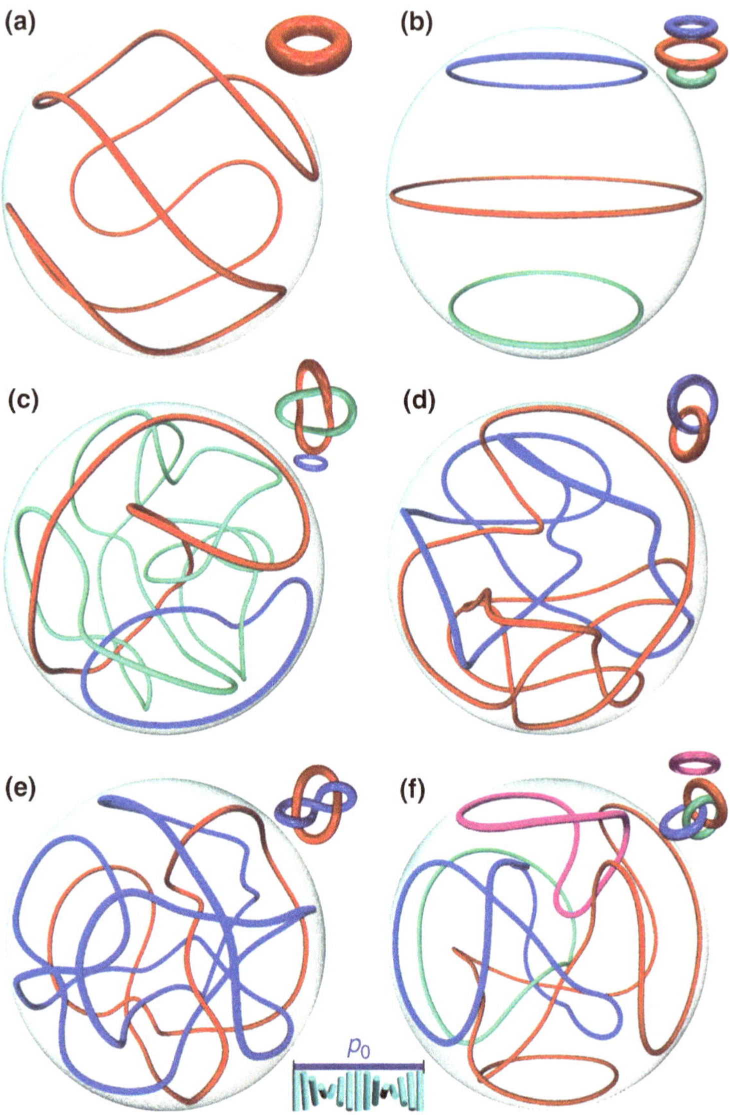

Fig. 7.9 Several different knots and links are obtained for $N = 5$. **a** A single loop is an unknot. **b** Three separated loops. **c** Solomon link with an extra loop. **d** Example of a Hopf link. **e** Example of a Whitehead link. **f** A three-component link denoted as 6^3_3 on the Rolfsen table with an additional unlinked loop. For the visualisation purposes, the disclination lines are presented as regions of reduced order parameter and colour coded for better visualisation. The insets show a schematic view of the corresponding knots and links. Reprinted by permission from Macmillan Publishers Ltd: D. Seč, S. Čopar, S.Žumer, Nat. Commun. **5**, 3057 (2014), copyright (2014)

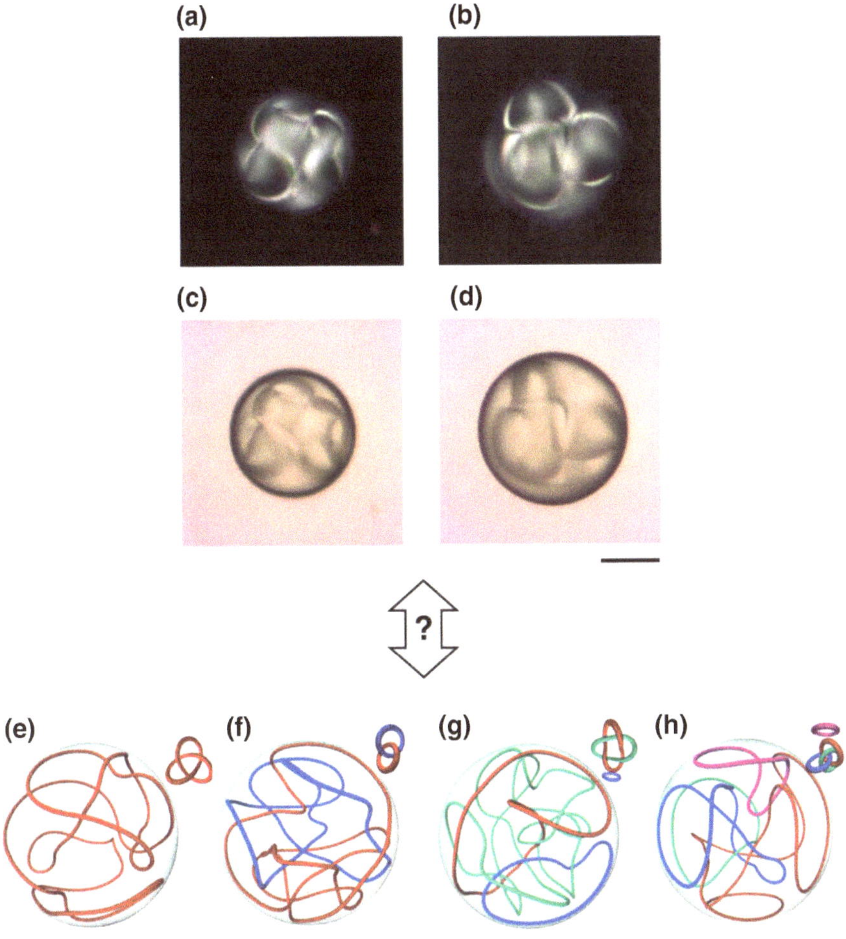

Fig. 7.10 Optical images of complex metastable states in chiral nematic droplets with perpendicular surface anchoring of liquid crystal molecules. In all cases the helical pitch $p = 55\,\mu\text{m}$. **a**, **c** Droplet diameter is $d = 105\,\mu\text{m}$, corresponding to $N = 3.8$. **b**, **d** Droplet diameter is $d = 125\,\mu\text{m}$, which correspond to $N = 4.5$. The images in **a**, **b** are taken between crossed polarisers. **b**, **d** Bright field imaging. Scale bar, $50\,\mu\text{m}$. *Bottom row* shows some illustration of numerically predicted topological states from Ref. [265], where $N = 5$. These are differently knotted and linked loops, as schematically presented in the *upper right* corners of each panel. Sketches are adapted from Ref. [265]. This figure is distributed under the terms of the Creative Commons Attribution 4.0 International License (http://creativecommons.org/licenses/by/4.0/). Authors: T. Orlova, S.J. Asshoff, T. Yamaguchi, N. Katsonis, E. Brasselet, Nat. Commun. **6**, 7603(2015)

either parallel or perpendicular surface anchoring. For strong chirality, they observed radial distribution of the helical axis with a radial disclination similar to the droplets with parallel surface anchoring. There was no supporting experimental evidence of the interfacial transition layer, where this structure should fit to the transitional interfacial surface anchoring. Kurik and Lavrentovich reported on the observations of an equatorial disclination and point surface singularity [267].

Several experimental studies were intrigued by the numerical predictions of knots and links in chiral nematic microdroplets. Orlova et al. [268] studied topological diversity in frustrated cholesteric droplets using polarisation microscopy. In the most interesting region of the chirality parameter $N = 5$, they observed indications of quite complex topological structures, which looked like linked and knotted disclination loops, shown in Fig. 7.10. It was concluded by Orlova et al. [268] that the polarisation microscopy alone does not give conclusive arguments for the unambiguous identification of the topological states of higher complexity. This is because of possible optical artifacts in optical microscopy, which is due to polarisation effects and lensing. Periodic birefringent structures are known to induce strong polarisation effects which depended of the ratio of the wavelength of light used and the helical pitch. In addition to these phase effects, these structures also act as birefringent lenses, which are able to focus light in quite unexpected manner. Polarisation effects and lensing depend on the birefringence and the characteristic length of distortion of the liquid crystal and are likely to cause strong optical artifacts, where twisted layers could appear as singular lines and so on. Fluorescent confocal microscopy was used to analyze the structure of chiral nematic droplets with perpendicular surface anchoring by Guo and Song [269] but no director reconstruction was performed, so only qualitative discussion of the 3D fluorescent intensity distribution within the droplet was possible.

7.4 Skyrmions and Torons in Chiral Nematic Microdroplets

Polarisation microscopy alone cannot give unambiguous answer to the existence of knots and links in chiral nematic droplets with perpendicular surface anchoring. Posnjak et al. [11] introduced a new method of reconstruction of the 3D director field from nematic droplets using fluorescent polarised confocal microscopy. They showed that in chiral nematic droplets of considerable size (tens of micrometers) no knots and links are stable but a variety of different point defects and nonsingular structures, such as skyrmions and torons are observed.

Whereas skyrmions and torons were reported in a number of experimental studies of chiral nematic liquid crystals confined to very thin homeotropic layers and are described in Chap. 8, the question of their stability in other geometries, such as microdroplets, remained relatively unexplored. The problem here is reliable reconstruction of the 3D nematic director in a spherical geometry, where many optical artifacts obscure the optical investigations and make them unreliable. Whereas simple topological defects like single points or single rings in liquid crystal microdroplets could be reliably observed in low-chirality microdroplets, it is much more difficult to analyse the topology of droplets containing highly chiral nematic liquid crystal. Besides polarising optical microscopy, which can give us information on a particular cross-section of the droplet, fluorescent confocal polarised microscopy is the best candidate for reconstruction of the 3D director in a droplet. In most cases, the director is reconstructed indirectly from the measured FCPM intensities, either by hand, by visually comparing the measured and the calculated FCPM intensities, or by observing the orientation of the in-plane director component on a selected intensitys iso-surface.

A new method of reconstruction of the 3D director field and topological defects through FCPM imaging was reported recently (Posnjak et al. [11]), which is using a numerical approach for director construction from fluorescent 3D images taken at different light polarisations. Let us remember that in a FCPM experiment, the light is collected from the focal region, which is diffraction-limited and forms a voxel of around 200 nm diameter with a depth of 500 nm. In this region, the vector of light polarisation is lying in the focal plane and the detected fluorescent intensity I_{FCPM} is proportional to the fourth power of the cosine of the molecular tilt θ with respect to the vector of polarisation

$$I_{FCPM} = I_{offset} + I_{norm} \cdot \cos^4 \theta \tag{7.2}$$

It is clear that FCPM cannot resolve the out-of-focal plane tilt, because the molecules tilted by $+\theta$ and θ with respect to the focal plane emit the same amount of fluorescent light. In other words, the FCPM cannot determine the correct sign of the z-component of the director (i.e. along the optical axis) and there are two possibilities for the orientation of the molecule in each measuring voxel. If the number of sampled voxels in the droplet is equal to M, there are M possible configurations. Because there are two possibilities for director orientation in each voxel, the liquid crystal droplet divided in M voxels is therefore similar to a system of M $\frac{1}{2}$ spins, which are interacting with each other. In a similar way, the neighbouring voxels interact elastically with each other, which brings in the analogy to the interactions of two neighbouring spins with two possible states for each spin. Minimization of the free energy of such a system of M coupled voxels is therefore formally equivalent to M interacting spins. Finding the minimal free energy of such a system is an optimisation problem, which was solved by Posnjak et al. [11] by using a simulated annealing algorithm, first used for optimising the placement and wiring of components in electronic systems. The second problem with the detected fluorescent intensity is its highly nonlinear dependence on the out-of plane molecular tilt θ. Most of the variation of the fluorescent intensity is concentrated around $\theta \approx 45°$, which makes it difficult to reliably reconstruct the director in those regions, where the molecules lie practically within the focal plane or are perpendicular to it.

Using the simulating annealing algorithm, 3D nematic director fields of chiral nematic droplets were fully and reliably reconstructed. In the experiments, the droplets were made by dispersing a mixture of very low birefringent liquid crystals in a glycerol, realising nearly perfect refractive index matching. The liquid crystal materials were doped with a small amount of fluorescent dye BTBP, which aligns with the radiative dipole parallel to the liquid crystal molecules. The fluorescent intensities were taken at four different polarisations of fluorescent excitations and the data were corrected for bleaching, deconvolution, abberations and background corrections. In non-chiral nematic droplets with perpendicular surface anchoring, a single point defect was observed in the centre of the droplet after director reconstruction. This was followed by increasing chirality of the liquid crystal, adding chiral dopants. The parameter which determines topological complexity of chiral nematic droplets is called the chirality parameter N and is equal to the diameter of the droplet divided by half of the intrinsic pitch of chiral nematic liquid crystal. Pitch is the distance over which the director rotates by 360°.

In the case of low-chirality, $N = 1.5$, one can see from the 3D reconstructed director profile that the original point defect is expelled from the centre of the droplet towards the surface, as shown in Fig. 7.11a, b. The droplet has a symmetry axis and if one examines the reconstructed director field in a plane perpendicular to this axis (Fig. 7.11b), one can see that the director twists along any diametrical line by almost 2π. This means that the local cross-section corresponds to a double twist cylinder and forms a 2-dimensional skyrmion, which is known as the Bloch-type skyrmion in chiral magnets. This 2D skyrmion is bound by a point defect on one side and connects smoothly to the opposite surface. The non-polarised transmission images of the droplets with the same type of structure as in Fig. 7.11a, b are shown in Fig. 7.11c, d. One can clearly see the defects while the twisted cholesteric layers appear as dark lobes, which is due to optical lensing effects.

By increasing the chirality parameter to $N = 3$, one can clearly see 3 collinear point defects, which are connected with two skyrmion-like profiles as shown in Fig. 7.11e, f. The skyrmion profile can be observed from the cross-section, taken along the dashed line in Fig. 7.11e and shown in Fig. 7.11f. The two skyrmions share a common hyperbolic point defect. By further increasing the chirality parameter, $N = 46$, similar skyrmion-like structures were found, but with additional cholesteric layers with a twist. This results in droplets with two point defects and a single disclination loop, as shown in Fig. 7.12a, b, with the central part of the droplet forming a typical double-twist cylinder. At this value of the chirality parameter one can observe up to five point defects and even more twists in the skyrmionic structures, as shown in Fig. 7.12c, d. The innermost layer marked with a dashed circle in Fig. 7.12c resembles a toron structure observed in thin homeotropic and unwound chiral nematic layers. In a droplet, the torons are nested along outer concentric layers instead of uniform director far-field in flat homeotropic cells.

The topology of a sphere with homeotropic anchoring forces the defect inside the sphere to have a total of $+1$ topological charge, which is realised as a single defect in the case of a non-chiral droplet. In chiral nematic droplets with homeotropic anchoring, multiple point (and eventually single loop) defects are stabilised by twisted cholesteric areas, which provide elastic repulsion forces, pushing the defects away from each other. Interestingly, the 3 and 5 point defect structures must have hedgehog defects of opposite topological charges which mutually compensate in pairs, leaving a total charge of $+1$ of the remaining, uncompensated point hedgehog. The

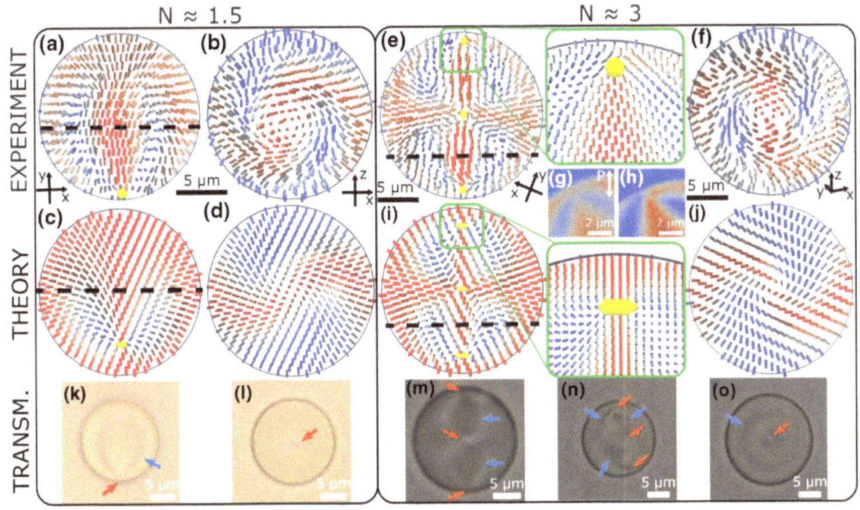

Fig. 7.11 **a** Cross-section of the reconstructed vector field in a droplet with a single point defect, expelled to the surface of the droplet, obtained at $N = 1.5$. **b** Cross-section of the droplet shown in **a**, taken along the *dashed line* in **a**. **c**, **d** transmission microscopy pictures of the same droplet as in **a**, **b** with different orientations of the symmetry axis. Location of the point defect is marked with *red arrow* and the cholesteric layers with *blue arrow*. **e** Three collinear point defects are observed in the reconstructed director of a droplet with increased chirality parameter $N = 3.0$. **f** Reconstructed director in a cross section taken along the *dashed line* in **e**. **g**, **i** Unpolarised transmission microscopy pictures of droplets with three point defects and different orientation of the symmetry axis. In **g**, the symmetry axis is in the focal plane, in **i** it is perpendicular to the focal plane, whereas in **h**, it is tilted out of the focal plane

topology is further complicated because the point defects are deformed by neighbouring twisted sections and do no form simple radial or hyperbolic hedgehogs. The signs of topological charges of these hedgehog cannot simply be determined by the geometry of the surrounding director field, but the director filed must be considered as a whole in order to assign the topological charge signs consistently.

In spite of large number of experiments performed on chiral nematic droplets with homeotropic surface anchoring, no signs of linked or even knotted defect loops were observed. These were predicted based on the LdG numerical simulations, performed on much smaller droplets of typical diameter 1 μm. In contrast, the experiments are performed on droplets with the diameter of tens of μm. A possible reason for this discrepancy between theory and experiments is a large difference in the diameters of theoretically analyzed and experimentally measured droplets. In smaller droplets, the structures are much more confined and the point defects may be forced by the surface and confinement to open into defect loops, which makes it possible to stabilise knots and links in chiral nematic droplets.

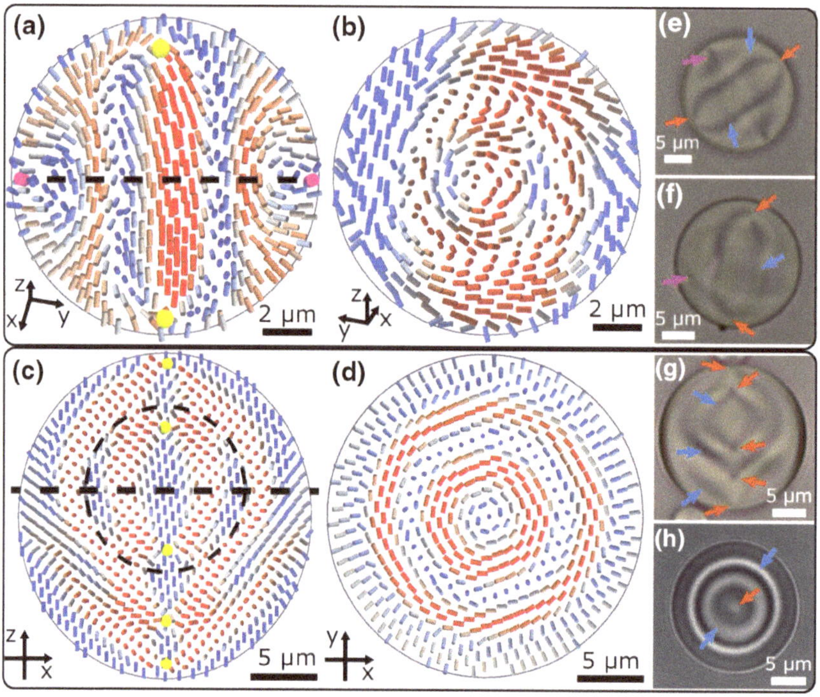

Fig. 7.12 Points and loops in droplets with higher chirality. **a**, **b** The director field, reconstructed from 3D FCPM images in two perpendicular cross-sections in a chiral nematic droplet with $N = 5$. Two point defects (*yellow dots*) and a disclination ring (*purple dots*) can be seen in **a**. The cross section shown in **b** includes the disclinations ring. **c** A different droplet with $N = 6$, showing a total of 5 point defects (*yellow dots*) with the region inside the *dashed circle* corresponding to a toron. In this cross-section, the elongation of the droplet along the z axis is an optical artefact due to refractive index mismatch between the liquid crystal and surrounding medium, as well as the birefringence of the LC. In this case the effect is pronounced because of the size of the droplet and the type and particular orientation of the director structure. **d** The reconstructed director projection on the $x - y$ plane. **e**, **h** Transmission microscopy pictures of droplets with different orientations of the symmetry axis. Panels **e** and **g** show droplets with the symmetry axis in the image plane, **f** and **h** with the symmetry axis perpendicular to the image plane. Locations of point defects are marked with *red*, disclination line with *purple* and cholesteric layers with *blue arrows*

7.5 Toroidal Nematics and Handlebodies

Spherical nematic droplets are easily obtained by simply mixing a nematic liquid crystal with another, immiscible fluid. This mixing results in a dispersion of nematic droplets in the continuous phase and the droplets are of perfect spherical shape due to the surface tension. In some extreme cases, when the surface tension is very small, like in water-based chromonic lyotropic nematics, spherical droplets are deformed into tactoid objects. A spherical droplet is topologically simple object with genus $g = 0$, which implies the formation of a single topological defect with charge -1

inside a nematic structure. This is due to the Gauss-Bonnet theorem, which dictates that the total topological charge on the bounding surface must be equal to the one half of Euler characteristic of that surface (O.D. Lavrentovich [43]). Let us remind that Euler characteristic χ is a topological invariant given by $\chi = 2(1 - g)$, where g is the genus of the surface, or the number of its handles. This theorem implies that in the ground state of a confined system there will be topological defects. In the case of a sphere, which has Euler characteristic $\chi = 2$, a single topological defect should be present in the sphere. Twisting, bending, stretching or deforming the sphere in some other manner does not change its Euler characteristic, which remains equal to 2.

However, the Euler characteristic of an object can be changed by adding handles to that object. Adding a handle to a sphere creates a torus and the Euler characteristic of a torus is $\chi = 0$. Creating nematic droplets with handles is a major experimental challenge, which was first resolved by the group of Fernandez-Nieves in 2013 [270]. To make nematic toroidal droplets, they injected a nematic liquid crystal through a tiny needle into a rotating bath containing a yield-stress material, which was a mixture of polyacrylamide microgels with glycerol, ethanol, polyvinyl alcohol and majority of ultra-pure water. The presence of polyvinyl alcohol guarantees degenerate tangential anchoring of liquid crystal at the interface to the external carrier fluid. During the rotation of the bath and injection of the liquid crystal, a ring of a liquid crystal closes into a torus immersed in the bath, as shown in Fig. 7.13a.

If such a nematic ring be formed in a simple liquid, it would spontaneously close and transform into a spherical droplet. However, the carrier fluid in a form of microgel has its own yield-stress, which prevents coalescence of a liquid crystal torus into a liquid crystal droplet. Once the torus has been formed, the elasticity of the surrounding continuous microgel phase provides the supporting force which overcomes the surface tension and prevents transformation of the toroidal droplet into a spherical droplet. Using this simple technique, nematic tori with very different aspect ratios $\xi = \frac{|R+a|}{a}$ ranging from 1 to 20 could be formed. Here, a is the tube radius and R is the inner radius of the torus. Top view of a nematic toroid is shown in Fig. 7.13b, c in brightfield (b) and between crossed polarisers (c). The structure of the nematic is completely smooth, with no observable topological defects. This is indeed expected for degenerate parallel surface anchoring, because the director can simply close into itself by looping once along the interior of the toroid.

However, when viewing such a nematic toroid from the side and between crossed polarisers, the images are quite surprising as shown in Fig. 7.13d–f. For convenience, the lower images in Fig. 7.13g, i give the bright field view of the same configuration as in Fig. 7.13d–f. By comparing Fig. 7.13d, g, i.e. when the plane of the toroid is coinciding with the direction of the polariser, the central part of the toroidal nematic particle is bright instead of being completely dark. Namely, if the interior of the nematic toroid is structured so that the nematic director is running in the plane of the toroid, then the molecules in the centre of Fig. 7.13d, g should be parallel to the polariser and the Fig. 7.13d should appear completely dark between crossed polarisers. Because this is clearly not the case, the interior of the nematic toroid should be twisted in the radial direction. This twist influences the propagation of light and makes the side view of the toroidal particle transmitting some light between

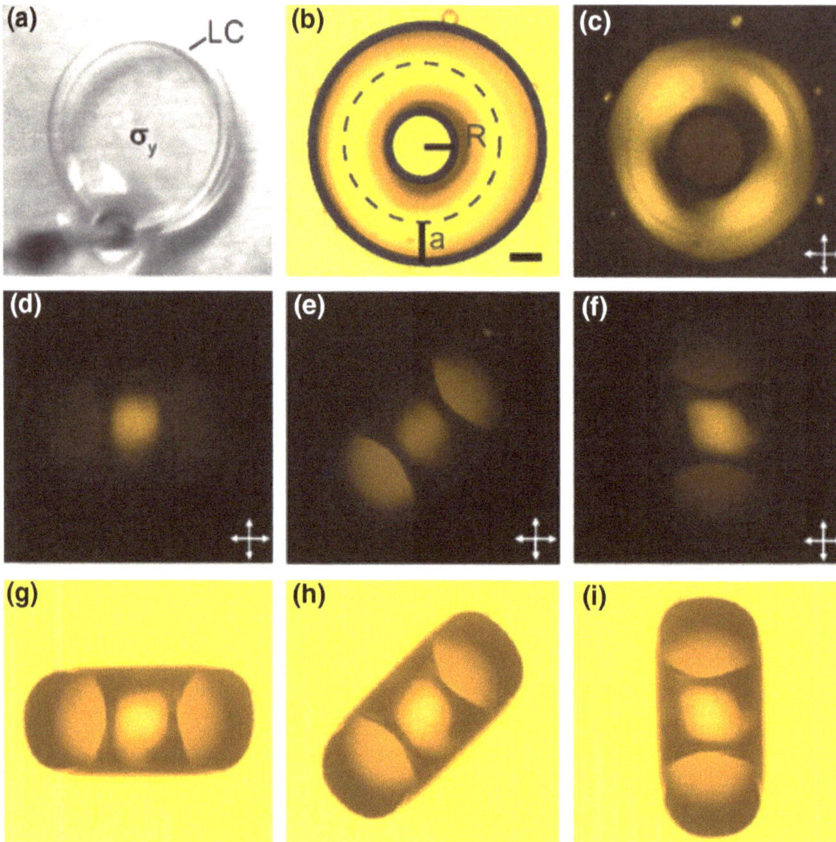

Fig. 7.13 Liquid droplets are forming tori. **a** A toroidal liquid crystal droplet is formed by injecting the liquid crystal into a material with yield stress. **b** Bright field image of a toroidal droplet of a nematic liquid crystal. The tube radius is a and the inner radius is R. **c** Crossed polarisers image of the same droplet. **d** Cross polarizers image taken from the side of the droplet with $\xi = 1.8$ when viewed at 0°. **e**, **f** The same droplet viewed at 45° and 90° with respect to the incident polarisation direction. The central part of the toroid remains bright for all orientations. **gi** Bright field images of the same droplet and orientation corresponding to panels **d**–**f**. Scale bar: 100 μm. Reprinted with PNAS permission from "Stable nematic droplets with handles" by E. Pairam, J. Vallamkondu, V. Koning, B.C. van Zuiden, P.W. Ellis, M.A. Bates, V. Vitelli, A. Fernández-Nieves, Proc. Natl. Acad. Sci. U. S. A. **110**(23), 9295 (2013)

crossed polarisers. This twist is also evident for different orientations of the toroidal particle as shown in Fig. 7.13e, f, h, i. Similar spontaneous twisting could be observed in twisted bipolar droplets and theoretical studies of DNA in toroidal geometries. Careful analysis of different nematic toroids reveals that this twisted configuration is stable for a very large interval of the aspect ratio ξ.

Spontaneous twisting of the interior of nematic toroids can be understood by considering the Frank elastic free energy of such a nematic, where the usual elastic

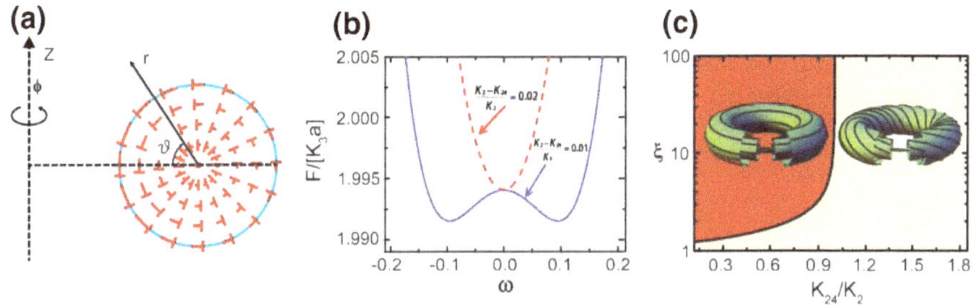

Fig. 7.14 Saddle-splay elasticity is important for spontaneous chiral symmetry breaking in nematic tori. **a** Illustration of the circular cross-section indicating the polar angle ϑ, radial separation r from the centre and the azimuth angle Φ. The director field is presented with nails, so one has the impression of the molecular tilt. The aspect ration of the torus is $\xi = 2$ and the configuration corresponds to a twisting strength $\omega = 0.49$. **b** Normalized free energy as a function of ω for two different values of $(K_2K_{24})/K_3$. There is a single minimum for $K_{24})/K_3 = 0.02$ that splits into two minima for $(K_2K_{24})/K_3 = 0.01$. **c** The phase diagram denoting the regions of non-twisted tori (*red shaded* area) and twisted tori (*yellow shaded* area). Reprinted with PNAS permission from "Stable nematic droplets with handles" by E. Pairam, J. Vallamkondu, V. Koning, B.C. van Zuiden, P.W. Ellis, M.A. Bates, V. Vitelli, A. Fernández-Nieves, Proc. Natl. Acad. Sci. U. S. A. **110**(23), 9295 (2013)

constants for splay (K_1), twist (K_2) and bend (K_3) are included and the saddle-splay deformation is considered as well with elastic constant K_{24}. By using an ansatz for the director field, the free energy of the toroid nematic can be analytically expressed as a function of the variational parameter ω, which is a measure for the degree of the radial twist. In the limit of large aspect ratio ξ, the Frank free energy can be expanded to quartic order in ω:

$$\frac{F}{\pi^2 K_3 a} \approx \frac{1}{\xi} + (4\frac{K_2 - K_{24}}{K_3}\xi - \frac{5}{4\xi})\omega^2 + \frac{1}{2}\xi\omega^4 \tag{7.3}$$

This simple form of the Frank free energy is actually very frequently meet in the Landau theory of phase transitions, where ω plays the role of the order parameter of the transition in question. If the coefficient of the quadratic term is positive, the Frank free energy will have a minimum at $\omega = 0$, as illustrated in Fig. 7.14b.

For $\xi = 0$, the internal structure of the toroid nematic will not be twisted. The twist will appear spontaneously when the coefficient of the quadratic term changes sign, and we shall have a transition from non-twisted to spontaneously twisted toroid neamtic. The analysis of the toroid nematic landscape given by the Eq. 7.3 shows the phase diagram, presented in Fig. 7.14c. It is the ratio of the saddle-splay elastic constant K_{24} to twist elastic constant K_2, which determines the phase boundary at a given aspect ratio of the nematic toroid. A more detailed numerical analysis of this spontaneous chiral symmetry breaking in toroid nematic was analysed by Koning et al. [271]. The comparison of the experimental measurement of the twist angle with theoretical prediction provide a robust and simple method to measure saddle-splay elastic constant K_{24}.

Notably, nematic toroids with degenerate planar surface anchoring have no defects in their ground state, but this changes in case if we add handles to this toroid. By adding handles, the Euler characteristic χ and hence the total topological charge decrease by -2 with each additional handle. The Poincare–Hopf theorem tells us about the appearance of additional charges by addition of handles, but it tells us nothing about the shape and location of these defects. This can be predicted by numerical modeling of the elastic free energy of nematic handlebodies, which is presented in Fig. 7.15a, b.

The result is interesting and quite expected for a double torus, where the theory predicts two different configurations of two defects, each with topological charge -1. The defects are located either at the innermost regions of the inner ring of each torus, as shown in Fig. 7.15a or in the outermost regions where the two tori meet, as shown in Fig. 7.15b. In both cases, the defects are located in local saddle geometry, where the Gaussian curvature is negative. Remember that the Gaussian curvature is the product of two principal curvatures at the selected point.

Nematic handle droplets were studied experimentally by forming individual tori, which were then merged together by the addition of the liquid crystal in the region between them. The top view of a typical double toroidal nematic is shown in Fig. 7.15c, d. When such a droplet is viewed along its side and between crossed polarisers, one can clearly observe the defect in the very centre of the droplets as shown in Fig. 7.15d. This defects were identified as -1 topological defects. The same technique could be used to form more complex nematic handlebodies, like triple tori shown in Fig. 7.15f, g. In this case, four point defects are formed, located in the region of negative Gaussian curvature.

Nearly simultaneously with the experiments of Pairam et al. [270], a similar study of topological properties of nematic handlebodies with degenerate planar anchoring were performed by Campbell et al. [272]. In this case, the nematic handlebodies were much smaller in size and were produced lithographically. These micrometer sized handlebodies were made by first producing silica handlebodies by photolithographic technique and then the hollow handlebody microstructures were obtained by replica methods. This produced nematic handlebodies of square, micrometer sized cross-section with variable genus from $g = 1$ to $g = 5$. There was a notable difference

► **Fig. 7.15** Genus $g = 2$ and $g = 3$ nematic droplets. **a, b** Numerical simulations of two different director profiles that show two surface defects (dark spots). Each defect has topological charge -1, and is located in regions of negative Gaussian curvature. **c** Bright field image of a double toroid with $g = 2$. *Dark spots* are the two surface defects. **d** The same torus viewed under cross-polarisers. **e** Crossed polarizers image of a $g = 2$ toroid. The image is focused to the region where the two tori meet, indicating the topological defect with charge $|s| = 1$. There is another $s = -1$ defect at the front of the double toroid. **f** Bright field image of $g = 3$ toroid with handles arranged in side-by-side fashion. **g** Another form of $g = 3$ toroid with the same topology as in **f**. **h** $g = 3$ toroid viewed between crossed polarizers and from the side. Note the presence of point defects in regions where the dark brushes meet. Scale bar is 100 μm. Reprinted with PNAS permission from "Stable nematic droplets with handles" by E. Pairam, J. Vallamkondu, V. Koning, B.C. van Zuiden, P.W. Ellis, M.A. Bates, V. Vitelli, A. Fernández-Nieves, Proc. Natl. Acad. Sci. U. S. A. **110**(23), 9295 (2013)

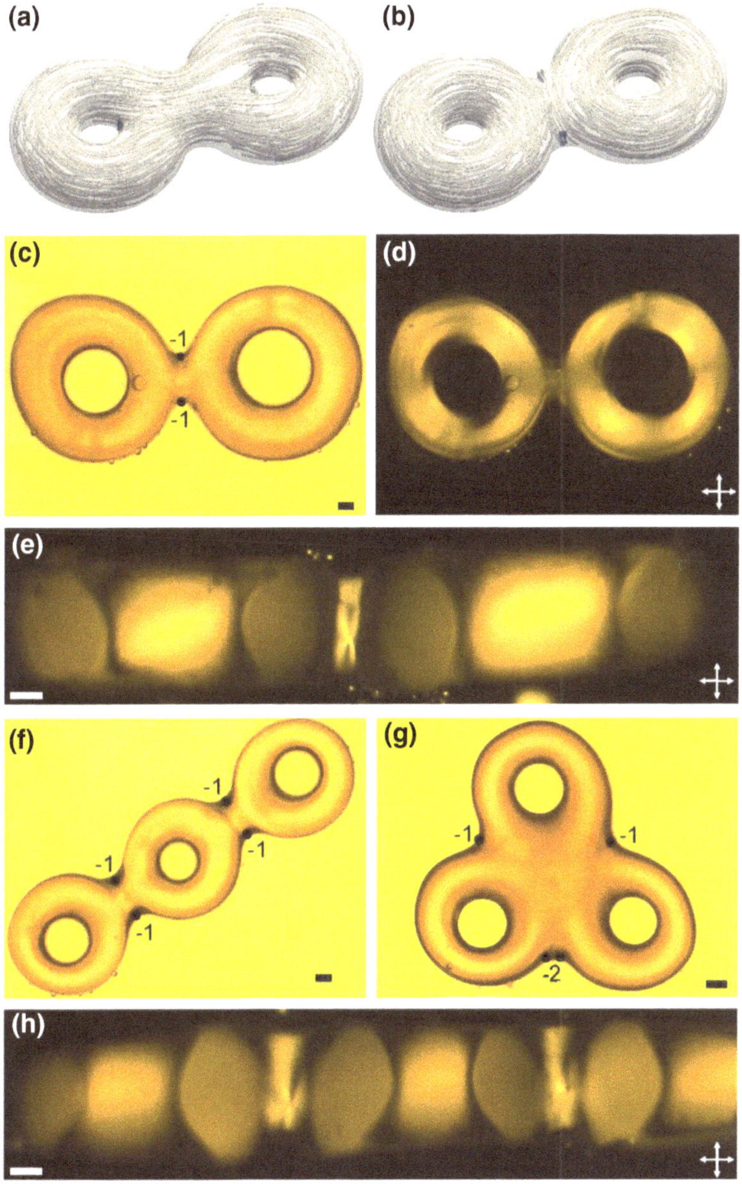
(a)
(b)
(c)
-1
-1
(d)
(e)
(f)
-1
-1
-1
-1
(g)
-1
-1
-2
(h)

between the double twisted nematic structures in circular cross-section, millimeter sized handlebodies and these much smaller handlebodies, which display no double twist nematic structure. This could be explained by slightly rounded square cross-section of the handle bodies which energetically does not favour the double twist as in the case of circular cross-section. Using three photon polarised fluorescent confocal microscopy, the 3D director distribution could be compared to theoretical predictions. The experiment showed that these micrometer sized nematic handlebodies host topologically stable half integer bulk defect lines, which are anchored at opposite side of handlebody surface. However, the small micrometer size of these handlebodies makes it difficult to obtain high resolution confocal imaging and reconstruction of 3D director field from the measured fluorescence.

The studies of nematic handlebodies with degenerate planar surface anchoring were followed by the studies of nematic handlebodies with perpendicular surface anchoring of the nematic liquid crystal by Tasinkevych et al. [273]. In contrast to topologically rather poor planar nematic handlebodies, these homeotropic nematic handlebodies display a large diversity of topological configurations including ones with linked and knotted half-integer defect lines. Nematic handlebodies were produced by replica method and were of micrometer size with genus g varying from 1 to 5 and the corresponding Euler characteristics $\chi = 2(1 - g)$ ranging from 0 to -8. The closed confining surfaces of handlebodies were chemically treated to impose strong homeotropic (normal) anchoring of liquid crystal molecules, so that the director locally aligns along the inner normal to the bounding surface. The topology of the confining surface dictates the existence of bulk defects of net topological hedgehog charge $m = \pm(1g)$ in the nematic liquid crystal. The existence of these topological defects assures the conservation of the topological charge and compensates the hedgehog charge of the ordering field on the inner closed confining surface of the nematic handle body, which is required by the Gauss-Bonnet and Poincare–Hopf theorems. This means that the nematic interior of a single torus with $g = 1$ is topologically uncharged, meaning that there are no defects present and the nematic toroidal structure is smooth as observed in the experiments. In a nematic handlebody with genus $g = 2$, which is made by merging together to nematic tori, the total hedgehog topological charge must be $m = \pm 1$. With increasing genus, nematic handlebodies should host an increasing number of topological defects. Here the sign of the topological charge depends on the choice of vector field direction.

The actual appearance of topological defects inside nematic handlebodies with homeotropic surface anchoring strongly depends on the energy of the director field embedding topological defects, more precisely on the elastic deformation, elastic constants, surface anchoring and the energy cost of melting the nematic order in the singular areas of defect cores.

For simple nematic toroids with $g = 1$, three different configuration of director field are predicted numerically as shown in Fig. 7.16: (i) a single disclination loop with winding number $s = 1$ (Fig. 7.16a), (ii) two disclination loops with winding numbers $s = 1/2$ (Fig. 7.16b) or (iii) non-singular, solitonic "escaped" director configuration containing no singular defects (Fig. 7.16e). The stability of each of these three structures was analysed numerically and it was found that they depend on

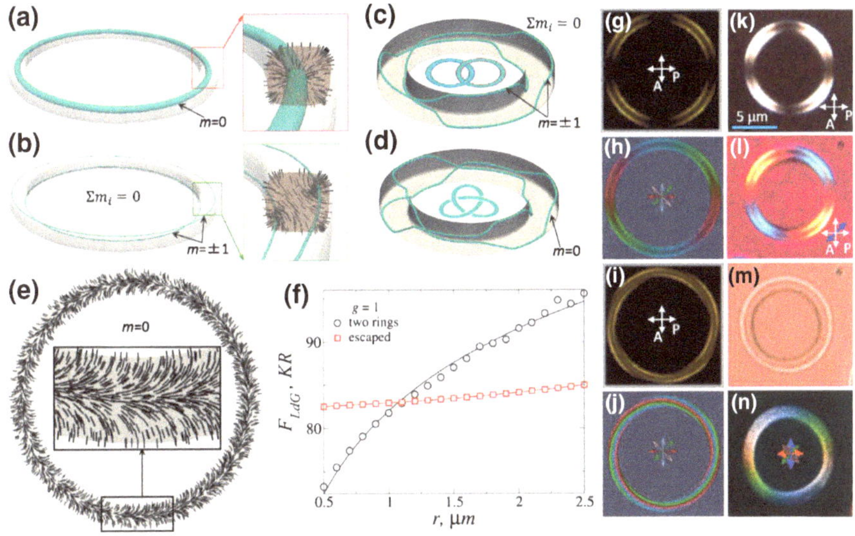

Fig. 7.16 **a** Numerical simulation of a single $s = 1$ disclination ring in a nematic toroid with $g = 1$. The ring is visualised by plotting the isosurfaces of reduced order parameter $S = 0.2$. **b** The second solution shows two $s = 1/2$ disclination rings, and is shown with order parameter isosurfaces of $S = 0.25$ shown in *blue*; $R = 2.5\,\mu$m, $r = 0.1R$, $a = 10$. The insets show details of the director field $\mathbf{n}(\mathbf{r})$ around the defects. **c** Numerical solution showing the Hopf link. **d** Trefoil torus knot of half-integer disclination loops with order parameter isosurfaces of $S = 0.3$, $R = 1\,\mu$m, $r = 0.2\,\mu$m, $\alpha = 6$. **e** Escaped director field $\mathbf{n}(\mathbf{r})$ in a plane of a torus for $R = 10\,\mu$m, $r = 0.1R$, $\alpha = 10$. **f** Dependence of the excess Landau de Gennes free energy (compared to uniform nematic), as a function of r at $R = 10r$ for the two rings and escaped structures. **g–j** Computer simulated images of nematic tori. **g** Crossed polarizers image of torus with two defect rings. **h** 3PEF-PM image of the torus with two defect rings. **i** Crossed polarizers image of a torus with escaped director $\mathbf{n}(\mathbf{r})$. **j** 3PEF-PM image of a torus with escaped director. **k** Optical micrograph of nematic torus between crossed polarizers. **l** The same torus as in **j** but with a *red* plate added. **m** Bright field image of the same torus as in **k** and **l**. **n** Experimental 3PEF-PM image, which is obtained by overlaying images with different polarizations of light, set to 0° (*red*), 45° (*green*), 90° (*blue*), and 135° (*pink*) and corresponding to **h**. Reprinted with PNAS permission from "Splitting, linking, knotting, and solitonic escape of topological defects in nematic drops with handles" by M. Tasinkevych, M.G. Campbell, I.I. Smalyukh, Proc. Natl. Acad. Sci. U. S. A. **111**(46),16268 (2014)

the torus major and minor radii, R and r, respectively. The shape of the cross-section also strongly influences the stability of each of these structures. The structures with a single $s = 1$ loop or two $s = 1/2$ loops are stable at small to intermediate values of the minor radius r. Both the single $s = 1$ and the pair of $s = 1/2$ disclination loops carry topological charges $m = 0$. Whereas these loops are unstable in a bulk nematic and annihilate, they are energetically stabilised by the surface confinement and perpendicular boundary conditions in nematic toroidal drops. The ground state with two equal $s = 1/2$ disclinations, shown in Fig. 7.16b exhibits repulsion between two equally topologically charged loops.

In addition to these three rather simple topological structures, numerical studies predict the existence of links and knots of s = 1/2 defect loops. Figure 7.16c shows a possible realisation of a Hopf link of two $s = 1/2$ defect loops, linked in the nematic torus. Figure 7.16d shows another topologically interesting structure, which is a trefoil knot of a single $s = 1/2$ defect loop, knotted inside the nematic toroid. Comparisons of theoretical and experimental micrographs of nematic tori with homeotropic surface anchoring are shown in Figs. 7.16g, n. The first set of panels (g, j) is generated from numerical simulations. Panels (g) and (h) show numerically calculated optical microscopy texture with two defect rings between crossed polarisers (g) and three photon fluorescence intensity (h). Panels (i) and (j) show numerically calculated microscopy textures of escaped nematic structure between crossed-polarisers (i) and three photon fluorescence intensity (j). The experimental obtained optical micrograph are shown in Fig. 7.16k, n. Panel k is a crossed-polarised micrograph and panel (l) is obtained with added red plate. Panel (m) is a micrograph in unpolarised light and panel (n) shows the three photon fluorescence intensity. The comparison of the image in panel (k) shows a close similarity to the theoretical simulation in panel (g), indicating the presence of two disclination lines, rather than a single escaped non-singular line, which gives quite different cross-polarised micrographs in panel (i).

Quite interesting structures were numerically predicted and experimentally observed in μm-sized nematic handlebodies with homeotropic surface anchoring. In contrast to $g = 1$ nematic toroids, the Poincare–Hopf index theorem requires existence of defects of non-zero net charge. In numerical separation, this is realised in the presence of multiple $s = 1/2$ defect loops, where the number of loops depends on genus g. in $g = 2$ nematic handlebodies, three half integer disclination loops are predicted running along the whole handlebody perimeter, whereas the others encircle the two holes of the handlebody. The existence of multiple $s = 1/2$ loops in nematic handlebodies is somehow similar to the predicted existence of knotted and linked loops in chiral nematic microdroplets. Here, the emergence of closed loops can be attributed to the smallness of the structures, which induce strong elastic deformation via strong surface anchoring. This imposed stress is then released by locally melting the liquid crystal and forming disordered liquid isotropic cores of extended defect loops, rather than concentrating the nematic disorder in hedgehog point defects. However the inherent problem with the smallness of these structures is the difficulty in the precise and reliable imaging and reconstruction of the director field in real nematic handlebodies.

7.6 Nematic Shells

The idea of producing nematic shells was first proposed by D.R. Nelson [274] as a way to produce colloidal particles with spatially oriented preferred bonding sites, which could be interesting for self-assembly of colloidal crystals with interesting photonic properties. It is discussed in Chap. 9: Photonic properties of nematic dispersions, that building photonic crystals out of colloidal crystals attracted enormous scientific interest for possible application in controlling light. Water-based colloids interact via Coulomb and van der Waals interactions, which are isotropic and result in a limited number of structurally different colloidal crystals. Directionality of colloidal interaction can be induced by grafting surfaces of colloidal particles with localised linkers, such as predetermined number of chemical groups, DNA strands, or kinesin molecules. However, it is difficult with this method to control the number and position of binding sites and the method is highly impractical. Controlled fabrication of colloidal particles which would act as tetravalent colloids is of particular interest. It was predicted (Ho et al. [275]) that assembly of tetravalent colloids into a diamond lattice structure and appropriate dielectric contrast would result in a photonic crystal with a very large photonic band gap. From a more general perspective, colloidal particles with a predictable 1-, 2-, 3-, or 4-fold valence would also allow creation of functionalized microobjects, thus mimicking molecular characteristics of organic chemistry.

Nelson proposed to cover a spherical colloidal particle or droplet with a thin layer of a nematic liquid crystal. This nematic phase could be formed of very different objects, such as orientational ordered triblock copolymers, gemini lipids, metallic or semiconducting nanorods or conventional thermotropic nematic liquid crystals. By covering a spherical particle with a thin orientationally ordered liquid, topological defects must be created in such a nematic shell, when the liquid crystal is forced to align parallel to the spherical surface. The formation of topological defects is inevitable, as it is not possible to cover the spherical surface smoothly, without having regions of vanishing order. The nematic shell, covering the spherical particle therefore creates a kind of "topological shell" with defects as topological inclusions. If such a shell with defects could be decorated with linker molecules concentrated in the regions of defects, one would obtain nanoscale binding sites, which would introduce a valance ability to micro-particles. The pair interaction of such valance colloids could then result in the hybridisation of these topological bonds between colloidal particles. If one is able to create a tetrahedral arrangements of topological defects in a nematic shell, such a colloidal particle would exhibit a 4-fold valence, similar to the sp_3 hybridised chemical bond associated with e.g. carbon, silicon and germanium.

The idea of these valanced colloidal particles is related to the paper by Lubensky and Prost, who have studied the ground states of more general order parameters with a p-fold symmetry on a sphere [276]. They pointed out that the ground state of a 2D nematic texture of a sphere consists of four $s = 1/2$ disclination at the vertices of the tetrahedral. This can be explained by the Poincare and Hopf theorem, which establishes that for nematic fields the total topological charge on a spherical surface must be 2. As result of the emergence of four defects each with topological charge $s = 1/2$, the total topological charge on a shell add up to $4 \times 1/2 = 2$. Nelson analysed the constellation of topological defects in a nematic shell by analysing the elastic free energy and found the repulsive interaction between the like-charged topological defects, which pushes defects as far apart from one another as possible. Besides the possibility of having four topological defects of strength 1/2, other

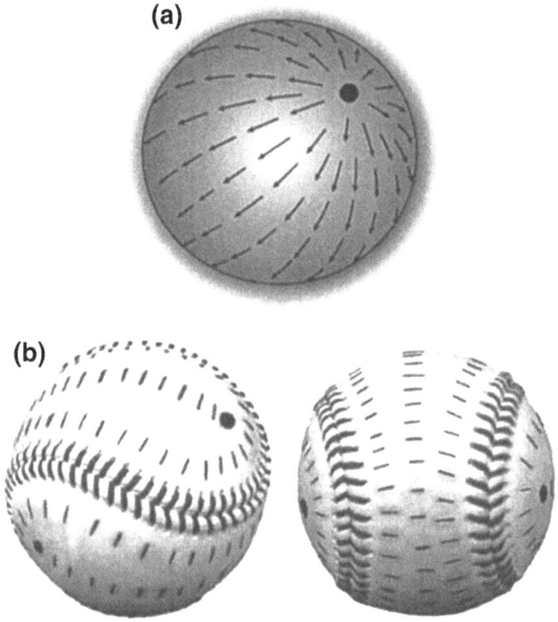

Fig. 7.17 **a** Vectorial field covering the surface of a sphere shows splayed configuration with two +1 vortices at each pole. **b** Splay configuration of a headless director field covering the sphere shows four disclinations, which are located at the vertices of a tetrahedron. Reprinted with permission from the article "Toward a Tetravalent Chemistry of Colloids", David R. Nelson, Nano Lett. **2**, 1125 (2002). Copyright (2002) American Chemical Society

combinations of defects with different strengths are also possible. The only requirement is that their sum topological charge is equal to 2. So in principle it is possible to have two topological defects with strength 1, as shown in Fig. 7.17a. This state is the state of pure splay for a vectorial order parameter on a sphere. However, for a headless vector on a sphere, this bipolar configuration has to be transformed for topological reasons to the tetrahedron configuration of four defects with 1/2 strength, shown in Fig. 7.17b.

The texture in Fig. 7.17b was predicted to be stable when the splay elastic constant K_1 is lower than the bend elastic constant K_3. In the opposite case, when $K_1 > K_3$, the bend texture of the headless nematic director is energetically preferred in the nematic shell, as shown in Fig. 7.18.

Topological defects in thin nematic shells were studied by Lopez-Leon et al. [277–280]. The spherical nematic shells were fabricated by producing double emulsions with a microcapillary device. Here we have two immiscible fluids such as water and the nematic liquid crystal, and the design of the microfluidic device enables to coat a water droplet with a thin nematic shell, immersed in a water environment. By using this method it is possible to fabricate reasonably small nematic shells, typically with the outer diameter in the range of 30–500 μm with a shell thickness of 1 μm.

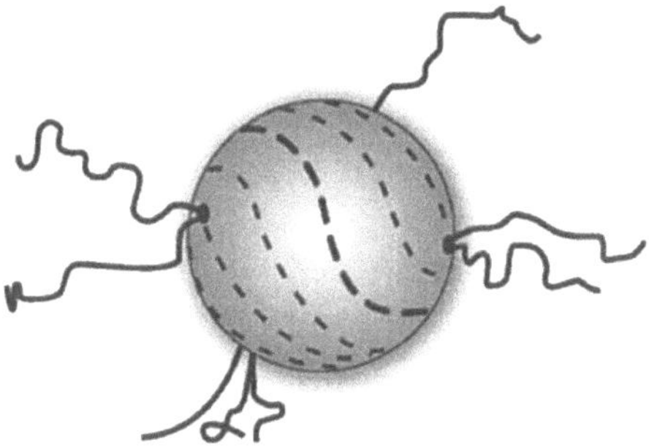

Fig. 7.18 When $K_1 > K_3$, the director field shows a bend-like structure on a sphere. For $K_1 < K_3$ the lower energy configuration is the splay texture, shown in Fig. 7.17b. Polymer linkers are emerging from the four disclinations. Reprinted with permission from the article "Toward a Tetravalent Chemistry of Colloids", David R. Nelson, Nano Lett. **2**, 1125 (2002). Copyright (2002) American Chemical Society

This thickness could be further lowered by increasing the volume of the inner drop once the double emulsion is formed. This is achieved by adding salt (up to 40 %) to the inner water droplet, which generates the osmotic pressure between the salty core and the external water. Water starts diffusing from outside through the partially permeable liquid crystal into the salty water core, which increases the volume of the inner droplet and therefore decreases the thickness of the nematic shell, which has to cover larger and larger water core. The nematic shells are imaged using optical microscopy and changing the focal plane of the microscope to precisely locate the position of topological defects. Examples of microscope images of nematic cells are shown in Fig. 7.19.

Three different constellations of topological defects were observed in thin nematic shells with a diameter of $R \approx 50\,\mu\text{m}$ and thickness of $h \approx 1\,\mu\text{m}$. Figure 7.19b, c shows the tetrahedron constellation of 1/2 topological defects in a thin nematic shell. Their positions were determined by focusing the microscope to different depths, therefore bringing into focus each of the four topological defects (Fig. 7.19b, c). The polarisation image in Fig. 7.19c clearly show that these are 1/2 defects because they display a single dark brush, accompanied by a bright brush. This means that on a circular path surrounding such defect, the liquid crystal molecules are rotated by π. The angular positions of defects were determined by measuring positions of all defects in a large number of shells with similar inner and outer diameters. The resulting distributions are Gaussian, as shown in Fig. 7.19e, f and the angles where these distributions are centred clearly show that defects are located on the vertices of a tetrahedron.

Besides this predicted tetrahedral structure, which is expected to be the ground state for nematic order on the surface of a sphere, two other constellations were observed, which involve higher charged defects of strength $s = 1$. The first structure is a constellation of three defects and in shown in Fig. 7.19g, h. Two of these defects have the winding number 1/2 and the third defect has a higher winding number of

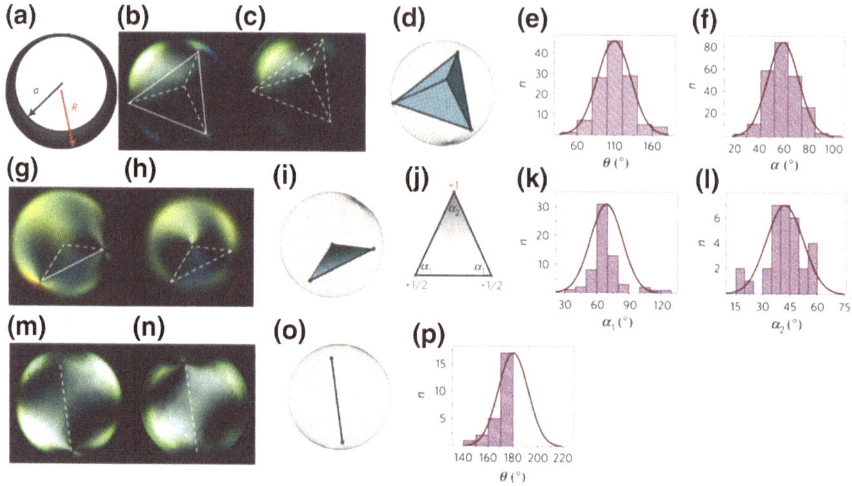

Fig. 7.19 Defects in *thin* nematic shells. **a** Schematic drawing of the shell geometry, showing displacement of the centers due to gravitation. **b**, **c** Four $s = 1/2$ defects are visible between crossed polarizers and the images are taken at different positions of the focal plane. **d** Presentation of defect constellation on a shell. **e** Histogram of the central angle between defects θ. **f** Histogram of the surface angle between defects, α. **g**, **h** Three defects are visible between crossed polarizers, the images are taken at different positions of the imaging plane. **i** Constellation of three defects in a nematic shell and **j** corresponding geometry. **k** Histogram of the surface angle α_1. **l** Histogram of the surface angle α_2. **m**, **n** Constellation with two defects, as seen between crossed polarizers at different imaging planes. **o** Schematic drawing of two defect constellation shown in **m**, **n**. **p** Histogram of the angle θ for two defect constellation. Reprinted by permission from Macmillan Publishers Ltd: T. Lopez-Leon, V. Koning, K.B.S. Devaiah, V. Vitelli and A. Fernandez-Nieves Nat. Physics **7**, 391 (2011), copyright (2011)

$s = 1$. These add up to the total topological charge of 2, which is consistent with the requirement of the Poincare–Hopf theorem. These three defects form an isosceles triangle, where the unequal angle is due to the single $s = 1$ defect, shown in Fig. 7.19j. The third constellation of defects in a thin nematic shell only has two $s = 1$ defects, as shown in Fig. 7.19m, n. They are located diametrically, as shown in Fig. 7.19o, which is confirmed by the measured angular distributions, having peak at 180°. The spatial constellation of defects is determined only by the elasticity of the nematic liquid crystal, if the thickness of the shell is uniform. However, in these water-liquid crystal-water emulsions, the difference between the densities of water and liquid crystal cause buoyancy forces, which displace the inner water droplet upwards. As a result, the nematic shell is always thinner on the upper part and thicker on the lower part, as illustrated in Fig. 7.19a. This buoyancy force allows for controlled tuning of the shell-thickness inhomogeneity, which is achieved using osmosis. Figure 7.20 presents the evolution of the defect constellation for different thickness of the nematic shell and three different types of the structures.

By increasing the nematic shell inhomogeneity, a continuous change in the defect arrangement was observed and the defects were always driven to the thinnest part of

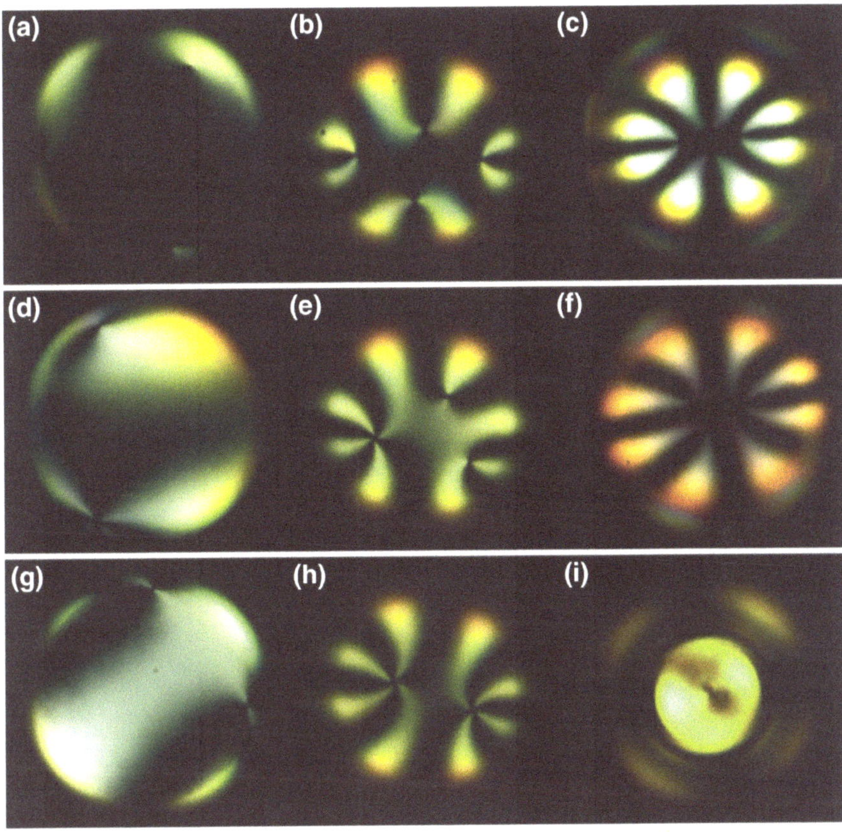

Fig. 7.20 Constellations of defects for different thickness inhomogeneity of a nematic shell. **a–c** Constellations of four defects between crossed polarizers at **a** $u/u_\circ = 1.6$, **b** $u/u_\circ = 7.8$ and **c** $u/u_\circ = 23.9$. **d–f** Nematic shells with three defects: **d** $u/u_\circ = 1.2$, **e** $u/u_\circ = 4.6$ and **f** $u/u_\circ = 6.8$. **g–i** Constellations of two defects in a nematic shell, **g** $u/u_\circ = 7.1$, **h** $u/u_\circ = 13.3$ and **i** $u/u_\circ = 53.7$. Reprinted by permission from Macmillan Publishers Ltd: T. Lopez-Leon, V. Koning, K.B.S. Devaiah, V. Vitelli and A. Fernandez-Nieves, Nat. Physics **7**, 391 (2011), copyright (2011)

the shell with increasing thickness inhomogeneity. Figure 7.20a, c shows the change in defect constellation for 4 $s = 1/2$ defects. Here the parameter u measures the difference in normal thickness of the nematic shell. Figure 7.20d, f shows the changing constellation of three topological defects, which are nicely presented in the intermediate inhomogeneity in Fig. 7.20e. In this panel, there are two $s = 1/2$ defects on the righthand side of the image and one $s = 1$ defect on the left. They are clearly differentiated by a number of dark brushes. Figure 7.20g–i present the change of constellation of two $s = 1$ defects. It should be noted that in all this constellations $s = 1/2$ defects are actually 1/2 disclination lines running from the outer to the inner surface. On the other hand, the $s = 1$ defect is a surface boojum, residing at the water-liquid crystal interface.

Theoretical studies of defects in nematic shells have shown that such a nematic coating could provide control of the valency and direction of the bonding sites using temperature, shape, elastic anisotropy or external fields. An interesting study was performed by Gharbi et al. [281] by incorporating small colloidal particles into the nematic shells. These particles are treated to produce strong homeotropic anchoring and when dispersed in a planar nematic shell, there are two force mechanisms which control the position of such an inclusion in the shell. First, the capillarity and the contact angle of the nematic liquid crystal on a colloidal inclusion with the diameter larger than the thickness of the shell. Capillary forces due to interfacial tension are much stronger than the elastic forces, produced by director distortion in the vicinity of the cell. The second force mechanism is the elastic distortion around a colloidal inclusion in the nematic shell.

The experiments were performed by producing nematic shell from water-LC-water double emulsions generated by microfluidic device. The addition of polyvinyl alcohol (PVA) ensured planar condition at the water-nematic liquid crystal interface. On the other hand, silica colloidal particles of $2\,\mu$m diameter were treated with DMOAP to produce strong homeotropic anchoring. When single bead of diameter $2\,\mu$m was inserted into a thin nematic shell of typical thickness of $23\,\mu$m, one could observe three topologically different nematic liquid crystal textures around the inserted microsphere, shown in Fig. 7.21.

In the first situation, shown in Fig. 7.21a, b, the colloidal inclusion induces uniform radial distortion and the microsphere acts as a pure $+1$ topological defect, surrounded by a 2D vector field. Under crossed polarisers, one can see four extinction brushes (Fig. 7.21a) and no accompanying topological defect. This means that the nematic director is radial, as shown in the schematic drawing accompanying Fig. 7.21b. In another situation, shown in Fig. 7.21c, d, the microsphere is accompanied by a topological defect of charge 1/2, whereas the sphere itself acts as a topological defect of charge $+1$. The 1/2 defect is clearly visible as a dark spot in unpolarised image in Fig. 7.21d, whereas the polarising image in Fig. 7.21c clearly supports the schematic drawing accompanying panel (d). The third situation in shown in Fig. 7.21e, f and the topological defect is of 1 charge, accompanying the microsphere with $+1$ charge. In these cases, the microsphere and the accompanying defect are thus topologically equivalent to a single structure with a net topological charge of $+1$, $+1/2$ and 0. Although this seems odd at the first sight, it does not violate the law of the net topological charge of a 2D shell, which should add up to $+2$. However, the conservation of the total topological charge does not prescribe or impose restrictions on the way this charge is obtained. In a very thin nematic shell, which presents a basically 2D nematic geometry, energetically stable defects have a charge of $\pm 1/2$. A homeotropic microsphere would repel $+1/2$ defect, but attract 1/2 defect. Further on, combinations with either one or two 1/2 defects are possible, which explains the topological characteristics of observed structures.

When the thickness of the shell is varying, the effects of the thickness non-uniformity of the shell become pronounced. In this case, the particles are usually located in thicker part of the shell, whereas defects are forced to be located in the thinner part of the shell, as shown in Fig. 7.22. The exact positioning of the spheres

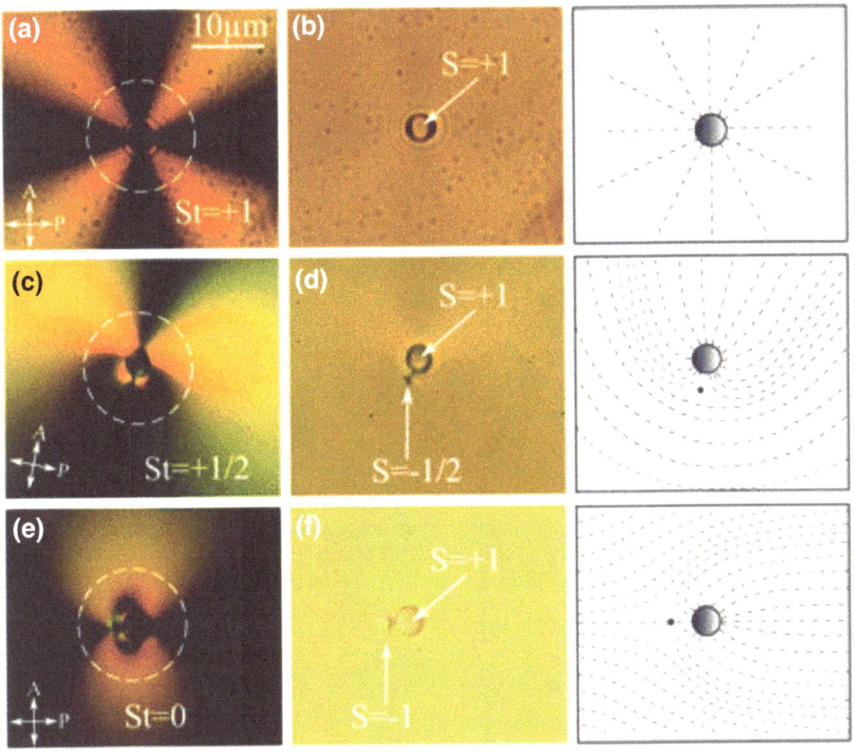

Fig. 7.21 Small colloidal particle with perpendicular surface anchoring of liquid crystal molecules in a $2 \sim 3\,\mu$m thin nematic shell. Three different director configurations are observed. **a** Crossed polarizers image of a bead with no visible defects. **b** Bright field image together with a schematic drawing of the director around the bead. **c** Cross polarizers image of another bead showing a defect where two dark brushes meet, therefore indicating a $-1/2$ defect. **d** Bright field image of the bead in **c**, together with schematics of the director. **e** Cross polarizers image of the third possible configuration: a bead is accompanied by a -1 defect due to the four dark brushes around the point defect. **f** Bright field image of the bead shown in **e** together with director schematics. Reproduced from M.A. Gharbi, D. Seč, T. Lopez-Leon, M. Nobili, M. Ravnik, S. Žumer, C. Blanc, Soft Matter **9**(29), 6911 (2013) with permission of The Royal Society of Chemistry

and defects in walls balance between the elastic and capillary forces and makes the analysis quite complicated. Beads and their associated defects appear at the thicker hemisphere of the shell, and this is due to capillary forces. They are trapped by the two interfaces delimiting the shell and they migrate towards the position where they fulfill the requirements of the Young equation. The rest of the defects organise themselves in the lower hemisphere of the shell where the nematic layer is thinner. The number of possible configurations in microspheres increases drastically with their number.

A lot of effort was devoted to understanding arrangements of topological defects in nematic shells, because this is the key advantage of shells, if they are to be used as

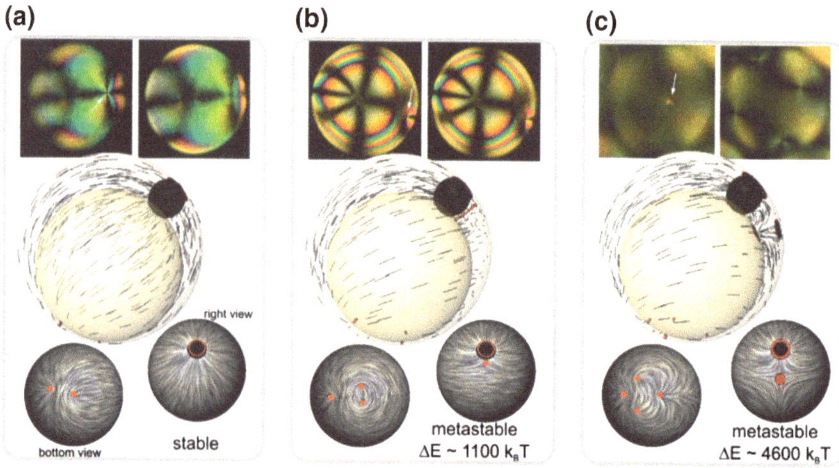

Fig. 7.22 Experimentally observed free $+1/2$ topological defect for a single bead in a nematic shell. The images in the *top line* are taken between crossed polarizers. The *white arrows* in *left panels* indicate positions of the beads. Defect is usually located in the thinnest part of the shell. The particle and disclinations bear an effective winding number **a** $s = +1$, **b** $s = +1/2$ and **c** $s = 0$. The simulated structures are shown *bellow* the experimental images, *black* spheres are the particles, *red dots* indicate defects, i.e. iso-surfaces of the nematic degree of order $S = 0.48$. Reproduced from M.A. Gharbi, D. Seč, T. Lopez-Leon, M. Nobili, M. Ravnik, S. Žumer, C. Blanc, Soft Matter **9**(29), 6911 (2013) with permission of The Royal Society of Chemistry

artificial "atoms" with their covalent bonds coinciding with the position of topological defects on a shell. It is clear that the main problem is the inhomogeneous thickness of the nematic shell, which is the consequence of density mismatch between the shell and the interior core. This causes gravitational effects and buoyancy, which inevitably results in different thicknesses of the upper and the lower part of the shell. In general, the repulsion between the pairs of boojums competes with the minimisation of the distance between the defects within a pair, which attract each other. If the inhomogeneity of the thickness is very large (Koning et al. [280]), the defects undergo a confinement transition to the thinnest part of the shell. There, they make a kind of de-confinement transition which maximises their mutual separation. Various experimental geometries were studied in nematic shells, such as the combination of the planar and perpendicular alignment on the inner and outer interface, behaviour at the nematic-isotropic phase transition and the shell structure at the nematic-smectic A transition (Liang et al. [282]). In conclusion, the inhomogeneous thickness of nematic shells due to density mismatch and buoyancy seems to be the major obstacle, which might prevent using the nematic shells as artificial atoms with spatially directed bonding sites. The other important problem is the minimum size attainable with a microfluidic technique, which is limited to several tens of μm diameters, whereas for microoptical applications, much smaller shells would be a great advantage.

Whereas simple nematic shells are quite easily produced in a microfluidic device, nematic shell handlebodies are much more difficult to produce in real experiments.

So far, there are theoretical predictions (Jesenek et al. [283]) on defect formation in toroid nematic shells. The existence of topological defects on handlebody nematic shells is governed by topological reasons and determined by the Euler-Poincare surface topological characteristic $\chi = 2(1g)$, where g is the genus of the surface, which is equal to the number of handles of the particular surface. For spherical topology, $g = 0$, whereas for toroid topology, $g = 1$, and the number of topological defects equals to 0. This means that from the topological point of view, there should be no topological defects stabilised on a thin toroid nematic shell. However, it is possible to keep the total topological charge to zero and still have pairs of opposite topological charges, which could be stabilised by some mechanism on the toroidal nematic shell. Here, the geometry of the shell supporting the metastable states of oppositely charged topological defect could dictate the number and stability of these pairs. A similar mechanism of geometrical stabilisation of mutually compensating topological charges was observed for Koch-star colloidal particles, discussed in Sect. 6.5. Creation and stabilization of a multitude of charged neutral pairs of topological defects on curved shells was studied by Mesarec et al. [284].

Numerical studies have indeed shown that the Gaussian curvature K of a surface, which is hosting these pairs of topological defects, could indeed stabilise topological states on toroid nematic shells, where several mutually compensating topological defects could coexist. It turns out that topological defects with positive winding numbers are attracted to regions where K attains its maximum value whereas low local Gaussian curvature of the shell attracts topological defects with negative winding number (Jesenek et al. [283]). Furthermore, regions which exhibit both positive and negative Gaussian curvature could induce unbinding of oppositely charged topological defect, therefore acting as a kind of dissociation site. Interestingly, local Gaussian curvature plays the role of an external field, whereas the topological charge od the defect attains the role of the electric charge. Gaussian curvature of handlebody nematic shells could have an interesting role in Kibble–Zurek mechanism production of topological charge. A comprehensive recent review of defect structures in nematic liquid crystal shells of different shapes was written by Mirantsev et al. [285]. Curvature of the nematic field and topological defects are discussed within the frame of structural forces and self assembly by F. Serra [286].

Chapter 8
Topological Particle-Like Structures in Chiral Nematics

Abstract This chapter describes some topologically interesting structures that can be observed in chiral nematics, such as skyrmions and torons in 2D chiral nematics. They can be considered as a kind of topological entities, forming more complex structures.

8.1 Strange Imperfections and Self-formed Structures in Chiral Nematic Liquid Crystals

In previous sections we have discussed dispersions of solid-like or liquid-like objects in nematic liquid crystals. In both cases, colloidal particles and the medium are distinct from each other and separated by the interface, either because of solid structure of colloidal particles or chemical immiscibility of one liquid in another. These particles (solid or liquid) may be of different shapes of topologies, and they induce a large variety of topological defects due to the surface anchoring of liquid-crystal ordering field.

In contrast to that, one can find in the literature on liquid crystals descriptions of interesting phenomena in a form of topological objects that are self-formed in a liquid crystal by a liquid crystal. They could be considered as a kind of "particles" that mutually interact and form more complex structures. One of the earliest examples is the spherulitic liquid crystal, first described in two different publications in 1994. The two publication appeared nearly simultaneously, one reporting on the "bubble domain texture" (Kawachi et al. [286]) and the other on the exactly the same phenomenon, this time named "spherulitic liquid crystals" (Haas and Adams [287]). In addition, a number of topological studies of chiral nematic liquid crystals were reported by Yves Bouligand in Journal de Physique in the 1973–1984 period [211, 217–219, 221, 288] and Pirkl et al. [289]. At that time, methods of observations of different textures in liquid crystals were limited to simple polarising microscope observations, as there were no laser tweezers or confocal fluorescent microscopes available yet. It has taken nearly 30 years to prove experimentally the internal structure of these particle-like objects, spontaneously formed in chiral nematic liquid crystals, which caused a revival of the experimental and theoretical topology of liquid crystals after the year 2010. This section will review various types of

I. Muševič, *Liquid Crystal Colloids*, Soft and Biological Matter,
DOI 10.1007/978-3-319-54916-3_8

topologically interesting particle-like structures in chiral nematic liquid crystals, such as the spontaneous formation of skyrmions and Torons in thin layers of chiral nematic liquid crystal with strong perpendicular anchoring at the confining surfaces.

8.2 Skyrmions in 2D Chiral Nematics

Skyrmions are inhomogeneous localised structures, which were first introduced as a new topological concept in 1960s by the British physicist Tony Hilton Royle Skyrme [290, 291]. Mathematically, a skyrmion is defined as a topological spatial feature of a particular physical field, which is preserved (self protected) under arbitrary continuous deformation. This mathematical concept attracted the attention of the physics community in 1962, when Skyrme found that it could explain how subatomic particles, such as neutrons and protons, could exist as discrete objects, shaped in a continuous nuclear field. These particles could be considered as stable geometric twists in an otherwise homogeneous background, resembling whirlpools in water. The idea went through ups and downs in nuclear theory, being obscured first by the idea of quarks and later by the string theory.

Quite unexpectedly, skyrmions were first observed as real physical objects in quantum Hall devices exhibiting unusual quantum effects in external magnetic fields, such as ultra-precise jumps in the electronic current. It was found that these subtle quantum effects could best be described in terms of topological features. This was a natural habitat for skyrmions, which were first predicted and then electronically detected in quantum Hall devices in the mid 1990s.

Physical realisation of skyrmions in condensed matter was described within a classical model of an isotropic ferromagnet by Belavin and Polyakov in 1975 [292]. They demonstrated that skyrmion localised states belong to metastable states of the ferromagnet and presented the analytical solutions for describing them. Skyrmions were then observed in magnetic systems by electron microscopy as swirling magnetisation in magnetic materials.

There are two basic types of magnetic skyrmions, which are shown in 2D images in Fig. 8.1. The arrows show the spin magnetisation in 2D, which forms two different axially symmetric structures. The first is the Bloch-type skyrmion, where the magnetisation twists by 180° as we move from the homogeneous far-field towards the centre of the skyrmion. It is therefore a helical-like object with no singularity and smoothly embedded into the uniform far-field background.

The second type of the skyrmion is shown in Fig. 8.1b and is called Neel-type skyrmion, with a cycloidal-like rotation of the spins from the periphery to the centre. The Neel-type skyrmions are expected to emerge in polar magnets and were recently observed in a magnetic semiconductor GaV_4S_8 [293].

Theory of skyrmions in liquid crystals was first developed by Bogdanov et al. [294] and later by Leonov et al. [295]. Analytical solutions for static 2D axially symmetrical localised states were found by minimising the free energy for the nematic liquid crystal. These objects have higher energy than the homogeneous state, hence, they can represent only local minima of the energy. They were found to be metastable

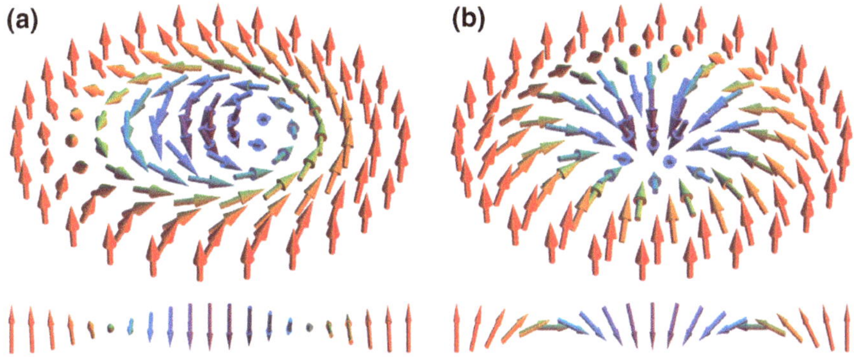

Fig. 8.1 Bloch- and Néel-type 2D skyrmions. **a** In a Bloch-type skyrmion, the spins rotate in the tangential planes when moving from the periphery to the core. This is therefore a twisted and smooth structure in 2D. **b** In a Néel-type skyrmion, the spins rotate in a radial direction from periphery to the core. This is a cycloid-like smooth structure, embedded in a uniform far-field background. Reprinted by permission from Macmillan Publishers Ltd: I. Kézsmárki, S. Bordács, P. Milde, E. Neuber, L. Eng, J. White, H.M. Ronnow, C. Dewhurst, M. Mochizuki, K. Yanai, H. Nakamura, D. Ehlers, V. Tsurkan, A. Loidl, Nat. Mater. **14**(11), 1116 (2015), copyright (2015)

and could be annihilated by the influence of applied internal electric or magnetic fields. The structure and the stability of these nematic skyrmions crucially depends on the values of the nematic elastic constant. While these idealised 2D skyrmions in nematic liquid crystal are interesting, in real materials, one has to consider the 3D skyrmion-like objects. It is also natural to consider that the skyrmion structures should be stabilised by the intrinsic chirality of liquid crystals, as the skyrmion structure implies spontaneous twist. Fukuda and Zumer predicted the existence of several types of periodic, 2D skyrmion lattices in very thin films of a blue phase liquid crystals [296]. These materials are highly chiral and the tendency to spontaneously twist is expected to stabilize the 2D skyrmion lattices.

The experimental evidence of the existence of 2D skyrmion structures, also called "baby-skyrmions" [297, 298], was first reported by Ackerman et al. [299]. They have studied laser-induced realignment of a cholesteric liquid crystal confined into a thin layer with homeotropic surface anchoring and they observed a variety of smooth (solitonic) and singular localised structures, which were formed after the localised treatment with laser tweezers. These structures were imaged in 3D using FCPM [300] and 3PEF-PM and were correspondingly classified based on their skyrmion number and singular defects accompanying various twist configurations.

One of the examples of smooth, solitonic, baby-skyrmion structures is shown in Fig. 8.2. It is a localised object, smoothly embedded in the far-field, homeotropically aligned, and unwound chiral nematic liquid crystal. It is axially symmetric, as can be seen from the FCPM vertical cross-sections in Fig. 8.2b–d. Here, the z-axis is directed perpendicularly to the confining surfaces. Figure 8.2a shows the in-plane fluorescence cross-section with an excitation polarisation directed along the y-axis. This image clearly shows the twist of the director by π as we move from the periphery of the

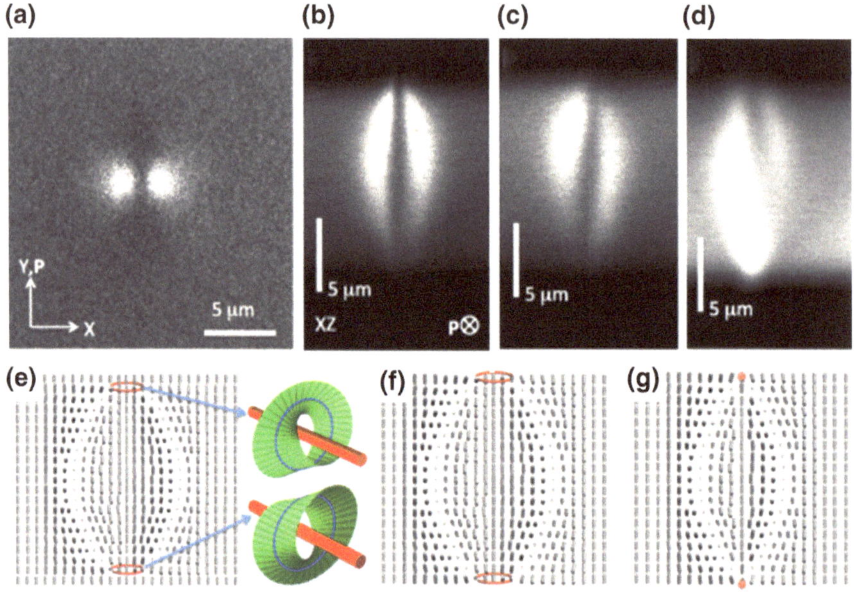

Fig. 8.2 Baby-skyrmions in a chiral nematic liquid crystal. **a** Fluorescence confocal polarized microscope image taken in the midplane of the cell. The direction of polarization is indicated with "P". **b**–**d** *x*–*z* vertical cross-sections of three different baby-skyrmion structures as imaged by FCPM with polarization indicated as "P". (e.g.) Schematic drawing of the nematic director. **e** Strong surface anchoring. **f**, **g** Weak surface anchoring. The twist-disclinations, which are forming loops, are shown in *red*. The director structure along these closed loops is forming a Möbius strip. The *red dots* in (**g**) are surface boojums. Reprinted figure with permission from P.J. Ackerman, R.P. Trivedi, B. Senyuk, J. van de Lagemaat, I.I. Smalyukh, Phys. Rev. E **90**(1), 012505 (2014). Copyright (2014) by the American Physical Society

object towards its centre. This twist is also evident from the vertical cross-sections in Fig. 8.2b, d, which is also showing the proposed structure, a baby-skyrmion presented in schematic drawings in Fig. 8.2e–g. This double twist cylinder (meaning it has a twist along the x- and the y-axis) terminates on the confining surfaces, where the director matches the vertical surface boundary conditions by small loops of singular twist disclinations.

8.3 Torons in 2D Chiral Nematics

In 1974, two independent publications reported on the observation of strange structures in chiral nematic liquid crystals confined to a thin layer with a strong homeotropic surface anchoring. The structures looked like a dispersion of colloidal particles in a uniform nematic background and were consequently named "bubble domain" or "Spherulitic liquid crystals" (Kawachi et al. [286] and Haas and Adams [287]). Both groups independently studied the behaviour of the MBBA liquid crystal doped with cholesteryl chlorite or oleyl carbonate, which induced chiral phase of MBBA with a predetermined helical pitch p. If such a chiral nematic mixture is introduced into a thin cell with homeotropic surface anchoring, one observes different structures for different thickness d of the layer. If the liquid crystal layer is much thicker than the pitch, one observes stripped domain texture, which consists of the cholesteric fingers topologically smooth structures with no singularities. However, if the ratio of the pitch to the thickness is in a narrow range, typically between $3 < p/d < 5$, the pitch is too large or the cell thickness is too small and the intermediate spherulitic or bubble-like texture is formed. In some cases this structure is induced by applying an external electric field, which induces some hydrodynamic instabilities, therefore triggering the onset of formation of bubble-like objects.

The first sketch of the possible structure of these bubbles was proposed by Kawachi et al. [286] and was followed by different studies with different suggestions of the internal structures of the bubbles. Pirkl et al. [289] proposed an alternative to the Steibs model by studying the experimental phase diagram for the appearance of these structures and proposing a topological model consisting of a double twisted toroidal structures accompanied by two hedgehog point defects located along the axes of the double twisted toroid structure. In the far field, this toroid, bubble-like structure is transformed smoothly into a uniformly aligned cholesteric liquid crystal, which is unwound by the presence of two confining walls with strong homeotropic surface anchoring. The first sketch of the topological structure, which was subsequently named "toron", is shown in Fig. 11. in Pirkl et al. [289].

Whereas the study of Pirkl et al. [289] did not give a direct evidence of the internal organisation of the bubble-like structure, the evidence was revealed by Smalyukh et al. in 2010 [185]. They were also studying strongly frustrated chiral nematic liquid crystals in a thin layer with strong homeotropic surface anchoring, where the thickness was chosen with a ratio of $0.75 < d/p < 1$. In such a strong confinement, chiral nematic liquid crystal is unwound by the strong homeotropic surface anchoring.

They were using laser tweezers with Laguerre–Gaussian beam, which was focused to the unwound chiral nematic liquid crystal. These Laguerre–Gaussian vortex beams belong to a family of beams with helical phase fronts and phase singularities. They are characterised by the topological charge l, comprising l-intertwined helical surfaces of the phase fronts. The integer charge value defines the number of twists the phase of the light makes in one wavelength and the Laguerre–Gaussian photons are carrying an orbital angular momentum of $m \cdot h$ per photon. When such a beam is focused into the bulk of an unwound chiral nematic liquid crystal and the power is increased, the electric field of light couples with the positive dielectric anisotropy of the liquid crystal and tends to rotate liquid crystal molecules perpendicular to the beam direction, i.e. along the oscillating electric field. At a lower power level, the Laguerre–Gaussian beam induces only a reversible elastic deformation, however, at a second threshold, this distortion transforms into a permanent, particle-like structure, as shown in Fig. 8.3.

The particle-like structures are of axial symmetry, and there are three different distinct structures, denoted by $T3$-1, $T3$-2 and $T3$-3, as shown in Fig. 8.3. The internal structure of these particle-like excitations can be reconstructed by using the fluorescent confocal polarising microscopy (FCPM) with circularly polarised light and

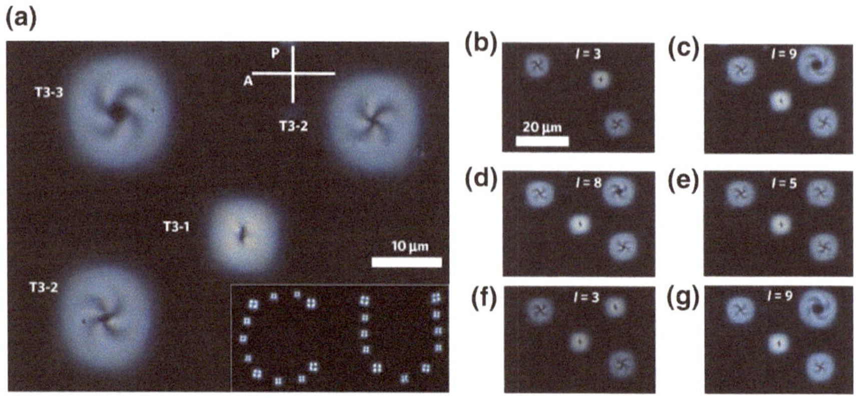

Fig. 8.3 Crossed polarizers optical micrographs of different torons, which are labelled, as explained in Fig. 8.5. **a** The smallest toron is T3-1, and is shown in the center of this micrograph. There are two torons of the type T3-2, which have the opposing winding. The largest structure is the T3-3 toron. The letters "CU" were written by optically inducing formation of torons. The orientation of the polarizer P and analyser A is also shown. **b**–**g** Time sequence of optical micrographs showing manipulation and transformation of individual torons using the infrared laser tweezers. Reprinted by permission from Macmillan Publishers Ltd: I.I. Smalyukh, Y. Lansac, N.A. Clark, R.P. Trivedi, Nat. Mater. **9**(2), 139 (2010), copyright (2010)

compared to computer simulations using Landau-de-Gennes theory. Example of 3D FCPM image of the structure, which is called a toron, is shown in Fig. 8.4a. This image was taken with circularly polarised probing light, where the electric field of the probing light is oscillating in the plane of the sample (x–y plane). Bright regions denote the areas where the director (and the radiative dipole moment of dye molecules) is also lying in the x–y plane. The structure therefore has a toroidal shape with inner dark region indicating vertical alignment of the director (i.e. along the z axis), which is also observed in the outside regions. This toroid-like liquid crystal structure is therefore smoothly embedded into the uniform and unwound chiral nematic liquid crystal. The cross-section in the equatorial plane is shown in Fig. 8.4c, clearly showing vertical alignment of liquid crystal molecules inside and outside of the toroid. The x–z vertical cross-section is shown in Fig. 8.4d and also supports toroid structure but with two singularities along the z axis of the toroid, where the direction of the director is obviously not well defined.

The director structure of the toron can be resolved with the help of numerical simulations and the results are shown in Fig. 8.4b. Red points indicate the points of escape of the director and are therefore cross-section of a red circle, positioned in the central cross-section of the toroid in a plane perpendicular to the axis of the toron. In contrast to red points, which belong to a non-singular defect, the two blue dots are point hedgehog defects of hyperbolic type. The presence of these two hedgehogs indicates that the toroid structure carries a topological charge of +2, which is attributed to the escape circle in the central region of the toron. This charge is compensated by the charge of the two hyperbolic hedgehogs, each carrying a topological charge of −1.

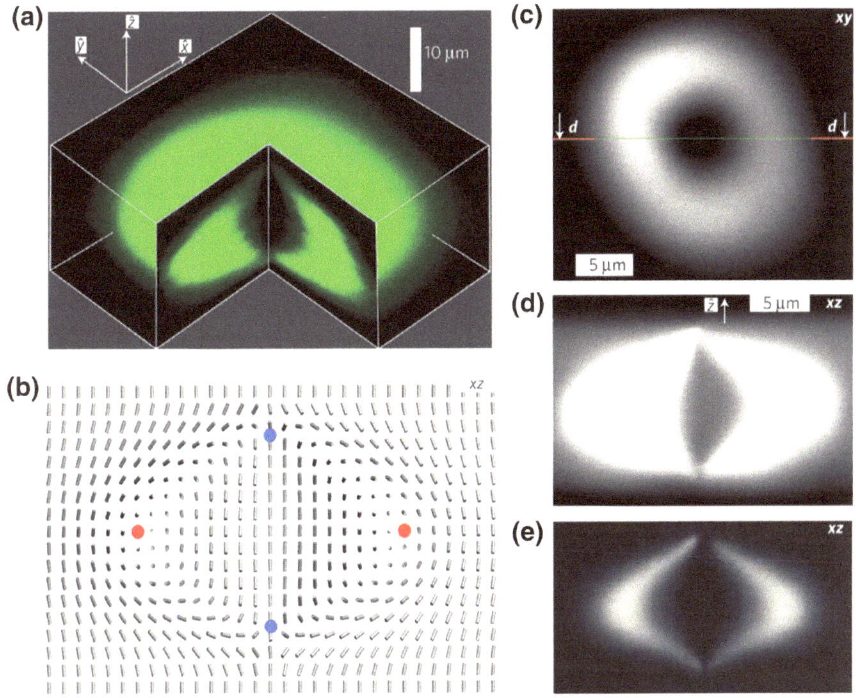

Fig. 8.4 Structure of torons, as elucidated by FCPM imaging and numerical simulations. **a** 3D representation of T3-1 toron obtained by FCPM imaging using circular polarisation of the fluorescence excitation beam. **b** Vertical cross section of the T3-1 toron obtained by numerical minimization. **c** FCPM cross section through the midplane intersecting the T3-1 toron. The *two red lines* indicate the positions of the vertical cross sections, shown in (**d**) and (**e**). **d** FCPM image of the cross-section indicated by the *green-red line* in (**c**). **e** Numerically calculated FCPM intensity cross section as shown in (**c**), using circularly polarized excitation beam and the director field shown in (**b**). The ratio of the cell thickness and the helical pitch is $d/p = 1$ in both the experiments and numerical simulations. Reprinted by permission from Macmillan Publishers Ltd: I.I. Smalyukh, Y. Lansac, N.A. Clark, R.P. Trivedi, Nat. Mater. **9**(2), 139 (2010), copyright (2010)

The general structure of a toron is shown in Fig. 8.5. There are three possible toron structures denoted with $T3$-1, $T3$-2 and $T3$-3. In all cases, the common central structure is shown in Fig. 8.5a. It is a double-twist cylinder looped on itself around the z-axis. In the mid plane of this toroid, the liquid crystal structure is skyrmion like, which can also be seen in the numerically generated director in Fig. 8.4b. When moving from left to right, the director rotates from the originally vertical direction through the horizontal direction, back to vertical direction in the centre of the toron. This is a 180° twist, which repeats on the other side of the toron, thus resulting in a total full 2π angle twist from one side of the toron to the other. The toron therefore contains a twist-escaped disclination ring, which is shown by the red line in Fig. 8.5a. This winding number +1 defect ring is equivalent to the +2 point defect, which was also discussed when analysing "bubblegum" topological defects accompanying

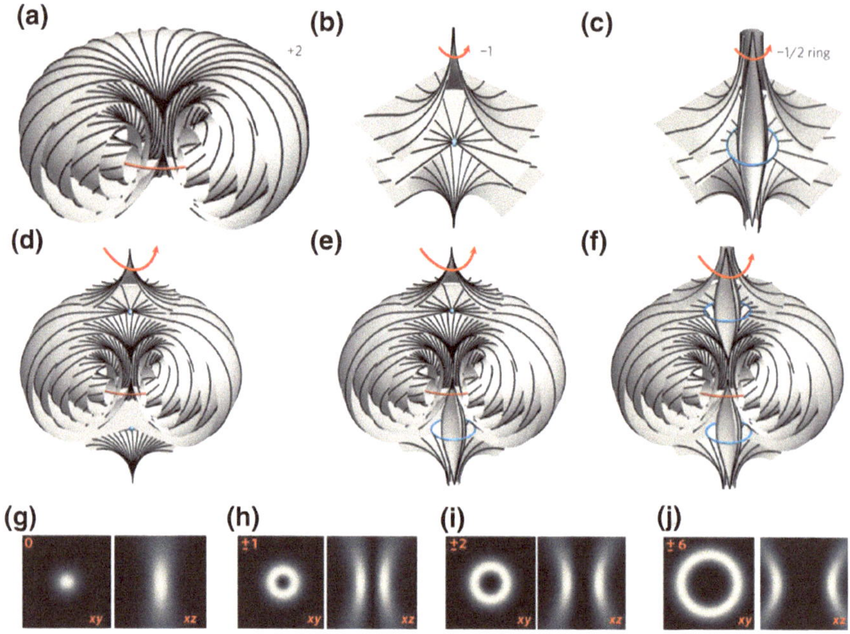

Fig. 8.5 Different types of torons can be generated by Laguerre–Gaussian beams. **a** Toron with escaped and non-singular disclination ring, similar to the bubble-gum colloidal structure with a vortex-like defect in Sect. 2.7, carrying the topological charge +2. *Red line* shows non-singular ring of strength +1. **b** Hyperbolic −1 point defect can open into a −1/2 disclination ring, shown in (**c**), therefore preserving the topological charge. **d** Toron of the type T3-1 with a constellation of two −1 hyperbolic point defects, neutralizing the +2 charge of the escape core ring of the toron. **e** Toron of the type T3-2 containing one hyperbolic point defect and a hyperbolic ring. **f** Toron of the type T3-3 with two $s = -1/2$ defect rings. **g–j** Distributions of light intensity for different Laguerre–Gaussian beams, carrying the topological charge marked with *red numbers*. The square cross-sections are 4 μm wide, different cross sections are indicated with *red letters*. Reprinted by permission from Macmillan Publishers Ltd: I.I. Smalyukh, Y. Lansac, N.A. Clark, R.P. Trivedi, Nat. Mater. **9**(2), 139 (2010), copyright (2010)

two microspheres with homeotropic surface anchoring. This double-twist cylinder therefore carries a topological charge of +2, which has to be somehow compensated by the opposite topological charges. This additional topological charges can appear in two different forms: either as a hyperbolic point defect carrying a charge of −1, shown in Fig. 8.5b or as small −1/2 ring, shown in Fig. 8.5c. These two charges appear in three different combinations with a double twist cylinder, as illustrated in Fig. 8.5d–f.

The toron T3-1 has two hyperbolic −1 point defects shown in Fig. 8.5d. The T3-3 toron is also up–down symmetric and has two −1/2 disclination rings shown by the blue lines in Fig. 8.5(f). The third type of toron, T3-2, is up–down asymmetric as it has a combination of a hyperbolic hedgehog and −1/2 disclination ring.

Chapter 9
Photonic Properties of Nematic Microdroplets

Abstract This chapter is an excursion from the nematic colloidal dispersions to the nematic emulsions, and we consider the structure and optical properties of micrometer-diameter nematic micro-droplets. The concept of liquid crystal photonics based on nematic dispersions is presented, as well as the structure and operation of tunable optical micro-cavities, omnidirectional liquid crystal micro-lasers and smectic liquid micro-fibers.

9.1 Photonics from Liquid Crystals

We have seen previously in Chap. 4 that the term "photonics" was invented by Yablonovitch and Gmitter [189] who published a theoretical paper entitled "Photonic band structure: The phase-centred-cubic case" in *Physical Review Letters* in 1989. By using the term "band gap" they establish clear physical correspondence between the well-known band-structure of electron levels in crystals and the band structure of dispersion relation for light, propagating in periodic dielectric structures. This is because the underlying differential equations are quite similar, and because the solutions of these equations must be periodic in space. Photonic properties of photonic crystals were discussed in Sect. 4.1 of this book.

Consequently, in case of the analysis of light propagation in periodic structures, the concept in the Brillioulin zone has to be introduced in the reciprocal, **k**-space. This is because of the spatial periodicity of the dielectric function of the material, which plays a similar role as the periodic crystal potential for the motion of electrons. As a direct consequence of the structural periodicity of the material, the corresponding dispersion relation, $\omega(\mathbf{k})$, which describes the relation between the frequency ω and the wave vector $\mathbf{k}$ of light modes in the material, must exhibit a band-like structure. The band structure of dispersion relation means there are some forbidden frequency gaps in this relation and the light within a certain frequency interval is not allowed to propagate inside the periodic matter and is therefore Bragg-reflected from the surface of that material. The concept of photonic bands in periodic and structured media is nowadays the pillar of emerging micro-photonic integrated circuits and architecture [301–305].

I. Muševič, *Liquid Crystal Colloids*, Soft and Biological Matter,
DOI 10.1007/978-3-319-54916-3_9

The concept of the photonic band gap has also been very well known in liquid crystals already since the discovery of cholesteric liquid crystals. When the cholesteric liquid crystal layer is illuminated by white light, it reflects an arrow band of light frequencies, and the layer appears vividly coloured. Even more interesting are the polarisation properties of the reflected light which turns out to be circularly polarised, and the handedness of this polarisation is the same as the handedness of the spherical cholesteric liquid crystal structure. The cholesteric liquid crystal is therefore an example of a spontaneously formed 1D photonic band gap material which acts as a perfect narrow band chiral mirror. There are other examples of spontaneously formed photonic structures in liquid crystals, and a notable example is the blue phase (BP) of liquid crystals. There are three different BP phases, all of them are chiral. They are all periodic in 3D, and they all imply a spontaneous appearance of a 3D lattice of topological defects. These topological defects appear emerge as structural imperfections which fill the space between the geometrically regular building blocks of the blue phases which are twisted cylinders. Because of the orientational periodicity of liquid crystal molecules forming the blue phases, they present 3D photonic structures which appear coloured when observed in white light.

A spatially periodic liquid-crystal phase that possesses a photonic band gap is an essential ingredient for lasers based on liquid crystals [306–329]. In a liquid-crystal laser, the spatially modulated structure provides an optical resonator, where the resonances are due to the spatial periodicity of the structure and corresponding Bragg reflections of light waves. In addition to having an optical resonator, which provides multiple passes of light through the resonator, an optical gain material has to be added to the resonator to amplify the light. In such a complex structure, there are two important physical mechanisms taking place when light with a given frequency is travelling along the direction of the spatial modulation. If the frequency of the light is within the forbidden frequency gap of the structure, the light will be Bragg-reflected back and forth within the resonator. However, if dye molecules are added and uniformly distributed within this spatially periodic structure, the light can be amplified by the process of stimulated emission. In order for this process to take place, the dye molecules must first be excited into higher electron energy levels, which involves illuminating the resonator using short light pulses. When these molecules emit fluorescent photons, they are Bragg-reflected by the surrounding periodic lattice and with a certain probability they create another identical photon through the process of stimulated emission from another excited dye molecule. The process of light amplification by stimulated emission in such distributed-feedback photonic structures will eventually result in lasing. This will take place when the light amplification exceeds the optical losses in the photonic structure, which plays the role of the distributed-feedback optical resonator. There are, therefore, two necessary components to form liquid-crystal lasers: the periodic structure, forming the optical resonator, and the dye molecules, playing the role of the optical amplification medium. Such a laser is usually triggered by an external pulsed-laser source, and the liquid-crystal laser emits the laser light along the direction of the periodic modulation of the refractive index.

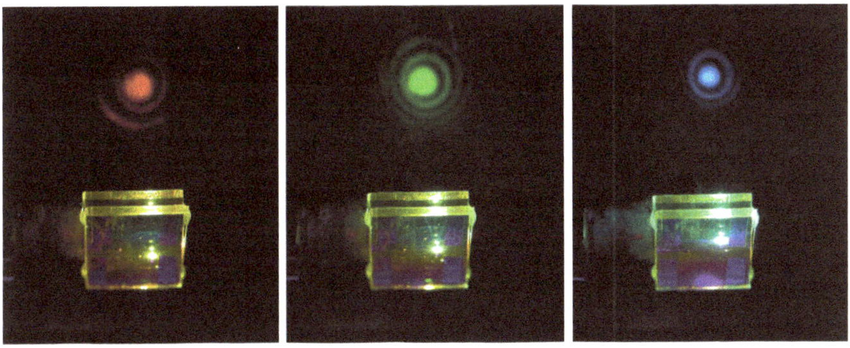

Fig. 9.1 Position-sensitive wavelength of the emitted laser light in a cholesteric liquid-crystal dye laser covering the full visible range. Two different cholesteric liquid crystals with two different dyes were injected into the glass cell from opposite sides. In this way, the pitch gradient was spontaneously achieved across the cell via the mixing of both liquid crystals. By pumping the dye laser at different positions, different wavelengths of the output light were selected. Image courtesy of Hideo Takezoe

Figure 9.1 shows an example of lasing in a cholesteric cell with a cholesteric liquid crystal and different fluorescent dyes. This cell is locally illuminated with short-wavelength excitation laser pulses (not visible in the images) and a cholesteric structure generates a strong laser beam, which propagates along the direction of the helical axis. The colour of the laser light is defined by the helical period of the structure and the corresponding position of the photonic band gap in the spectrum. Because in Fig. 9.1 the pitch depends on the position inside the cell, different laser colours can be generated by exciting different spots on the cell. In all cases, liquid-crystal lasers are dye lasers, and have all the advantages and disadvantages associated with such lasers.

Lasers based on liquid crystals have been well known for a long time [151, 306]; their basic advantage is their spontaneous formation of periodic structures. The first dye-doped cholesteric liquid-crystal lasers were realized independently by Kopp et al. [42, 307] and later by Taheri et al. [308]. Later on, the lasing was realized in a series of liquid-crystal structures and phases, such as the chiral ferroelectric smectic phase [319], liquid-crystal polymers [310, 318, 323, 325, 326], chiral nematic elastomers [309], cholesteric glasses [324, 328], an intermediate phase between the chiral nematic phase and the smectic A phase [316], and blue phases I and II [311, 329]. The liquid-crystal dye lasers described in these studies exhibit a very small thickness of the order of ten micrometres, but have large lateral dimensions, as they are fabricated in centimetre-sized glass cells filled with liquid crystals. The actual active area of such a laser is in fact relatively small, as the lasing takes place only from the region of the liquid crystal, which is illuminated with an external pumping laser. This area is small: of the order of tens of micrometres. One obvious question is how to reduce the dimensions of such lasers to micrometre-sized objects?

A natural way of producing micrometre-sized lasers from liquid crystals is to disperse the liquid crystal in another liquid in which it is immiscible [10, 330]. Then,

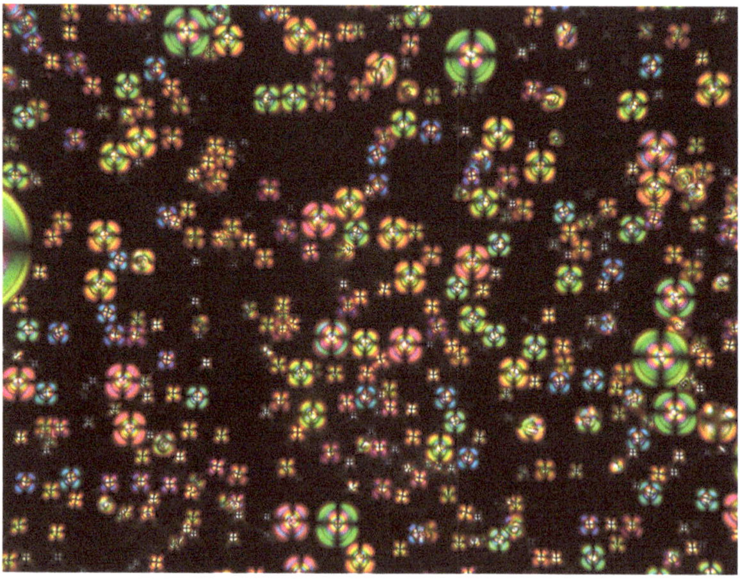

Fig. 9.2 Millions of microdroplets of a nematic liquid crystal in a thin layer of PDMS, as seen under crossed polarizers. Image courtesy of M. Humar

because of the chemical incompatibility, the liquid crystal will spontaneously form micrometre-sized droplets with a perfect spherical shape, as shown in Fig. 9.2. It was recently demonstrated that such small droplets of a nematic liquid crystal, or a cholesteric liquid crystal, dispersed in an immiscible fluid could be used as optical microcavities that are tuneable with an electric field [241] and microlasers [331]. If the nematic liquid crystal is replaced by a chiral nematic liquid crystal with a helical period in the range of optical frequencies, we can obtain an unusually structured droplet that acts as a three-dimensional Bragg-onion microresonator. In this case, the liquid-crystal molecules at the interface are oriented parallel to the interface, and the helical structure is formed all the way to the centre of the droplet, where a topological defect is situated.

These appealing properties of immiscible liquids and soft matter have triggered the idea of soft matter photonics [332–336], where basic photonic elements such as photonic crystals, wave guides, optical microcavities, and microlasers should be assembled entirely of soft matter microdroplets. This Chapter is devoted to the description of these basic elements; they are all self-grown from dispersion of various liquid crystals in various liquids and prove to be fascinating elements for microphotonics.

9.2 Whispering Gallery Mode Resonances in Nematic Microdroplets

Probably the first liquid matter optical microelement was the liquid optical microresonator, studied by Ashkin and Dziedzic in 1977 [337]. Considering a simple spherical microdroplet of liquid being levitated in air, one finds that light could be trapped inside such a microsphere by total internal reflection. Because the index of reflection of the liquid droplet is much higher than the refractive index of air, the light can circulate inside the microsphere with a series of consecutive total internal reflections at the liquid-air interface. If the circulating light reaches the point of origin with the same phase after one circulation, the resonance condition for this light mode is fulfilled. This kind of optical resonances are called Whispering Gallery Modes (WGM) or morphology-dependent resonances. Although the sphere is a geometrically perfect object sustaining optical resonances, these resonances could be supported in bodies with other shapes. For example, if one considers a transparent cube made of high-refractive index material, it can be immediately recognised that WGM resonances are possible in this case as well, but now with only four consecutive total internal reflections at each surface of the cube. Similarly, morphology-dependant optical resonances could be found in other objects, including the topologically interesting tori.

In optical resonators, optical modes are confined to very small volumes [338], and as a consequence, local electric fields could be very high. If the material is loss-less, and has a high refractive index, the quality factor of such a resonator could be very high, and Q-factors of the order of 10^{10} have been reported. Optical microresonators are extremely interesting for application as laser sources [151, 339, 340], and all-optical switches [338]. Their tunability is one of the most desirable properties, and it was reported in literature that optical resonances could be tuned by size, shape, temperature, or external field [338, 341–343]. The largest tunability can be obtained by the mechanical deformation of the resonator, but this method is slow, inaccurate, and not practical for real applications. Solid-state optical microresonators can be tuned electrically, however, the range of tunability is usually quite small [342, 343]. A different method of tuning is controlled heating and temperature-induced changes of the refractive indices and dimensions [344], but this is energetically very costly.

Liquid optical microresonators are appealing because of their perfect spherical shape, dictated by the surface tension. Whereas optical resonators based on isotropic liquids are difficult to tune, it was demonstrated in 2009 by Humar et al. [241] that small droplets of a nematic liquid crystal in an external fluid or polymer matrix is a low-loss microresonator, which can be tuned efficiently by an external electric field. Nematic microdroplets are prepared by mixing a small amount (several percents by weight) of a fluorescently-labelled nematic liquid crystal, such as E12 or 5CB, in another fluid. This could be water, glycerine, fluorinated oil, polydimethylsiloxane (PDMS), or another polymer. The methods of preparation of this kind of liquid-crystal dispersions are very well known from the previous studies of polymer-dispersed liquid crystals (PDLCs). In contrast to submicron-sized liquid-crystal droplets used

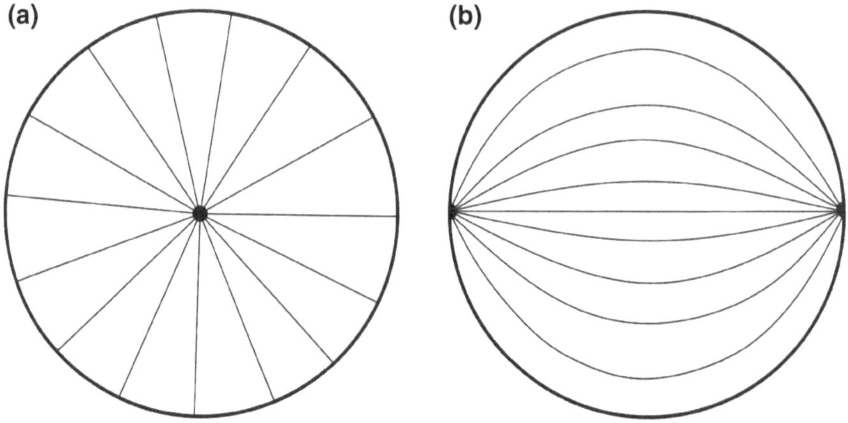

Fig. 9.3 **a** When the nematic LC molecules are aligned perpendicularly at the interface to the external fluid, the droplet exhibits a "radial" nematic structure with the radial point defect in the centre. The defect carries the topological charge of +1 and has the winding number +1. **b** For planar surface anchoring of nematic liquid crystal molecules, the droplet is "bipolar" and there are two surface hedgehog defects at each pole. Each of them carries 1/2 fractional topological charge and has the winding number of +1/2

in PDLC materials, droplets with diameters of tens of micrometers are used for optical microresonators. The reason for such large droplets is in the curvature of the surface and the associated leaking of light, circulating in such an optical resonator. Smaller droplets with larger surface curvature emit more light from the internal optical modes, and their Q-factors are correspondingly low.

An important aspect is the surface alignment of the liquid crystal, forming the droplet. Two characteristic optical resonators made of a nematic liquid crystal are possible, as shown in the schematic drawing in Fig. 9.3. They are different in the type of surface alignment of nematic liquid crystal molecules and the resulting topological defects.

Let us first discuss the optical properties of a radial nematic microdroplet, embedded in an external medium, which has a refractive index lower than each of the refractive indices of the nematic. An example of such a droplet in polymer is shown in unpolarised light and between crossed polarisers in Fig. 9.4a and b, respectively.

When viewed between crossed polarisers, one can see from Fig. 9.4b a uniaxial dark cross, which clearly indicates radial organisation of the director field inside the droplet. Sometimes this uniaxial cross is slightly curved, which indicates that the interior of the droplet is not purely radial but is bent and twisted. This happens when the splay elastic constant of the liquid crystal is large, and the energy is minimised by spontaneously bending and twisting the director field, thereby transforming some splay energy into the energetically more favourable bend-twist energy. There is a topological defect in the centre of the droplet, which, interestingly, appears to be strongly fluctuating when the droplet is observed under the microscope.

Optical resonance properties of nematic microdroplets are most easily revealed by labelling the droplet with a fluorescent dye. This is done by dispersing a small

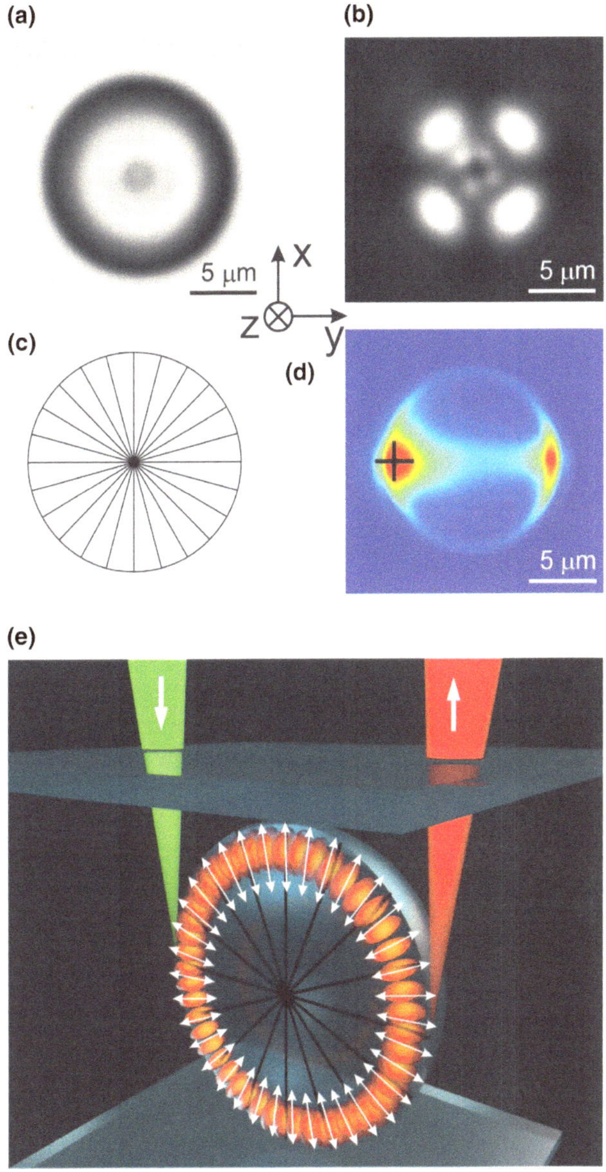

Fig. 9.4 Light in the liquid-crystal microdroplets. **a** Microdroplet of nematic liquid crystal E12 in PDMS. **b** The same droplet under crossed polarisers. **c** Structure of the director field inside the microdroplet with perpendicular surface anchoring. The *lines* representing the director field merge in the *centre*, where a radial hedgehog defect is located. **d** Light intensity under illumination by a strongly focused beam of the Ar^+ laser tweezers, which is illuminating the *left* side rim of the droplet indicated by the *black cross*. Note the strong intensity on the other side, indicating the circulation of the light inside the droplet. **e** Schematic view of WGMs in a liquid-crystal droplet with the electric field oscillating in the radial direction. *Green* beam is the excitation light, *red* beam indicates the detected light

amount of dye into the nematic liquid crystal prior to making a dispersion of droplets. The dye molecules are usually rod-like or platelet-like, and they usually tend to order with respect to the surrounding liquid crystal molecules. Rod-like fluorescent dyes tend to line their long axis parallel to the nematic director, this is the case for the Nile red and BTBP fluorescent dyes.

The incorporation of a small number of fluorescent molecules into the nematic host provides the generation of light homogeneously and throughout the whole volume of the nematic liquid crystal. This is done by using a laser light with the proper excitation wave length, which causes excitation of molecular levels and a subsequent emission of fluorescent light. If the dye is illuminated with a very short light pulse (such as a nanosecond laser source), the typical fluorescent decay time is in the order of nanoseconds to microseconds.

Having the "internal" light source in the form of fluorescent molecules, fluorescent photons could occupy available light modes in the nematic liquid crystal droplet. This means that WGM resonances are occupied by the emitted fluorescent photons, and light starts circulating inside the nematic microcavity, according to the resonant conditions. Because of the inherent leakiness of the WGM resonances and because of additional light leakage, induced by fluorescent molecules, some light from these resonances will leave the nematic resonator. Hence, it should be observed under an optical microscope when the fluorescence in the nematic droplet is excited. This is indeed observed in the experiment where the nematic droplet is locally illuminated by the focus light from the Ar^+ laser tweezers, as illustrated in Fig. 9.4d. When the left side of the droplet is illuminated locally, one observes a rather bright rim at the equator of the droplet and the strong source of light on the opposite pole of the droplet. This can easily be explained by light circulation in WGM resonances and the corresponding leakage of light, which is the strongest on the opposite side of the illuminated spot. The observed fluorescent ring and strong fluorescent emission from the ring clearly demonstrate this light is due to the WGM resonances.

The analysis of the polarisation properties of the emitted light shows that the electric field is directed radially. The WGM resonances in the radial nematic droplet are therefore transverse magnetic mode resonances (TM), which is consistent with the internal structure of the droplet. One should recall that the nematic liquid crystal is highly birefringent; the extraordinary index of refraction is sensed by light which is polarised along the long molecular axis, as shown in Fig. 9.5. The ordinary index of refraction is sensed by light which is travelling along the long molecular axis or is polarised perpendicularly to the long molecular axis. Usually, the extraordinary refractive index is larger than the ordinary one, which means that the refractive index contrast is the largest for the TM WGMs. These WGMs have their electric field oscillating radially and are therefore sensing a higher dielectric constant. The Q-factor for TM modes is therefore higher than for the transverse electric modes (TE), and these resonances are stronger and sharper.

In principle, one should be able to observe two sets of resonances in a radial nematic microdroplet, corresponding to TE and TM resonant modes. However, this depends on the diameter of the droplet and the associated leakage of WGM resonances, which is stronger for smaller droplets with a strongly curved surface.

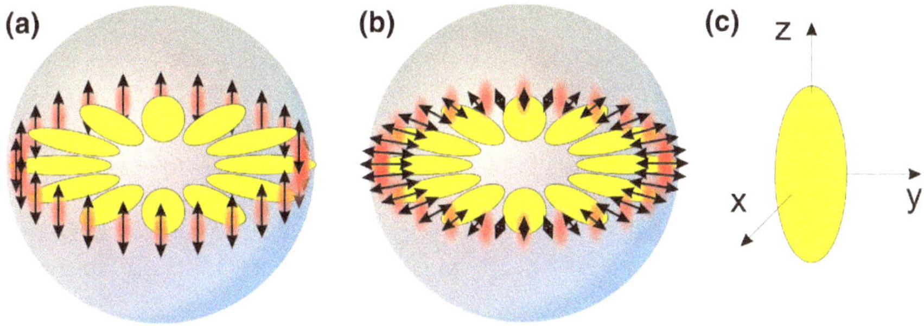

Fig. 9.5 Schematic view of the electric field oscillations in **a** transverse electric (TE) and **b** TM WGMs. The prolate objects represent the liquid-crystal molecules, pointing with their exis in the radial direction. **c** The reference frame with the representation of the uniaxial director tensor of the nematic liquid crystal with eigenvalues $\varepsilon_{\parallel}$ and $\varepsilon_{\perp}$

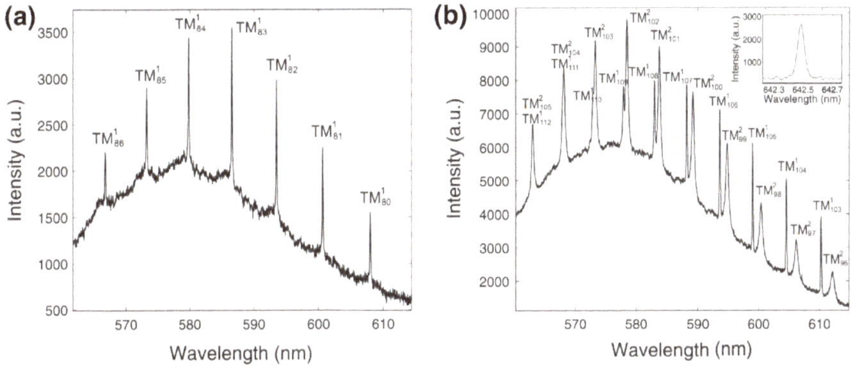

Fig. 9.6 Spectrum of light circulating in a liquid-crystal droplet. **a** A single set of WGM resonances is observed in a 10.1 μm droplet of E12 in PDMS, corresponding to the lowest radial modes ($n = 1$) with TM polarisation. **b** In a larger radial birefringent droplet (12.6 μm) second radial modes appear with $n = 2$. The *inset* shows details of a WGM spectral line in a 53 μm-diameter E12 droplet. The linewidth is approximately 0.055 nm, and the liquid crystal cavity Q-factor is of the order of ~12 000. This spectrum was measured using a high-resolution micro-Raman spectrophotometer

The spectrum of light, emitted from a 10 μm-diameter droplet of the nematic liquid crystal E12 in PDMS is presented in Fig. 9.6a. There is a broad fluorescent background and a series of very sharp intensity peaks which correspond to WGM resonances. For this rather small size of the droplet, only one polarisation of WGM resonances is supported by the droplet resonator, and this set of modes corresponds to the fundamental WGM TM modes with $n = 1$. Here, n is the radial number of the mode, which determines the number of maxima in the radial intensity distribution inside the sphere. For ordinary liquids, one usually observes both sets of modes (TE and TM) due to the liquids' optical isotropic properties. However, nematic liquid crystals are highly selective in this respect, and only the modes sensing the extraordinary index of refraction are stable in smaller droplets. In somewhat larger droplets, such as the 12.6 μm droplet of E12, shown in Fig. 9.6b an additional set of WGM resonances is observed. It turns out that this additional set corresponds to higher radial TM modes with $n = 2$. We expect that TE modes would appear in even larger droplets as well.

The mode numbers in Fig. 9.6a, b were determined by fitting the experimentally obtained spectra to the numerically calculated resonant frequencies of the radial anisotropic dielectric microspheres. As the mode numbers are very sensitive to the material parameters, the angular mode number l could be determined with the precision of ± 1.

The sharpness of the WGM resonant peaks is directly related to the Q-factor of the microresonator; high quality resonators support narrower resonant lines, and by measuring the widths of the resonances, one is able to determine the Q-factor of the liquid-crystal cavities. An example of the high-resolution spectra of WGM resonances in a 53 μm-diameter droplet of E12 is shown in the inset to Fig. 9.6b. The linewidth of the resonance is very low, 0.055 nm, and the Q-factor of the nematic liquid crystal cavity is of the order of 12 000. This is quite a surprising number considering that liquid crystals are very well known for their high light scattering cross-sections. Light scattering in nematic liquid crystals occurs due to the orientational fluctuations of the NLC molecules. Because of high birefringence (typical value 0.1) and the softness of the liquid-crystal orientational fluctuations, light scattering is very strong from the thermally induced collective fluctuations of the NLC molecules. The scattering is so strong that in a bulk nematic liquid crystals it can be easily observed under an optical microscope.

Rather large Q-factors of the nematic microcavities indicate low scattering losses due to nematic fluctuations. This can in fact be understood by considering the distribution of the electric field of the WGM resonances, which is concentrated at the interface between the nematic liquid crystal and the surrounding fluid. Due to strong orientational anchoring at this interface, the amplitude of the orientational fluctuations of NLC molecules seems to be quite low at a position of the maximum of the WGM field. This is in fact confirmed when the droplet is observed under the microscope: one can easily observe strong fluctuations of the NLC in the centre of the droplet, including strong positional fluctuation off the defect itself. However, there are no observable thermal fluctuations off the center of the droplet, which means that either the amplitude of the fluctuations is smaller, or the frequencies of these fluctuations are much higher. Higher frequencies of relaxation modes in nematic liquid crystals also lead to a lower amplitude of the mode and lower scattering cross-section [1].

As the nematic liquid crystals are well known for their large optical response to an external electric field, one expects that the frequencies of WGM resonances should be sensitive to the applied external field [241]. The effect of the field is very easily analysed by observing the droplet under the microscope with crossed polarisers, as shown in Fig. 9.7a and b. In zero electric field, one can observe a twisted dark cross on the texture of the droplet. However, this twisted cross straightens into a rather broad dark cross when an external field of only 2.6 Vμm^{-1} is applied. The comparison of the two images in Fig. 9.7a and b clearly shows that a rather small electric field has a strong influence on the internal structure of a NLC droplet. In

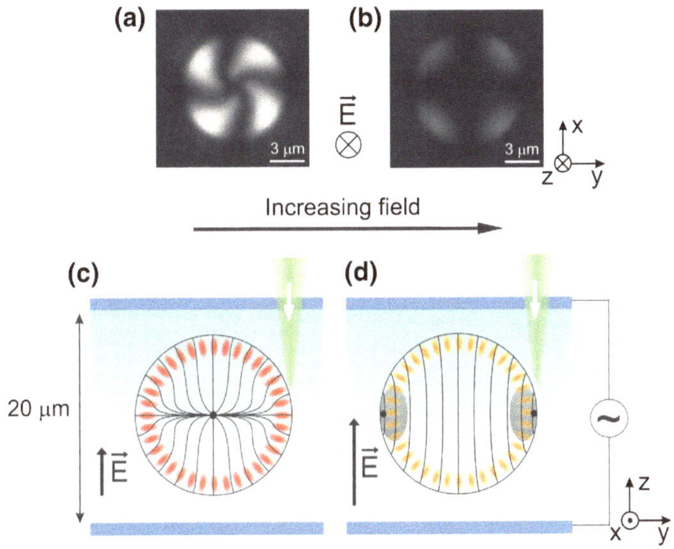

Fig. 9.7 Liquid crystal microresonator in an external electric field. **a** Crossed-polariser micrograph of a 9.3 μm droplet of E12 in PDMS at 1.9 $V_{RMS}\mu m^{-1}$. **b** The same droplet at 2.6 $V_{RMS}\mu m^{-1}$. **c**, **d** Schematic drawing of the alignment of NLC molecules at different levels of the external electric field

this case, the nematic liquid crystal E12 with positive dielectric anisotropy was used, making it very straightforward to understand what happens to the internal structure of the originally radial nematic droplet when the external electric field is applied. This is shown in Fig. 9.7c and d.

In zero field, the droplet of E12 in PDMS has a radial structure with a point topological defect in the centre. Even a small electric field of 1.9 $V_{RMS}\mu m^{-1}$ induces distortion of the nematic director inside the droplet. As the molecules tend to align along the electric field, the point defect becomes surrounded with a disc-region of a highly splayed nematic liquid crystal. At even larger fields, this point defect opens into a small ring, located in the equatorial plane of the droplet. As the electric field increases, the ring opens and allows the uniform nematic to penetrate through the ring. Finally, at very large fields, most of the interior of the droplet is oriented along the electric field and the ring is completely opened and pushed to the equator of the droplet. This defect transformation under the application of an external field is rather well known and expected, and it preserves the topological charge of the original point defect.

The electric-field-induced transformation of the nematic liquid-crystal microdroplet has an important impact on the resonant frequencies of the WGMs. By following the circular path of the WGM light, one can see from Fig. 9.7d that the electric field of light senses an inhomogeneous refractive index when circulating. By

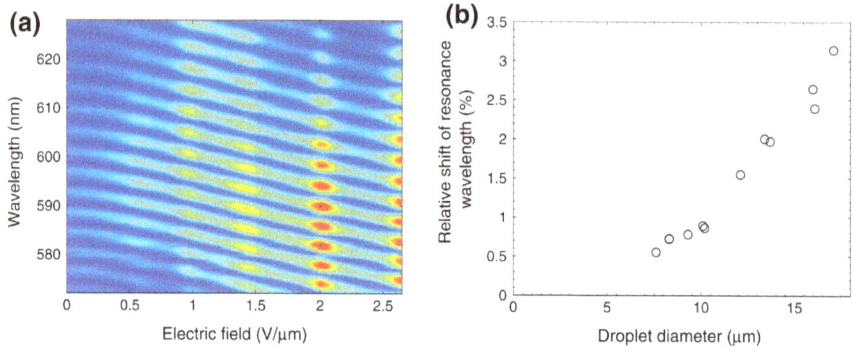

Fig. 9.8 Electric-field-induced shift of TM WGM resonances. **a** Wavelength shift in a 16-μm-diameter microresonator, filled with E12 nematic liquid-crystal and SPP-106 fluorescent dye. The *colour scale* indicates the intensity of detected light. Note the enhancement of the resonances at discrete wavelengths. **b** Range of tunability of E12 microresonators as a function of droplet diameter, determined at 2.6 $V_{RMS}\mu m^{-1}$

considering the electric field of the TM mode which is pointing radially inwards one can see that in the vicinity of the defect ring, the refractive index is lower, which causes the shortening of the overall optical path for this mode. This reduction of the optical path also reduced the resonant wavelength for resonant WGMs, which indicates that WGM resonances should shift into the blue region with increasing electric field. This expected decrease of the wavelength of the TM WGM resonances in a radial nematic liquid-crystal microresonator is indeed observed in the experiments and is shown in Fig. 9.8.

In a case of a very small field there is also very little influence of the field; between 0.5 V/μm. The wave length shift increases nearly linearly with increasing electric field. This blue-shift has no hysteresis and is completely reversible. The magnitude of the electric-field-induced WGM shift depends on the diameter of the droplet, but is as high as 20 nm at 2.6 V/μm in 17 μm diameter radial microdroplets of E12, as illustrated in Fig. 9.8b. This value is one to two others of magnitude larger that any mechanism of tuning in solids. The experimentally measured blue-shift in Fig. 9.8a has not reached the saturation regime yet. One could estimate the maximum possible WGM shift for a droplet by calculating the decrease of the optical path length between the initial radial configuration and an asymptotic configuration at very high fields, where the entire interior of the droplet is uniformly aligned along the field. This can be achieved by integrating the effective index of refraction around the circumference of the droplet, where simultaneously the rotation on the local axis is taken into account. For the nematic liquid crystal E12 the maximum range of tunability is as high as 6.7% or 41 nm at 600 nm.

We should note as a final comment of this section that the configuration of the nematic liquid crystal inside the droplet could not only be changed by applying external field (such as electric or nematic), but is also very sensitive to the chemical changes at the interface of the droplet with the medium outside the droplet. This could most easily be realised by placing nematic droplets in water with added surfactant molecules, which prefer to bind at the liquid crystal-water interface. Depending on the concentration of the specific surfactant one could therefore reach different configurations of the nematic droplet: radial or bipolar. This change in the internal configuration of the droplet has a similar effect on the spectrum of WGM resonances as the external electric field described in this Section. This gives us an exciting opportunity to use a structural transition in nematic microdroplets as a sensing mechanism, where the read-out is simply the WGM optical spectra emitted from such a droplet. One could therefore create freely floating or even actively driven microsensors which are excited by light, and also the read-out is taken by collecting the emitted light from the microdroplet. An example of such a mechanism will be described in Sect. 9.3.

9.3 Nematic Microdroplets as WGM Microlasers

In a WGM microresonator, the modes could be considered as light circulating inside the microdroplet by consecutive total internal refractions at the interface. If the light amplification medium is added inside such a resonator, one might obtain a WGM microlaser, where the light is amplified by stimulated emission during the circulation of resonant modes. Whereas this kind of experiments have been demonstrated for isotropic liquid droplets, the first experiments on WGM lasing in nematic microdroplets were performed in 2011 by Humar and Musevic [242]. Droplet of a NLC 5CB which were doped with 0.1 wt% fluorescent dye 7-dyethilamino-3,4-benzophenoxazin-2-1 (Nile red) which serves as the light amplification medium and is uniformly distributed inside the nematic microdroplet. These microdroplets were produced by mechanical mixing of the dye-doped 5CB and the water solution of Sodium Dodecyl Sulphate (SDS). Depending on the molar concentration of the SDS, the 5CB nematic microcroplets obtain different internal configuration. At high concentration of SDS, exceeding 2.0 mM, the 5CB droplets are of a radial configuration. For zero molar concentration of SDS, the anchoring of 5CB on water is weak and planar, and one obtains practically uniformly aligned nematic liquid crystal inside the microdroplets of the diameter of tens of μm.

In a typical lasing experiment one is using a pulsed source of a short wavelength light, which excites the electronic levels of the fluorescent gain medium (pumping light). In the case of Nile red fluorescent molecules, a frequency-doubled Nd-YAG nanosecond pulsed laser emitting 532 nm pulsed light is used. The laser is slightly focused through the optical microscope onto the sample, forming several tens of μm diameter illuminated region of fluorescently labelled liquid crystal. At a low level of the pumping light the sample emits usual fluorescence, which under the microscope appears as a uniformly redish-glowing light, emitted after the electrons of the molecules fall back to the ground state. In this regime, there is not enough photons produced by stimulated light emission and the emitted spectrum is broadband and incoherent. However, if one increases the level of the pumping light, the threshold

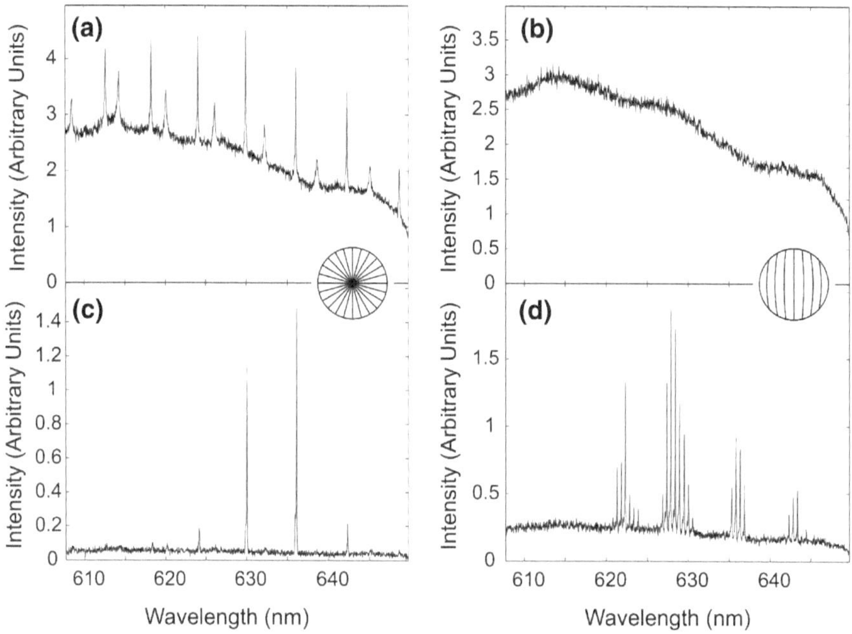

Fig. 9.9 Fluorescence and lasing spectra of radial and bipolar nematic microdroplets. **a** At the SDS concentration of 4 mM the 5CB droplet is in the radial configuration with a point defect in the centre. The *lines* inside the droplet indicate molecular orientation. The fluorescence shows characteristic spectrum of WGMs. **b** In pure water with no SDS added, the 5CB droplet is in the bipolar configuration. No modes are visible in the fluorescence spectra of the microdroplet. **c** Above the lasing threshold of 0.25 mJ/cm^2, lasing of the WGMs is clearly observable in the radial droplet configuration. **d** Above the lasing threshold of 0.7 mJ/cm^2 several groups of lasing modes are clearly recognisable in the spectrum of 5CB microdroplet with bipolar director configuration

of lasing is eventually reached. In this regime, each emitted photon produces another photon by stimulated emission, and this process is amplified through the periodic light circulation inside the spherical microresonator. In contrast to the incoherent light emitted by fluorescence emission below the lasing threshold, this light is coherent and strongly monochromatic. Under an optical microscope this is observed by the characteristic laser-speckles, which could only be formed by the interference of coherent light, emitted from the sample.

Whereas the transition from fluorescence to lasing is observed qualitatively by the naked eye through the optical microscope, this process is also analysed by taking the spectra of light emitted from the nematic liquid-crystal micro-droplet at different levels of the pumping light. This is shown in Fig. 9.9 for two configurations of the 13.7 μm 5CB nematic microdroplet.

At high concentration of SDS solution, the configuration of 5CB is radial, and at low pumping levels, one can see the characteristic spectra of WGM resonances shown in Fig. 9.9a. One can see two different sets of resonances, corresponding to different

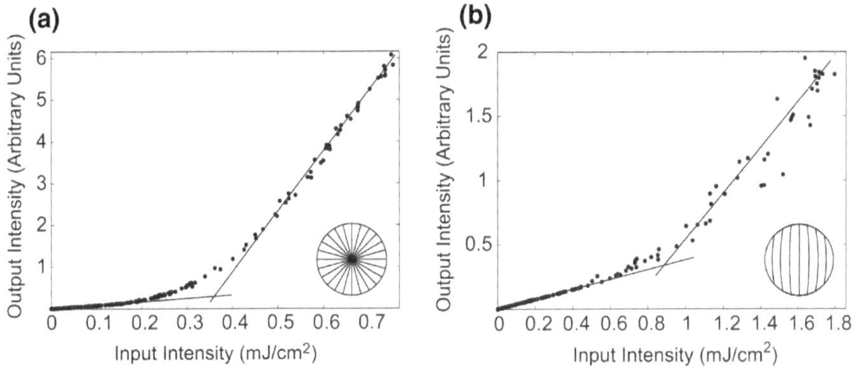

Fig. 9.10 The intensity of the laser line as the pump laser intensity is increased, for **a** radial 5CB nematic microdroplet and, **b** bipolar 5CB nematic microdroplet, both 13.7 μm in diameter. The *lines* are drawn as a guide for the eye. In both cases a clear threshold for lasing is observed

radial numbers, and these resonant peaks are of practically uniform height. However, the strength of these peaks becomes quite different when the lasing threshold is exceeded, as shown in Fig. 9.9c. One can see from this Figure that only two of the WGM resonant lines are strongly enhanced, whereas the rest of the resonances are practically gone. Furthermore, one could see in comparison to Fig. 9.9a that there is very low and broad fluorescent background, as seen in Fig. 9.9a below the threshold.

Above the lasing threshold the fluorescent background is strongly depressed and some of the lines are strongly enhanced. The mechanism of light amplification therefore "picks" and amplifies some of the resonant lines which become very strong. Whereas in a radial nematic liquid-crystal microresonator, the WGM resonances and WGM lasing lines are nearly equally spaced single peaks, this situation is quite different for the nearly uniformly aligned liquid-crystal droplet in pure water, shown in Fig. 9.9b, d. Below the lasing threshold there is practically no detected WGM resonances, but above the threshold there is a series of sharp lasing lines, forming a characteristic band-like structure. Figure 9.10 shows the measured light intensity of the spectral lines (including fluorescent background) with increasing pumping intensity. In both cases, one can observe a characteristic "knee" in the output intensity, which is typical for the threshold of lasing. This threshold of lasing is rather low for radial nematic droplets (0.25 mJ/cm^2) and is approximately three times higher than the lasing threshold for the bipolar configuration (0.7 mJ/cm^2)

WGM microlasers made of liquid crystals are therefore sensitive to the chemical conditions at their surfaces, and this could be used as a way to detect the presence of targeted molecules, floating in the surrounding liquid and binding to the microlasers' surfaces [345, 346]. As this process of binding could influence the anchoring condition of LC molecules and therefore the interior structure of the droplet as well, this kind of microlasers could be used as sensors of targeted molecular species. WGM microlasers are therefore wireless, optically addressed, and optically read-out

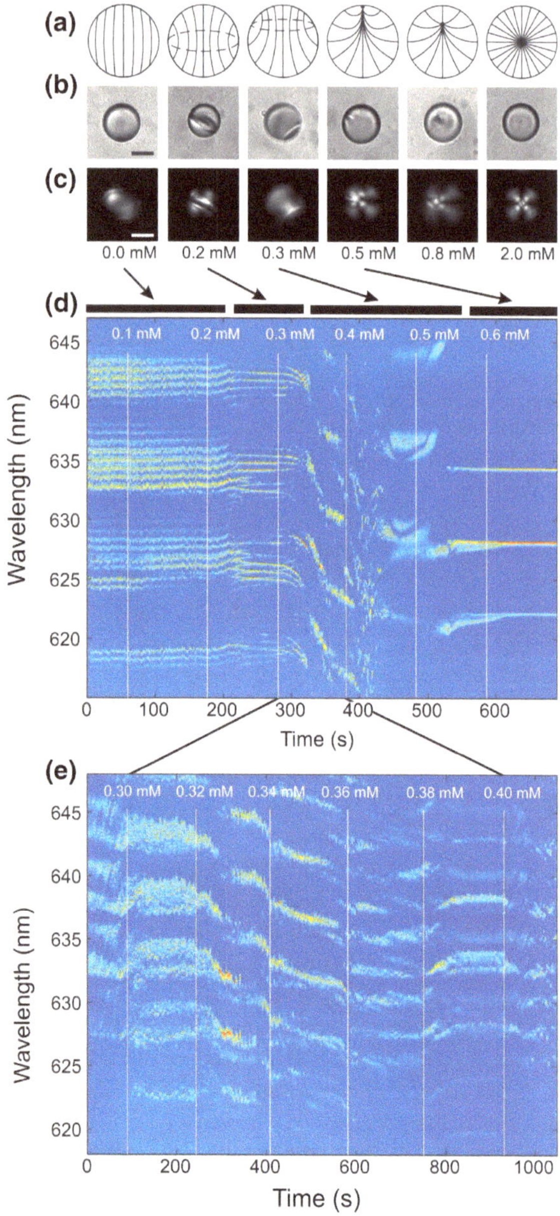

Fig. 9.11 Changes of the structure of a small droplet of the nematic liquid crystal at increasing concentrations of SDS. **a** The *lines* represent the orientation of the long axes of the LC molecules. The *dots* are point defects, where the orientation is not defined. In pure water, the LC molecules align parallel to the water-LC interface and the structure is bipolar. By increasing the SDS concentration, the surface anchoring of LC molecules gradually changes towards the perpendicular molecular orientation, obtained at 2.0 mM of SDS and beyond. **b** Non-polarised optical microscope images of 17 μm diameter microdroplets of 5CB in water and SDS. The "inner" ring is observable at 0.2 mM of SDS. The point defect evoles at the surface and sinks into the center at 0.8 mM concentration of SDS. Scale bar 10 μm. **c** The same images as in (**b**), taken between crossed polarisers. **d** The spectrum of laser light, emitted from a 13 μm 5CB droplet in water with various concentrations of SDS added. **e** Part of the lasing spectrum in the "chaotic" regime of intermediate SDS concentrations (0.3 – 0.4 mM) of a 16 μm droplet

microsensors, where the shift of the lasing line gives the information on the chemical conditions of their interface.

An example of such a sensing process is observed when a liquid crystal droplet is immersed in the water solution of SDS with varying concentration. For a low SDS concentration, the surface anchoring of LC molecules is planar, whereas at a certain threshold concentration, a transition to the radial structure is expected and observed. This process is shown in Fig. 9.11 where panel (b) shows a series of nonpolarised optical micrographs of a 5CB microdroplet in water with varying SDS concentration from 0 to 2.0 mM. For zero SDS concentration the droplet is practically homogeneous, as shown in the upper schematic panel (a). Here, the two surface boojums are virtual ones. They are situated outside of the LC microdroplet. With increasing SDS concentration, a ring defect is observed in the second panel at the 0.2 mM concentration, which then gradually transforms into a radial point hedgehog located in the centre of the microdroplet at 2.0 mM SDS concentration. The changes induced by different surfactant concentrations are completely reversible since the surfactant molecules residing on the interface are always in thermodynamic equilibrium with the surrounding water solution and can therefore adsorb and desorb from the interface.

When the droplets are excited and lasing starts, one can analyse the spectrum of the emitted light at different SDS concentrations, which is shown in Fig. 9.11d and e. It is clear from from the recorded spectra that the changes of the lasing spectra are quite dramatic and show two characteristic regimes. At low SDS concentration up to 0.3 mM, there is a typical band-structure lasing spectrum consisting of several bands with finely spaced narrow lasing lines. This band-like structure is characteristic of the bipolar configuration of the nematic microdroplet and is due to the presence of the two surface boojum defects. Each line of this band corresponds to a stable trajectory of circulating light with a plane of trajectory inclined with aspect to the line connecting both boojums. On the other hand, the spectrum is very simple above 0.6 mM SDS concentration where only a series of sharp equidistant lasing lines are observed, characteristic of the radial nematic droplet. In-between the spectrum is rather chaotic and extremely sensitive to the variations of the SDS concentration. This is understandable because it corresponds to the intermediate regime, where dramatic changes into the interior structure of the droplets are taking place at even the smallest variation of the SDS concentration.

9.4 3D Microlasers from Cholesteric Liquid Crystal Droplets

WGM microlasers, described in the previous section, could be realised form both normal, isotropic, and nematic LC. The difference is in the spontaneous order of NLC, which brings new phenomena into the interplay when the LC is confined to a droplet: internal structure, topology, and response to external fields. Whereas nematic liquid crystals therefore bring substantial novelty to WGM microlasers and their tuning, interesting phenomena are observed when microdroplets of chiral nematic liquid crystals are considered as optical microresonators and optical microcavities in general. In the following, we shall consider a spherical microdroplet of a chiral nematic liquid crystal, perfectly shaped into a spherical form by the action of the surface

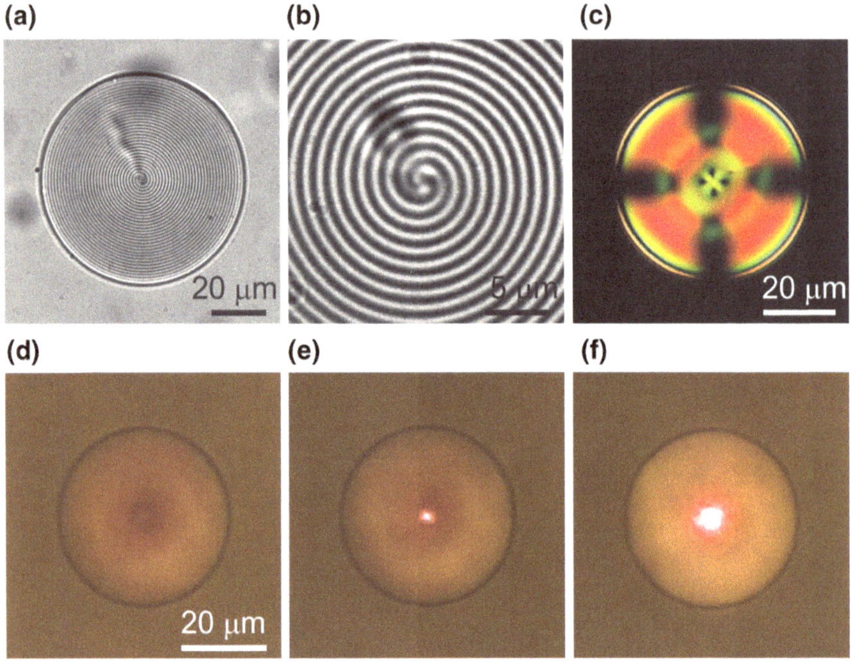

Fig. 9.12 **a** A typical cholesteric droplet with a pitch $p = 2.2\,\mu\text{m}$ in glycerol. The *light* and *dark* concentric shells are due to the spatial variation of the refractive index of the cholesteric liquid crystal in the radial direction. **b** Close up of the centre of the cholesteric droplet, when viewing in the direction parallel to the disclination line. **c** Cholesteric droplet with PBG in the visible range of light, viewed under crossed polarisers and white-light illumination. **d–f** Omnidirectional (3D) lasing in a cholesteric droplet illuminated by laser pulses ($\lambda = 532\,\text{nm}$) and a weak white background illumination. **d** Below the lasing threshold ($1.6\,\text{mJ/cm}^2$), a bright spot of radiating monochromatic light can be observed in the centre of the droplet. **f** Lasing becomes very intense at a high pump power ($12\,\text{mJ/cm}^2$). Image courtesy of Matjaz Humar

tension. The liquid-crystal molecules at the interface will have planar anchoring on the external medium (solid or fluid), which imposes an interesting organisation of the chiral nematic liquid crystal in the interior of such a droplet.

Microdroplets of chiral liquid crystals with planar surface anchoring have been considered in the past for purely theoretical reasons because of rich topology. It turns out that the chiral nematic liquid crystal organises itself in several fascinating structures, with the helical direction pointing from the surface towards the centre of the droplet. One therefore obtains a layered structure, also called an onion-like structure with the layer periodicity equal to the helical period the chiral nematic liquid crystals. As it is not possible to fill a spherical object with a chiral nematic liquid crystal without defects, there are two most probable structures and two different kinds of topological defects which appear in chiral nematic microdroplets with planar surface anchoring [258]. The first is the diametrical spherical structure (DSS) with a disclination line extending from one to the other surface through the centre of

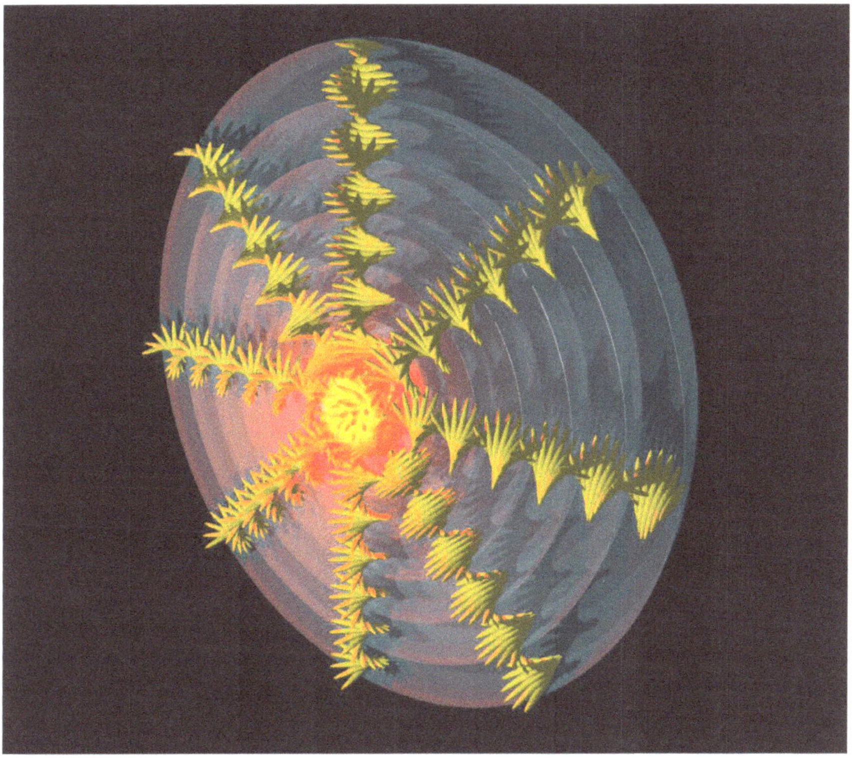

Fig. 9.13 The schematic view of the arrangement of CLC molecules in a cholesteric microdroplet with parallel anchoring of the LC molecules at the surface. The helical structure of the liquid crystal originates from the centre of the droplet and gives rise to concentric shells of constant refractive index. This dielectric structure is optically equivalent to the well-known Bragg-onion optical microcavity. Image courtesy of Matjaz Humar

the droplet. The second is the most common radial spherical structure (RSS) with a disclination line extending from the surface to the centre of the droplet. An example of such a droplet is shown in Fig. 9.12a, where the helical period is visible as a series of concentric dark lines filling the droplet uniformly and completely. One can also see a broad and irregular λ line extending from the centre towards the surface. This is the topological defect line of the double helix of two intertwined cholesteric λ cholesteric lines. Figure 9.12b shows details of the structure in the centre of the droplet, where one can see the origin of the cholesteric layers formation and winding. In both cases, the helical period was of the order of 2.2 μm in order to make the cholesteric structure inside the droplet observable with an optical microscope.

A chiral nematic microdroplet with planar surface anchoring and perfect onion-like structure is an interesting optical object due to the optical anisotropy of the medium which is combined with the chirality of the material. This kind of materials are known to selectively reflect light and therefore act as chiral mirrors for light. If such a chiral mirror is wrapped into a sphere as in the case of our chiral nematic droplet, it shows new and surprising optical properties, such as the chirality-selective reflector, which plays an important role in laser trapping of such chiral nematic microdroplets with laser tweezers. More importantly, such a chiral nematic microdroplet acts as a special kind of optical cavity, which results in interesting polarization-dependent optomehanics [175, 347, 348]. If we consider molecular distribution in a chiral nematic droplet with planar surface anchoring, as shown in Fig. 9.13, one can immediately see that in terms of optics, this structure is the so-called Bragg-onion microcavity. Such a microcavity is formed of a series of nested and concentric layers of transparent dielectric material with consecutively higher and lower refractive index. Because of this regular variation of the index of refraction as one is moving from the centre to the surface of such a resonator, it acts as a Bragg spherical microcavity. In terms of optics, such a microcavity will Bragg-reflect any light with a frequency in the forbidden Bragg region towards the centre of the cavity.

If optical gain is added into the Bragg-onion microcavity, made of chiral nematic microdroplets, an interesting lasing device is obtained. In such a device, light amplification takes place for photons which are Bragg-reflected from the Bragg resonator radially inwards or outwards. This means that above that lasing threshold, where the optical gain exceeds losses, such a lasing device should emit light in all directions equally. This is therefore a 3D omnidirectional microlaser which emits coherent light in all directions.

Such a 3D microlaser was indeed realised by Humar and Muševič in 2010 [331] by taking chiral nematic microdroplets of chiral nematic material with a sufficiently short helical period. Fluorescent dye was added into this kind of droplets to act as an optically amplifying medium, and the helical period (or Bragg-reflection interval) was selected to match the maximum emission of the fluorescent dye. When such a droplet of fluorescently labelled chiral nematic liquid crystal is illuminated with a pulsed light, exciting the fluorescent, one observes lasing, as shown in Fig. 9.12d–f. At low pumping levels the chiral nematic microdroplet shows weak fluorescence, uniformly distributed within the volume of the droplet, as shown in Fig. 9.12d. By increasing the pumping level, one observes a tiny and very bright spot of monochromatic light emerging only from the centre of the droplet. By increasing the intensity of the pumping light, the spot remains a spot but becomes much stronger. If the droplet is then rotated and observed from a different direction, one can always see the same picture: a single bright spot of monochromatic light emerging from the centre of the droplet. There is a simple and unique explanation of this phenomenon: a chiral nematic microdroplet emits laser light in all directions, and the presence of characteristic speckles in the bright spot is the proof of the coherence of the emitted light.

The threshold characteristics for 3D lasing of the chiral nematic microdroplets were measured by increasing the intensity of the pumping pulses. The emitted spectral line intesity was also measured, as shown in Fig. 9.14. A typical threshold behaviour is observed with a threshold energy of 20 nJ for a 1 ns pumping pulse, uniformly illuminating a circular area of the sample of 40 μm diameter.

The line width of the emitted laser line is typically 0.1 nm, which is comparable to the line widths of lasing in thin planar chiral nematic layers or liquid crystal

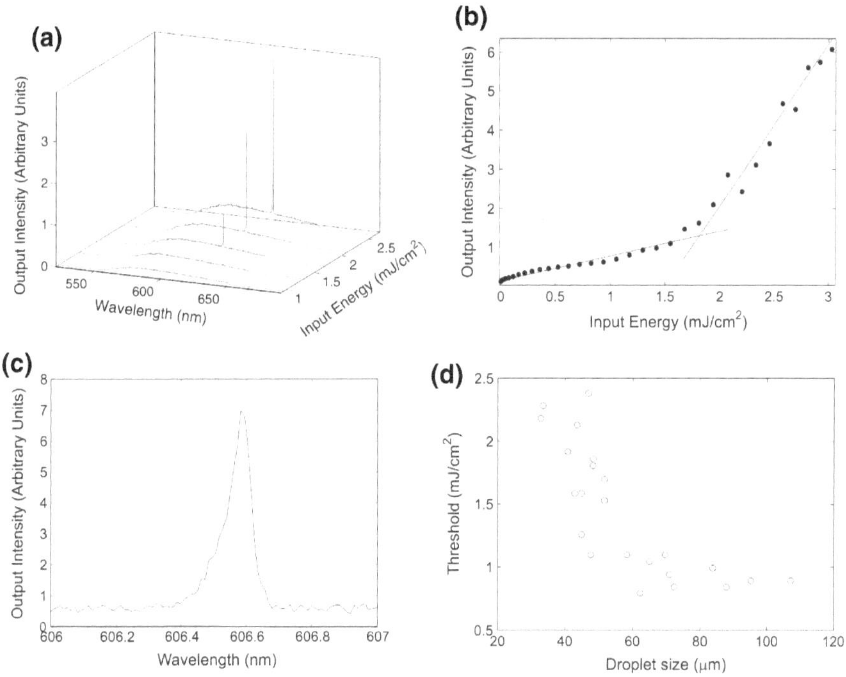

Fig. 9.14 Lasing characteristics of a single droplet of dye-doped CLC. **a** The spectra of light emitted from the center of the CLC microdroplet at different energies of the pumping pulse. **b** The radiated laser-light intensity as a function of the input-pulse energy density. The threshold for lasing is clearly seen at ~1.8 mJ/cm^2. **c** Magnified lasing spectrum showing a laser linewidth of ~0.10 nm. **d** The threshold for lasing as a function of the diameter of the CLC microdroplet. All the spectra were measured using an imaging spectrometer with a 0.05 nm resolution (Andor, Shamrock SR-500i) and cooled EM-CCD camera (Andor, Newton DU970N). Image courtesy of Matjaz Humar

blue phase. With decreasing diameter the threshold for lasing increases, as shown in Fig. 9.14d, and the smallest droplets that were lasing were made of high-birefringence liquid-crystal droplets of 15 μm diameter. The typical emitted average power of such a 3D microlaser is up to 0.05 mW at a 200 Hz repetition rate. Another phenomenon observed in these LC spherical lasers is the thermal fluctuations of the intensity, which might be related to the strong, low-frequency orientation fluctuations of the liquid crystal within the microdroplet.

Because the 3D microlaser is made of chiral nematic liquid crystals and exhibits helical internal organisation, one expects that the output light would be circularly polarised. The reason for this is that the Bragg reflector is reflecting only circularly polarised light of the same handedness as the helical structure of the chiral nematic liquid crystal. The other helicity of light is not reflected back and forth, and there is no resonance condition for this polarisation. It turns out to be quite difficult to determine the polarisation properties of the emitted light, and no clear proof of the

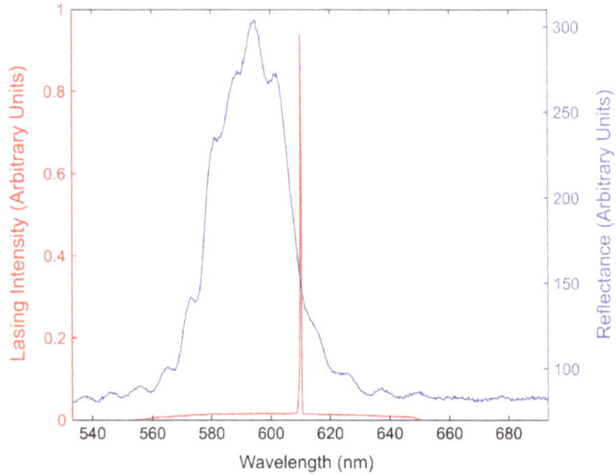

Fig. 9.15 Lasing spectrum of a single CLC droplet compared to the reflection spectrum of a 30 μm planar cell filled with the same CLC mixture. The reflection spectrum was measured for light propagating along the helix of the CLC. Image courtesy of Matjaz Humar

circular nature of 3D emitted light was convincingly presented. On the other hand, such a laser shows the characteristic band-edge lasing, which is shown in Fig. 9.15.

In Fig. 9.15, the lasing line of the 3D microlaser is superposed to the reflection spectra of the same chiral liquid crystal in a planar cell. The lasing line is clearly positioned at the red edge of the reflection band, which is due to the highest density of photon states, at the band edges characteristic for chiral liqiud crystal microlasers.

The spatial uniformity of lasing intensity was measured by inserting the 3D microlaser into a glass capillary, which allowed for wide-angle measurements of the emitted light intensity [331]. This turns to be quite uniform and confirms a simple observation when rotating the sample with 3D microlasers. Although all lasers that we know emit light into a given direction, this is a simple consequence of the construction of their optical resonator. This is usually a 1D resonator made of reflective mirrors or Bragg gratings, which allow for photon reflection and optical amplification via stimulated emission in 1D. The laser is made to emit light along a given direction by meeting one of the reflective elements slightly transmissive. However, other types of lasers are known, such as tori, which emit light within a plane, containing the equator of the tori. The microlaser emits light in full solid angle for a spherical resonator with radial Bragg reflectors. Although this laser seems odd at the first glance, it may have clear advantages in some special cases, such as internal source holography, sensing, or imaging. In this case, the volume is uniformly emitted from an internal point source of coherent light, which is reflected by the surrounding medium and collected with an outside detector. The Bragg-onion laser is therefore a natural candidate for such microscopic applications.

As a final note, we should comment on the range of wave-lengths which could be emitted by such a laser and its tuning. The range of possible wave-lengths, emitted by such a laser is determined by two limiting factors: the natural helical period of the chiral nematic LC and the range of light emission of the light amplifying medium. As far as the helical pitch is concerned, chiral nematic liquid crystals show an extremely large tuning range, accessible by chemical synthesis from 200 nm to a micrometre helical period. Besides using chemical tuning, helical period can also be made tunable by temperature or external electric fields. Both temperature and external fields are known to have a strong influence on the helical period of the chiral nematic liquid crystals. Light amplification in 3D microlasers could be achieved by using various fluorescent materials, such as dye molecules, quantum dots, or even individual atoms (such as Erbium), which are attached to the molecules forming the liquid-crystal phase. There are various advantages in using a particular fluorescent source. Organic dyes are very well known for their high quantum yield, they also dissolve very well and at high molecular concentration in liquid crystals. This makes them a highly efficient fluorescent source, which provides for low threshold microlasing. However, their disadvantage is bleaching and molecular instability. A typical 3D microlaser based on fluorescent organic dyes will therefore bleach with time because there is no regeneration of the fluorescent material. In this respect, quantum dots are superior materials, but their disadvantage is their limited solubility in nematic liquid crystals. One way to solve this problem is to chemically attach quantum dots to the molecules that constitute the liquid-crystal phase. Finally, organo-metallic complexes in a form of rod-like molecules incorporating fluorescent Er atoms are known to form the nematic phase. In this case, the emitted light is rather narrow, which has a strong constraint on the properties of the optical resonator, as it has to be tuned to this emitted wavelength.

3D microlasers have been investigated in different aspects and forms, such as band-edge random lasing in paintable laser emulsions [349], and random sphere microlasers [350, 351]

9.5 Wave Guiding and Lasing in Smectic a Liquid Crystal Fibres

A straightforward way to make dispersions of liquid-crystal material a carrier immiscible liquid is simply mechanically mixing a small amount of LC in this medium. This produces spherical droplets of LC, because the surface tension tends to minimise the surface. The elasticity of the liquid-crystal material forming the droplet is in most cases irrelevant, as the surface energy contribution is much larger than the elastic energy, stored in deformed liquid crystals. In some special cases, such as chromonic liquid-crystal droplets of non-spherical shapes are observed, known as tactoids. In this case, the internal elasticity of the confined liquid crystal imposes force to the interface and deforms it into a non-spherical, tactoidal shape. This happens when the surface energy is low and the elastic distortion of the interior of the droplet becomes important. Another example of non-spherical liquid crystal microdroplets are nematic droplets threaded by cellulose fibres [352].

However, there is a special case showing unusual solubility of smectic-A liquid crystals in contact with the water solution of some surfactants, such as cetyl ammonium bromide (CTAB) [353, 354]. If a droplet of smectic-A material is exposed to such a surfactant solution, an interesting phenomenon of the instability of this smectic-A surface is observed. The surface wrinkles and occasionally ejects tiny jets of the smetic-A liquid crystal, which self-forms into tiny liquid fibers. This phenomenon of smetic-A fiber growth is observed only at surfactant concentrations which are much higher than the critical micellar concentration of the given surfactant. For CTAB in water, this is approximately 1 mM at 25 °C. Below this concentration, CTAB is uniformly dispersed in water and starts forming micelles above this critical concentration. The growth of smectic-A fibers into the CTAB dispersion is similar to processes of growth of tubular-like structures at the interface of lyotropic lamellar liquid crystals and water. These myelin figures are tubular-like structures consisting of cylindrically bent water-surfactant bi-layers. In contrast to lyotropic myelin figures which have poor optical contrast (i.e. the difference in the two optical indices), smectic-A fibers have very large optical contrast, as the refractive index of lyotropic myelin figures is 1.33 and the refractive index of smectic-A materials is 1.5 and above). This makes smectic fibers in water an appealing candidate for optical wave guides, but their internal structure remains open. One could actually conjecture that the only possible organisation of smectic-A material in these fibers is in a series of coaxial smectic layers, rolled-up into hollow tube-like structures. A fiber would therefore be organised as a radial stacking of coaxial smectic layers, ending with perpendicular molecular orientation at the interface to the external water and forming a line topological defect in the centre of the fiber. Such a coaxial organisation of smectic fibers is indeed confirmed by simple optical observations of microfibers between crossed polarisers, as shown in Fig. 9.16.

Figure 9.16a shows a smectic fiber made of 8CB LC in water solution of C16TAB with added fluorescent Nile red [354]. The structure is extremely homogeneous under crossed polarisers and is obviously highly birefringent. There are two hemispherical endings at each end of this fiber and there is clearly a line connecting both hemispheres through the centre of the fiber. Using red plate imaging (the inset to Fig. 9.16a), one is able to determine the orientation of the LC molecules in the fiber. The orientation is clearly radial, as seen by blue and yellow parts of the fiber. More precisely the arrangement of LC molecules can be determined by Fluorescent Confocal Polarising Microscopy (FCPM), which is collecting polarised light from the fluorescent molecules, dissolved in the LC. In the case of Nile red molecules, it is known they behave as molecular rods, which align with the rod-like molecules in the smectic-A liquid crystal. Their radiative dipole moment, which is pointing along their long axis is therefore locally aligned with the surrounding liquid-crystal molecules and determines the polarisation of locally emitted light. If such a 8CB fluorescently labelled microfiber is scanned with confocal polarising microscope, one can clearly determine the direction of fluorescent molecules within the imaging plane, which also determined the local direction of the 8CB molecules. Such a FCPM cross-section is shown in Fig. 9.16b and c for two different polarisations of the emitted light. In panel (b), the polarisation of the collected light is along the axis of the fiber, and one can see that light with this polarisation is emerging only from the hemispherical end of the fiber. If the same fiber is now scanned for perpendicular light polarisation, the bright regions are distributed along the surface of the fiber. It

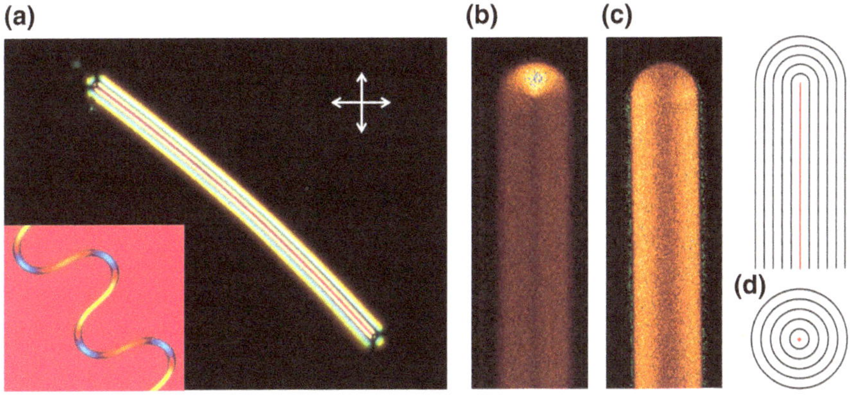

Fig. 9.16 Polarising microscopy and FCPM images of smectic-A microfibers of 8CB (doped with 0.01 wt.% of Nile *red*) in aqueous C16TAB solution. **a** 8CB microfiber (diameter 10 μm) between crossed polarisers demonstrating the strong birefrigence of the smectic fiber. The inset shows a thinner fiber (diameter 3 μm), which has adopted a bend S-like shape, between crossed polarisers with the addition of a *red* (λ) wave plate. The alternating *yellow/blue* colour sequence indicates the the LC molecules, and thus the local optical axis, are always aligned perpendicular to the surface of the fiber. **b** FCPM image of a smectic-A 8CB microfiber (diameter 35 μm) with the polarisation of the exciting light along the axis of the microfiber. High fluorescence intensity is observed at the spherical tip of the fiber. **c** FCPM image of the same fiber with the polarisation of the exciting light perpendicular to the axis of the microfiber. High fluorescence intensity is observed along the cylindrical body of the fiber. The distribution of the fluorescence intensities in both images indicates that the Nile red molecules, and thus the LC molecules, are oriented perpendicular to the surface of the fiber. **d** Schematic drawings of the coaxial arrangement of the smectic layers in a microfiber. *Top* cross section parallel to the fiber axis. *Bottom* cross section perpendicular to the fiber axis. The *red line* or *dot* indicates the topological line defect. The rod-like LC molecules (not shown in the drawings) are oriented perpendicular to the layer planes and the fiber surface

is straightforward to determine the schematic structure of the fiber, which is shown in panel (d). The smectic layers are rolled into smectic tubes which fit nicely one into another and form a concentric series of smectic layers. The end of the fiber is naturally completed by a hemispherical arrangement of smectic layers which are distorted and curved but are well known from smectic-A structures. And finally, there must be a topological defect line in the centre of each fiber, where the molecule's direction is ill-determined and therefore singular.

One can see from Fig. 9.16a that the fibers are very uniform in diameter and that there is practically no structural defect observable throughout the length of the fiber. The reason for this is the smectic layering which perfectly defines the diameter of the fiber as the layers are obviously not ruptured at any point on the surface of the fiber. Regarding their optical properties, there are two interesting features of smectic microfibers. First, they are of high optical quality, and they are highly birefringent with the optical axis pointing radially outwards, and there is a topological line defect in the core. Because of this topological line defect, one expects these microfibers to likely support Gaussian–Laguerre electromagnetic waves. It was demonstrated in several experiments with light propagation through topological defects in liquid

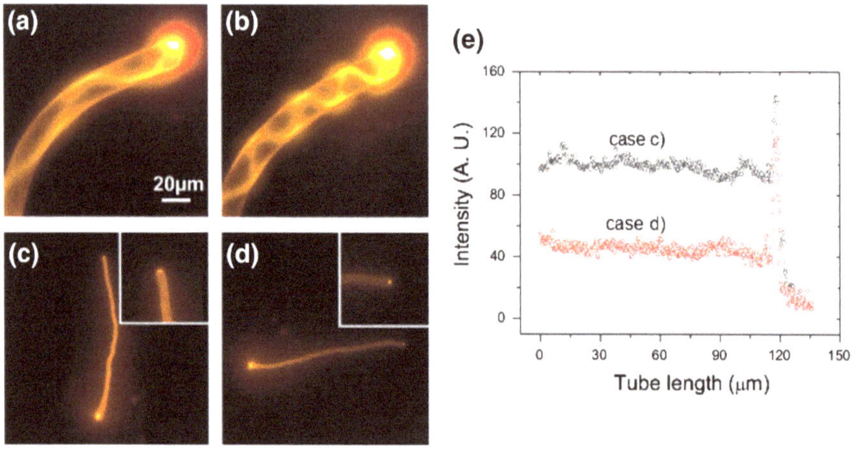

Fig. 9.17 Light guiding through 8CB smectic-A microfibers doped with 0.01 wt % of Nile red. The concentration of C16TAB surfactant in the surrounding aqueous medium is 20 mM. The power of the applied continuous Ar+ laser beam for fluorescence excitation is 1 mW. **a–b** A focused Ar^+ beam is positioned at different points at the hemispherical cap of a 20 μm thick fiber. The spiral-shaped trajectories of the guided light are clearly visible because of the fluorescence. **c–d** Light guiding by a thin (2 μm) 8CB microfiber. The fluorescence is excited at the *lower* (**c**) or *left* (**d**) end of the microfiber using different polarisations of the Ar^+ beam. The *insets* show the respective other end of the fiber at a higher magnification, the bright spot at the fiber end indicating the leaking of the guided light. The angle between the major axis of the microfiber and the polarisation of the Ar^+ beam is 0 in (**c**) and 90° in (**d**). The corrugated appearance of the thin fiber is a result of some random bending and the small thickness of the fiber. There is no indication for an axial variation of the fiber thickness. **e** The intensity I of the emitted fluorescent light as a function of position (i.e. length) along the thin microfiber in (**c, d**) remains fairly constant. Note the intensity peak at the end of the microfiber, corresponding to the bright spot shown in the insets in (**c, d**)

crystals that the Gaussian, plane wave beam picks up the topological defect from the matter and transforms the light wave into topologically nontrivial waves. Topological defects of the matter are therefore imprinted into the electromagnetic waves as phase singularities. Second, both ends of the fiber are kept with birefringent hemispheres which act as perfect spherical lenses.

The guiding of light by smectic fibers is observed by using the technique of fluorescent labelling of the fiber [354]. Without this labelling it is very difficult to couple light into the fiber and observe the its propagation along the fiber. By adding fluorescent molecules, which align themselves along the local orientation of the liquid crystal, it is possible to create light inside the fiber simply by shining a slightly focused excitation beam towards the selected part of the fiber. As fluorescent light is produced at the illuminated volume, it propagates in all directions and some of it is accepted by the cylindrical fiber and is guided along the fiber by the total internal reflection (TIR) at the interface between the fiber and the surrounding water. During this light propagation fluorescent light is emitted all along the guiding light, which makes it visible to the external observer.

An example of light generation and propagation in a smectic-A fiber with a rather large diameter is shown in Fig. 9.17a–d. In these images, a focused beam of the Ar^+ laser tweezers was positioned on the fluorescently labelled 8CB microfiber, and intense fluorescent spiralling pattern of light was observed along the fiber. The changes in this fluorescent pattern were clearly visible if the tweezers were positioned on another spot. In thinner 8CB microfibers, as shown in Fig. 9.17c, d, the fluorescent light is emitted nearly uniformly across and along the fiber and the bright spot is observed on the other side, where the light leaks out of the fiber. Figure 9.17e shows the intensity of the emitted fluorescent light, measured all along the fiber. No substantial intensity drop was observed and the losses are lower than 5% over 130 μm over the length of the fiber. The amount of generated light is the highest

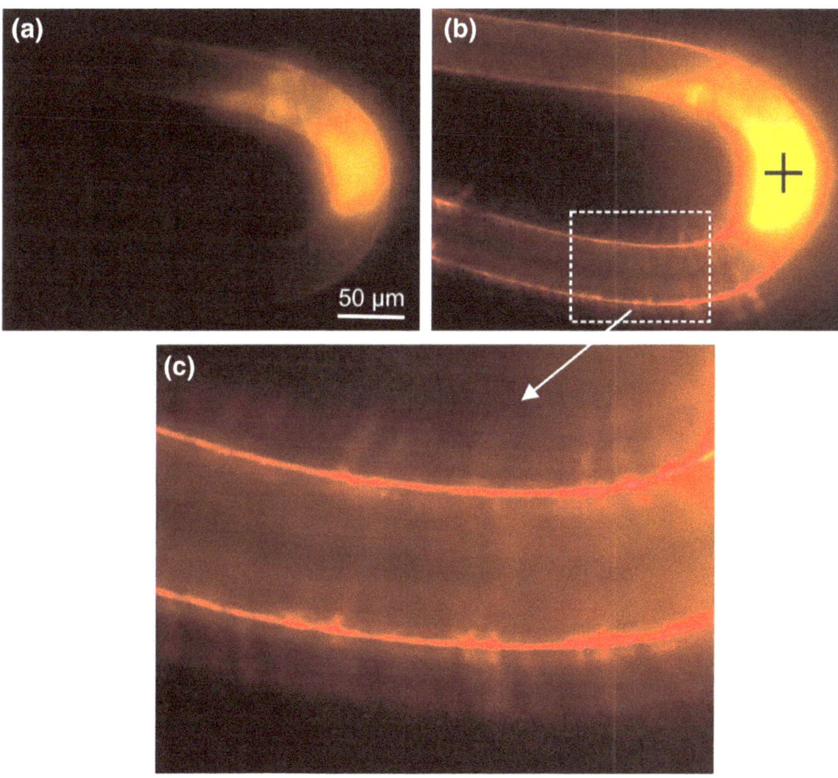

Fig. 9.18 An example of lasing from a large, 50 μm-diameter Nile red-doped 8CB microfiber in a 100 mM C16TAB water solution. The power of the pumping laser is below the lasing threshold in (**a**) and above the threshold in (**b**). In both cases the pumping pulsed laser beam is illuminating a 20 μm diameter region encircling the *black cross*. The polarisation of the pumping light is along the tangent to the fiber at this position. Note the very distant laser speckles in (**b**), shining from the surface of the fiber at hundreds of μm separation (see enlarged section shown in (**c**)). Thinner fibers (thickness of a few μm) showed essentially the same behaviour

when the polarisation of the fluorescent excitation beam matches the major axis of the tube.

Smectic-A fibers are perfectly transparent optical objects of cylindrical symmetry, which obviously support wave-guiding of light along their axis. However, if one considers a cross section of the fiber, it is clear that such fibers could also support Whispering Gallery Modes resonances (WGMs). In this case, light is not transported along the axis (as in wave guiding) but is circulating in a direction perpendicular to this axis along the circular cross section of the fiber. This kind of resonances are well known from microphotonics and are called "morphological resonances". In our case, the diameter of the fiber is very uniform, i.e. within one smectic layer (1 nm), and one expects rather large spatial coherence of the WGMs in the lasing regime. For this purpose, the fiber is locally illuminated with a pulsed pumping beam (532 nm) which excites a rather large part of the fiber, as shown in Fig. 9.18a. In this image, fluorescent light is primarily emitted from the illuminated region, but the light is also fluorescently emitted from the surrounding, where it was partially excited by the light guided through the fiber. When the intensity of the pumping beam is increased, a clear difference can be noticed at some threshold value of the density. This regime is shown in Fig. 9.18b, where it clearly observable that there is a thin sheet of monochromatic red light apparently shining from the surface of the fiber. By looking at the enlarged part of the fiber far away from the illuminated region, one can clearly see interference speckles shining from the tiny light sheet at the interface of the fiber.

Considering lasing from the WGM resonances, there is a simple explanation for this sheet of speckles. In WGM resonantors with an optical gain, the light is amplified by stimulated emission by light circulating within the resonator. In our case, the light is circulating along the circumference of the circular cross section of the fiber; due to fluorescent dyes, this light is amplified by the stimulated emission of photons and the fiber therefore acts as a cylinder WGM laser. Such a laser emits light from the WGM modes localised at the interface of the exterior. When viewed under the microscope, the most intensity is emitted from the interface. This is therefore the reason for the strong monochromatic light emanating from the surface of the fiber. Spectral properties of light emitted from this interfacial WGM layer show lasing lines, similar to the WGM resonances in spherical droplets. However, the number of resonances is higher compared to WGM lasers made of LC droplets. The observed threshold of WGM lasing is comparable to the threshold for lasing of WGMs in spherical nematic droplets and therefore confirms the same underlying physical mechanism.

References

1. P.G. de Gennes, *The Physics of Liquid Crystals* (Clarendon Press, Oxford, 1974)
2. R. Blinc, B. Žekš, *Soft Modes in Ferroelectrics and Antiferroelectrics* (North-Holland, Amsterdam, 1974)
3. J.C. Tolédano, P. Tolédano, *The Landau Theory of Phase Transitions* (World Scientific, Singapore, 1987)
4. P.M. Chaikin, T.C. Lubensky, *Principles of Condensed Matter Physics* (Cambridge University Press, Cambridge, 1997)
5. M. Ravnik, Colloidal structures confined to thin nematic layers. Ph.D. thesis. Faculty of Mathematics and Physics (University of Ljubljana, Slovenia, 2009)
6. P.P. Karat, *Electric and Magnetic field effects in liquid crystals*, Raman Research Institute, Bangalore (1977)
7. Th. Rasing, I. Muševič (eds.), *Surfaces and Interactions of Liquid Crystals* (Springer, Heidelberg, 2004)
8. M. Nikkhou, M. Škarabot, I. Muševič, Eur. Phys. J. E **38**(3), 1 (2015)
9. M. Humar, Liquid-crystal microdroplets as optical microresonators and lasers. Ph.D. thesis (Jozef Stefan International Postgraduate School, Ljubljana, Slovenia, 2012)
10. P. Poulin, D. Weitz, Phys. Rev. E **57**(1), 626 (1998)
11. G. Posnjak, S. Čopar, I. Muševič, Sci. Rep. **6**, 26361 (2016). doi:10.1038/srep26361
12. A. Yariv, P. Yeh, *Optical Waves in Crystals* (Wiley, New York, 1984)
13. R.M.A. Azzam, N.M. Bashara, *Ellipsometry and Polarized Light* (North-Holland, Amsterdam, 1987)
14. I.C. Khoo, *Liquid Crystals: Physical Properties and Nonlinear Optical Phenomena* (Wiley, New York, 1995)
15. D.W. Berreman, J. Opt. Soc. Am. **62**(4), 502 (1972)
16. N.D. Mermin, Rev. Mod. Phys. **51**(3), 591 (1979)
17. D.R. Nelson, *Defects and Geometry in Condensed Matter Physics* (Cambridge University Press, Cambridge, 2002)
18. M.I. Monastyrsky, *Riemann, Topology, and Physics*, 2nd edn. (Birkhäuser, Boston, 2008)
19. M.I. Monastyrsky (ed.), *Topology in Condensed Matter*, vol. 150 (Springer Series in Solid-State Sciences (Springer, Heidelberg, 2006)
20. D.J. Bishop, P.L. Gammel, D.A. Huse, C.A. Murray, Science **255**(5041), 165 (1992)
21. N. Mermin, T.L. Ho, Phys. Rev. Lett. **36**(11), 594 (1976)
22. V. Ruutu, V. Eltsov, A. Gill, T. Kibble, M. Kruslus, Y.G. Makhlin, B. Placais, G. Volovik, W. Xu et al. Nature **382**(6589), 334 (1996)
23. C.N. Weiler, T.W. Neely, D.R. Scherer, A.S. Bradley, M.J. Davis, B.P. Anderson, Nature **455**(7215), 948 (2008)

I. Muševič, *Liquid Crystal Colloids*, Soft and Biological Matter,
DOI 10.1007/978-3-319-54916-3

24. A. Wachowiak, J. Wiebe, M. Bode, O. Pietzsch, M. Morgenstern, R. Wiesendanger, Science **298**(5593), 577 (2002)
25. O. Tchernyshyov, G.W. Chern, Phys. Rev. Lett. **95**(19), 197204 (2005)
26. S.B. Choe, Y. Acremann, A. Scholl, A. Bauer, A. Doran, J. Stöhr, H.A. Padmore, Science **304**(5669), 420 (2004)
27. A. Patani, M. Schlindwein, Q. Shafi, J. Phys. A **9**(9), 1513 (1976)
28. E. Brasselet, N. Murazawa, H. Misawa, S. Juodkazis, Phys. Rev. Lett. **103**(10), 103903 (2009)
29. P.J. Ackerman, Z. Qi, I.I. Smalyukh, Phys. Rev. E **86**(2), 021703 (2012)
30. L. Allen, M.W. Beijersbergen, R. Spreeuw, J. Woerdman, Phys. Rev. A **45**(11), 8185 (1992)
31. T.W.B. Kibble, J. Phys. A **9**(8), 1387 (1976)
32. W.H. Zurek, Phys. Rep. **276**(4), 177 (1996)
33. M. Kléman, Points, lines and walls in liquid crystals, magnetic systems and various orderedmedia. Wiley (1983)
34. M. Kléman, Rep. Prog. Phys. **52**(5), 555 (1989)
35. T. Lubensky, D. Pettey, N. Currier, H. Stark, Phys. Rev. E **57**(1), 610 (1998)
36. M. Kléman, O.D. Lavrentovich, Philos. Mag. **86**(25–26), 4117 (2006)
37. H. Stark, Phys. Rep. **351**(6), 387 (2001)
38. S. Čopar, S. Žumer, Phys. Rev. E **85**(3), 031701 (2012)
39. S. Čopar, Topology and geometry of defects in confined nematics. Ph.D. thesis. Faculty of Mathematics and Physics (University of Ljubljana, Slovenia, 2012)
40. S. Čopar, Phys. Rep. **538**(1), 1 (2014)
41. G. Posnjak, S. Čopar, I. Muševič, Nature Communications **8**, 14594 (2017)
42. O. Lavrentovich, Liq. Cryst. **24**(1), 117 (1998)
43. B. Senyuk, Q. Liu, S. He, R.D. Kamien, R.B. Kusner, T.C. Lubensky, I.I. Smalyukh, Nature **493**(7431), 200 (2012)
44. P. Oswald, P. Pieranski, *Nematic and cholesteric liquid crystals: concepts and physical properties illustrated by experiments, The Liquid Crystals Book Series* (CRC Press, Boca Raton, 2005)
45. P. Poulin, H. Stark, T. Lubensky, D. Weitz, Science **275**(5307), 1770 (1997)
46. J. Joanny, Science **275**(5307), 1751 (1997)
47. P. Cladis, M. Kléman, P. Piéranski, C.R. Acad, Sci. Paris **273–279**(6), 275 (1971)
48. J. Rault, C.R. Acad, Sci. Ser. B **272**, 1275 (1971)
49. P. Poulin, V. Raghunathan, P. Richetti, D. Roux, J. Phys. II **4**(9), 1557 (1994)
50. V. Raghunathan, P. Richetti, D. Roux, F. Nallet, A. Sood, Mol. Cryst. Liq. Cryst. **288**(1), 181 (1996)
51. F. Brochard, P.G. de Gennes, J. Phys. **31**(7), 691 (1970)
52. M. Kreuzer, T. Tschudi, R. Eidenschink, Mol. Cryst. Liq. Cryst. **223**(1), 219 (1992)
53. E. Terentjev, Phys. Rev. E **51**(2), 1330 (1995)
54. O. Kuksenok, R. Ruhwandl, S. Shiyanovskii, E. Terentjev, Phys. Rev. E **54**(5), 5198 (1996)
55. H. Stark, Eur. Phys. J. B **10**(2), 311 (1999)
56. X. Wang, Y.K. Kim, E. Bukusoglu, B. Zhang, D.S. Miller, N.L. Abbott, Phys. Rev. Lett. **116**(14), 147801 (2016)
57. R. Pratibha, N. Madhusudana, Mol. Cryst. Liq. Cryst. **178**, 167 (1990)
58. O. Mondain-Monval, J. Dedieu, T. Gulik-Krzywicki, P. Poulin, Eur. Phys. J. B **12**(2), 167 (1999)
59. Y. Gu, N.L. Abbott, Phys. Rev. Lett. **85**(22), 4719 (2000)
60. J. Loudet, P. Poulin, Phys. Rev. Lett. **87**(16), 165503 (2001)
61. J. Loudet, O. Mondain-Monval, P. Poulin, Eur. Phys. J. E **7**(3), 205 (2002)
62. C. Völtz, Y. Maeda, Y. Tabe, H. Yokoyama, Phys. Rev. Lett. **97**(22), 227801 (2006)
63. S. Ramaswamy, R. Nityananda, V. Raghunathan, J. Prost, Mol. Cryst. Liq. Cryst. **288**(1), 175 (1996)
64. B. Lev, P. Tomchuk, Phys. Rev. E **59**(1), 591 (1999)
65. J.i. Fukuda, B.I. Lev, H. Yokoyama, J. Phys.: Condens. Matter **15**(23), 3841 (2003)
66. J.i. Fukuda, H. Stark, M. Yoneya, H. Yokoyama, Phys. Rev. E **69**(4), 041706 (2004)

67. V. Pergamenshchik, V. Uzunova, Condens. Matter Phys. **13**(3), 33602 (2010)
68. V. Pergamenshchik, V. Uzunova, Phys. Rev. E **83**(2), 021701 (2011)
69. V. Pergamenshchik, V. Uzunova, Eur. Phys. J. E **23**(2), 161 (2007)
70. V. Pergamenshchik, V. Uzunova, Phys. Rev. E **76**(1), 011707 (2007)
71. S. Chernyshuk, O. Tovkach, B. Lev, Phys. Rev. E **85**(1), 011706 (2012)
72. P. Poulin, V. Cabuil, D. Weitz, Phys. Rev. Lett. **79**(24), 4862 (1997)
73. M. Škarabot, M. Ravnik, S. Žumer, U. Tkalec, I. Poberaj, D. Babič, N. Osterman, I. Muševič, Phys. Rev. E **76**(5), 051406 (2007)
74. C. Noel, G. Bossis, A.M. Chaze, F. Giulieri, S. Lacis, Phys. Rev. Lett. **96**(21), 217801 (2006)
75. K. Takahashi, M. Ichikawa, Y. Kimura, J. Phys.: Condens. Matter **20**(7), 075106 (2008)
76. N. Kondo, Y. Iwashita, Y. Kimura, Phys. Rev. E **82**(2), 020701 (2010)
77. K. Izaki, Y. Kimura, Phys. Rev. E **88**(5), 054501 (2013)
78. O.P. Pishnyak, S.V. Shiyanovskii, O.D. Lavrentovich, Phys. Rev. Lett. **106**(4), 047801 (2011)
79. I. Smalyukh, O. Lavrentovich, A. Kuzmin, A. Kachynski, P. Prasad, Phys. Rev. Lett. **95**(15), 157801 (2005)
80. J. Kotar, M. Vilfan, N. Osterman, D. Babič, M. Čopič, I. Poberaj, Phys. Rev. Lett. **96**(20), 207801 (2006)
81. R. Ruhwandl, E. Terentjev, Phys. Rev. E **56**(5), 5561 (1997)
82. M. Vilfan, N. Osterman, M. Čopič, M. Ravnik, S. Žumer, J. Kotar, D. Babič, I. Poberaj, Phys. Rev. Lett. **101**(23), 237801 (2008)
83. S.J. Kim, J.H. Kim, Soft Matter **10**(15), 2664 (2014)
84. M. Škarabot, M. Ravnik, S. Žumer, U. Tkalec, I. Poberaj, D. Babič, N. Osterman, I. Muševič, Phys. Rev. E **77**(3), 031705 (2008)
85. K. Takahashi, M. Ichikawa, Y. Kimura, Phys. Rev. E **77**(2), 020703 (2008)
86. R. Ruhwandl, E. Terentjev, Phys. Rev. E **55**(3), 2958 (1997)
87. U. Ognysta, A. Nych, V. Nazarenko, I. Muševič, M. Škarabot, M. Ravnik, S. Žumer, I. Poberaj, D. Babič, Phys. Rev. Lett. **100**(21), 217803 (2008)
88. U. Ognysta, A. Nych, V. Nazarenko, M. Škarabot, I. Muševič, Langmuir **25**(20), 12092 (2009)
89. U.M. Ognysta, A.B. Nych, V.A. Uzunova, V.M. Pergamenshchik, V.G. Nazarenko, M. Škarabot, I. Muševič, Phys. Rev. E **83**(4), 41709 (2011)
90. S.V. Burylov, Y.L. Raikher, Phys. Rev. E **50**(1), 358 (1994)
91. S. Burylov, Y.L. Raikher, Phys. Lett. A **149**(5–6), 279 (1990)
92. D. Andrienko, M.P. Allen, G. Skačej, S. Žumer, Phys. Rev. E **65**(4), 041702 (2002)
93. D. Andrienko, M. Tasinkevych, P. Patricio, M.P. Allen, M.T. da Gama, Phys. Rev. E **68**(5), 051702 (2003)
94. D. Andrienko, G. Germano, M.P. Allen, Phys. Rev. E **63**(4), 041701 (2001)
95. D.L. Cheung, M.P. Allen, Phys. Rev. E **76**(4), 041706 (2007)
96. D.L. Cheung, M.P. Allen, Phys. Rev. E **74**(2), 021701 (2006)
97. D.L. Cheung, M.P. Allen, Langmuir **24**(4), 1411 (2008)
98. C.J. Smith, C. Denniston, J. Appl. Phys. **101**(1), 014305 (2007)
99. U. Tkalec, M. Škarabot, I. Muševič, Soft Matter **4**(12), 2402 (2008)
100. A. Eremin, P. Hirankittiwong, N. Chattham, H. Nádasi, R. Stannarius, J. Limtrakul, O. Haba, K. Yonetake, H. Takezoe, Proc. Natl. Acad. Sci. U. S. A. **112**(6), 1716 (2015)
101. M. Nikkhou, M. Škarabot, S. Čopar, M. Ravnik, S. Žumer, I. Muševič, Nat. Phys. **11**(2), 183 (2015)
102. F.R. Hung, O. Guzmán, B.T. Gettelfinger, N.L. Abbott, J.J. de Pablo, Phys. Rev. E **74**(1), 011711 (2006)
103. M. Conradi, M. Ravnik, M. Bele, M. Zorko, S. Žumer, I. Muševič, Soft Matter **5**(20), 3905 (2009)
104. B. Senyuk, I.I. Smalyukh, Soft Matter **8**(33), 8729 (2012)
105. J.C. Loudet, A.G. Yodh, B. Pouligny, Phys. Rev. Lett. **97**(1), 018304 (2006)
106. C.P. Lapointe, T.G. Mason, I.I. Smalyukh, Science **326**(5956), 1083 (2009)
107. J. Dontabhaktuni, M. Ravnik, S. Žumer, Soft Matter **8**(5), 1657 (2012)
108. J.S. Evans, C.N. Beier, I.I. Smalyukh, J. Appl. Phys. **110**(3), 033535 (2011)

109. Q. Liu, Y. Cui, D. Gardner, X. Li, S. He, I.I. Smalyukh, Nano Lett. **10**(4), 1347 (2010)
110. G.M. Koenig, B.T. Gettelfinger, J.J. de Pablo, N.L. Abbott, Nano Lett. **8**(8), 2362 (2008)
111. A.R. Tao, D.P. Ceperley, P. Sinsermsuksakul, A.R. Neureuther, P. Yang, Nano Lett. **8**(11), 4033 (2008)
112. B. Senyuk, J.S. Evans, P.J. Ackerman, T. Lee, P. Manna, L. Vigderman, E.R. Zubarev, J. van de Lagemaat, I.I. Smalyukh, Nano Lett. **12**(2), 955 (2012)
113. C. Blanc, D. Coursault, E. Lacaze, Liq. Cryst. Rev. **1**(2), 83 (2013)
114. W.L. Barnes, A. Dereux, T.W. Ebbeson, Nature **424**, 824 (2003)
115. J. Pendry, D. Schurig, D. Smith, Science **312**, 1780–1782 (2006)
116. W. Cai, U.K. Chettiar, A.V. Kildishev, V.M. Shalaev, Nat. Photon **1**(4), 224 (2007)
117. M. Škarabot, I. Muševič, Soft Matter **6**(21), 5476 (2010)
118. A. Ryzhkova, I. Muševič, Phys. Rev. E **87**(3), 032501 (2013)
119. A. Ryzhkova, M. Škarabot, I. Muševič, Phys. Rev. E **91**(4), 042505 (2015)
120. M. Ravnik, G.P. Alexander, J.M. Yeomans, S. Žumer, Proc. Natl. Acad. Sci. U. S. A. **108**(13), 5188 (2011)
121. X. Wang, D.S. Miller, E. Bukusoglu, J.J. de Pablo, N.L. Abbott, Nat. Mater. (2015)
122. C. Williams, P. Piéranski, P. Cladis, Phys. Rev. Lett. **29**(2), 90 (1972)
123. U. Tkalec, M. Ravnik, S. Žumer, I. Muševič, Phys. Rev. Lett. **103**(12), 127801 (2009)
124. R.B. Meyer, Mol. Cryst. Liq. Cryst. **16**(4), 355 (1972)
125. A. Mertelj, D. Lisjak, M. Drofenik, M. Čopič, Nature **504**(7479), 237 (2013)
126. N.A. Clark, Nature **504**(7479), 229 (2013)
127. Q. Zhang, P.J. Ackerman, Q. Liu, I.I. Smalyukh, Phys. Rev. Lett. **115**(9), 097802 (2015)
128. Q. Liu, P.J. Ackerman, T.C. Lubensky, I.I. Smalyukh, Proc. Natl. Acad. Sci. U. S. A. **113**(38), 10479 (2016)
129. V. Jampani, M. Škarabot, S. Čopar, S. Žumer, I. Muševič, Phys. Rev. Lett. **110**(17), 177801 (2013)
130. V.S.R. Jampani, Chiral nematic colloidal interactions and photonic properties of nematic colloids. Ph.D. thesis. Faculty of Mathematics and Physics (University of Ljubljana, Slovenia, 2013)
131. I. Gvozdovskyy, V.S.R. Jampani, M. Škarabot, I. Muševič, Eur. Phys. J. E **36**(9), 97 (2013)
132. A. Nych, U. Ognysta, I. Muševič, D. Seč, M. Ravnik, S. Žumer, Phys. Rev. E **89**(6), 062502 (2014)
133. J. Lintuvuori, K. Stratford, M. Cates, D. Marenduzzo, Phys. Rev. Lett. **105**(17), 178302 (2010)
134. J. Lintuvuori, A. Pawsey, K. Stratford, M. Cates, P. Clegg, D. Marenduzzo, Phys. Rev. Lett. **110**(18), 187801 (2013)
135. N. Hijnen, T.A. Wood, D. Wilson, P.S. Clegg, Langmuir **26**(16), 13502 (2010)
136. E.G. Rawson, A. May, Appl. Phys. Lett. **8**(4), 93 (1966)
137. A. May, E. Rawson, E. Hara, J. Appl. Phys. **38**(13), 5290 (1967)
138. A. Ashkin, Phys. Rev. Lett. **24**(4), 156 (1970)
139. A. Ashkin, J. Dziedzic, Appl. Phys. Lett. **19**(8), 283 (1971)
140. A. Ashkin, J. Dziedzic, Appl. Phys. Lett. **24**(12), 586 (1974)
141. A. Ashkin, J. Dziedzic, Science **187**(4181), 1073 (1975)
142. A. Ashkin, Sci. Am. **226**, 62 (1972)
143. A. Ashkin, Phys. Rev. Lett. **40**(12), 729 (1978)
144. A. Ashkin, Science **210**(4474), 1081 (1980)
145. A. Ashkin, J. Dziedzic, J. Bjorkholm, S. Chu, Opt. Lett. **11**(5), 288 (1986)
146. A. Ashkin, J. Dziedzic, T. Yamane, Nature **330**(6150), 769 (1987)
147. G. Roosen, C. Imbert, Phys. Lett. A **59**(1), 6 (1976)
148. A. Ashkin, Biophys. J. **61**(2), 569 (1992)
149. K. Svoboda, S.M. Block, Annu. Rev. Biophys. Biomol. Struct. **23**(1), 247 (1994)
150. J.E. Molloy, M.J. Padgett, Contemp. Phys. **43**(4), 241 (2002)
151. A.E. Siegman, *Lasers* (University science books, Sausalito, 1986)
152. N. Osterman, Comput. Phys. Commun. **181**(11), 1911 (2010)
153. S.M. Block, D.F. Blair, H.C. Berg, Nature **338**, 514 (1989)

154. S.M. Block, L.S. Goldstein, B.J. Schnapp, Nature **348**, 348 (1990)
155. M. Friese, T. Nieminen, N. Heckenberg, H. Rubinsztein-Dunlop, Nature **394**(6691), 348 (1998)
156. S.C. Kuo, M.P. Sheetz, Science **260**, 232 (1993)
157. J.E. Curtis, B.A. Koss, D.G. Grier, Opt. Commun. **207**(1), 169 (2002)
158. D.G. Grier, Nature **424**(6950), 810 (2003)
159. P.Y. Chiou, A.T. Ohta, M.C. Wu, Nature **436**(7049), 370 (2005)
160. K. Dholakia, Nature Mater. **4**(8), 579 (2005)
161. I. Muševič, M. Škarabot, D. Babič, N. Osterman, I. Poberaj, V. Nazarenko, A. Nych, Phys. Rev. Lett. **93**(18), 187801 (2004)
162. M. Škarabot, M. Ravnik, D. Babič, N. Osterman, I. Poberaj, S. Žumer, I. Muševič, A. Nych, U. Ognysta, V. Nazarenko, Phys. Rev. E **73**(2), 021705 (2006)
163. B. Lev, A. Nych, U. Ognysta, S. Chernyshuk, V. Nazarenko, M. Škarabot, I. Poberaj, D. Babič, N. Osterman, I. Muševič, Eur. Phys. J. E **20**(2), 215 (2006)
164. I.I. Smalyukh, A.V. Kachynski, A.N. Kuzmin, P.N. Prasad, Proc. Natl. Acad. Sci. U. S. A. **103**(48), 18048 (2006)
165. S. Samitsu, Y. Takanishi, J. Yamamoto, Nat. Mater. **9**(10), 816 (2010)
166. M. Škarabot, Z. Lokar, I. Muševič, Phys. Rev. E **87**(6), 062501 (2013)
167. O.D. Lavrentovich, Soft Matter **10**(9), 1264 (2014)
168. J.i. Hotta, K. Sasaki, H. Masuhara, Appl. Phys. Lett. **71**(15), 2085 (1997)
169. Y. Iwashita, H. Tanaka, Phys. Rev. Lett. **90**(4), 045501 (2003)
170. S. Juodkazis, M. Shikata, T. Takahashi, S. Matsuo, H. Misawa, Appl. Phys. Lett. **74**(24), 3627 (1999)
171. S. Juodkazis, S. Matsuo, N. Murazawa, I. Hasegawa, H. Misawa, Appl. Phys. Lett. **82**(26), 4657 (2003)
172. H.F. Gleeson, T.A. Wood, M. Dickinson, Philos. Trans. R. Soc. A **364**(1847), 2789 (2006)
173. T.A. Wood, H.F. Gleeson, M.R. Dickinson, A.J. Wright, Appl. Phys. Lett. **84**(21), 4292 (2004)
174. Y. Yang, P. Brimicombe, N. Roberts, M. Dickinson, M. Osipov, H. Gleeson, Opt. Express **16**(10), 6877 (2008)
175. G. Cipparrone, A. Mazzulla, A. Pane, R.J. Hernandez, R. Bartolino, Adv. Mater. **23**(48), 5773 (2011)
176. G. Tkachenko, E. Brasselet, Nat. Commun. **5**, 4491 (2014)
177. R. Trivedi, D. Engström, I. Smalyukh, J. Opt. **13**(4), 044001 (2011)
178. T. Lee, R.P. Trivedi, I.I. Smalyukh, Opt. Lett. **35**(20), 3447 (2010)
179. R.P. Trivedi, T. Lee, K.A. Bertness, I.I. Smalyukh, Opt. Express **18**(26), 27658 (2010)
180. I. Smalyukh, A. Kuzmin, A. Kachynski, P. Prasad, O. Lavrentovich, Appl. Phys. Lett. **86**(2), 021913 (2005)
181. I.I. Smalyukh, D.S. Kaputa, A.V. Kachynski, A.N. Kuzmin, P.N. Prasad, Opt. Express **15**(7), 4359 (2007)
182. L. Lucchetti, L. Criante, F. Bracalente, F. Aieta, F. Simoni, Phys. Rev. E **84**(2), 021702 (2011)
183. L. Criante, F. Bracalente, L. Lucchetti, F. Simoni, E. Brasselet, Soft Matter **9**(22), 5459 (2013)
184. A. Martinez, H.C. Mireles, I.I. Smalyukh, Proc. Natl. Acad. Sci. U. S. A. **108**(52), 20891 (2011)
185. I.I. Smalyukh, Y. Lansac, N.A. Clark, R.P. Trivedi, Nat. Mater. **9**(2), 139 (2010)
186. J.C. Crocker, D.G. Grier, J. Colloid Interface Sci. **179**(1), 298 (1996)
187. J. Loudet, P. Hanusse, P. Poulin, Science **306**(5701), 1525 (2004)
188. T. Turiv, I. Lazo, A. Brodin, B.I. Lev, V. Reiffenrath, V.G. Nazarenko, O.D. Lavrentovich, Science **342**(6164), 1351 (2013)
189. E. Yablonovitch, T. Gmitter, Phys. Rev. Lett. **63**(18), 1950 (1989)
190. I. Muševič, M. Škarabot, U. Tkalec, M. Ravnik, S. Žumer, Science **313**(5789), 954 (2006)
191. M. Humar, M. Škarabot, M. Ravnik, S. Žumer, I. Poberaj, D. Babič, I. Muševič, Eur. Phys. J. E **27**(1), 73 (2008)
192. M. Ravnik, S. Žumer, Liq. Cryst. **36**(10–11), 1201 (2009)

193. A. Nych, U. Ognysta, M. Škarabot, M. Ravnik, S. Žumer, I. Muševič, Nat. Commun. **4**, 1489 (2013)
194. M. Škarabot, M. Ravnik, S. Žumer, U. Tkalec, I. Poberaj, D. Babič, I. Muševič, Phys. Rev. E **77**(6), 061706 (2008)
195. O. Guzmán, E.B. Kim, S. Grollau, N.L. Abbott, J.J. de Pablo, Phys. Rev. Lett. **91**(23), 235507 (2003)
196. T. Araki, H. Tanaka, Phys. Rev. Lett. **97**(12), 127801 (2006)
197. S. Žumer, Keystone, Colorado, July pp. 2–7 (2006)
198. M. Ravnik, M. Škarabot, S. Žumer, U. Tkalec, I. Poberaj, D. Babič, N. Osterman, I. Muševič, Phys. Rev. Lett. **99**(24), 247801 (2007)
199. S.Z. Lin, X. Wang, Y. Kamiya, G.W. Chern, F. Fan, D. Fan, B. Casas, Y. Liu, V. Kiryukhin, W.H. Zurek, C.D. Batista, S.W. Cheong, Nat. Phys. **10**(13), 970 (2014)
200. M. Ravnik, S. Žumer, Soft Matter **5**(22), 4520 (2009)
201. S. Čopar, T. Porenta, V.S.R. Jampani, I. Muševič, S. Žumer, Soft Matter **8**(33), 8595 (2012)
202. S. Čopar, S. Žumer, Proc. R. Soc. A **469**(2156), 20130204 (2013)
203. K.S. Chichak, S.J. Cantrill, A.R. Pease, S.H. Chiu, G.W. Cave, J.L. Atwood, J.F. Stoddart, Science **304**(5675), 1308 (2004)
204. J.P. Sauvage, C. Dietrich-Buchecker (eds.), *Molecular Catenanes, Rotaxanes and Knots: A Journey Through the World of Molecular Topology* (Wiley, Weinheim, 1999)
205. P.G. de Gennes, *Scaling Concepts in Polymer Physics* (Cornell University Press, New York, 1979)
206. L. Faddeev, A. Niemi, Nature **387**, 58 (1997)
207. H. Kedia, I. Bialynicki-Birula, D. Peralta-Salas, W.T. Irvine, Phys. Rev. Lett. **111**(15), 150404 (2013)
208. M.R. Dennis, R.P. King, B. Jack, K. O'Holleran, M.J. Padgett, Nat. Phys. **6**(2), 118 (2010)
209. D. Kleckner, W.T.M. Irvine, Nat. Phys. **9**(4), 253 (2013)
210. D. Han, S. Pal, Y. Liu, H. Yan, Nat. Nanotechnol. **5**(10), 712 (2010)
211. Y. Bouligand, J. Phys. **35**(3), 215 (1974)
212. U. Tkalec, M. Ravnik, S. Čopar, S. Žumer, I. Muševič, Science **333**(6038), 62 (2011)
213. V.S.R. Jampani, M. Škarabot, M. Ravnik, S. Čopar, S. Žumer, I. Muševič, Phys. Rev. E **84**(3), 031703 (2011)
214. C.C. Adams, *The Knot Book: An Elementary Introduction to the Mathematical Theory of Knots* (American Mathematical Society, Providence, 2004)
215. S. Čopar, U. Tkalec, I. Muševič, S. Žumer, Proc. Natl. Acad. Sci. U. S. A. **112**(6), 1675 (2015)
216. S. Čopar, N.A. Clark, M. Ravnik, S. Žumer, Soft Matter **9**(34), 8203 (2013)
217. Y. Bouligand, J. Phys. **34**(11–12), 1011 (1973)
218. Y. Bouligand, J. Phys. **35**(12), 959 (1974)
219. Y. Bouligand, B. Derrida, V. Poenaru, Y. Pomeau, G. Toulouse, J. Phys. **39**(8), 863 (1978)
220. Y. Bouligand, M. Kléman, J. Phys. **40**(1), 79 (1979)
221. Y. Bouligand, F. Livolant, J. Phys. **45**(12), 1899 (1984)
222. M. Nikkhou, M. Škarabot, I. Muševič, Phys. Rev. E **93**(6), 062703 (2016)
223. I. Chuang, R. Durrer, N. Turok, B. Yurke, Science **251**(4999), 1336 (1991). doi:10.1126/science.251.4999.1336
224. M.J. Bowick, L. Chandar, E.A. Schiff, A.M. Srivastava, Science **263**(5149), 943 (1994). doi:10.1126/science.263.5149.943
225. M. Cavallaro Jr., M.A. Gharbi, D.A. Beller, S. Čopar, Z. Shi, R.D. Kamien, S. Yang, T. Baumgart, K.J. Stebe, Soft Matter **9**(38), 9099 (2013)
226. B. Senyuk, M.B. Pandey, Q. Liu, M. Tasinkevych, I.I. Smalyukh, Soft Matter **11**(45), 8758 (2015)
227. A. Martinez, T. Lee, T. Asavei, H. Rubinsztein-Dunlop, I.I. Smalyukh, Soft Matter **8**(8), 2432 (2012)
228. A. Martinez, M. Ravnik, B. Lucero, R. Visvanathan, S. Žumer, I.I. Smalyukh, Nat. Mater. **13**(3), 258 (2014)
229. K. Robbie, D. Broer, M. Brett, Nature **399**(6738), 764 (1999)

230. S. Donaldson, *Riemann Surfaces* (Oxford University Press, Oxford, 2011)
231. T. Machon, G.P. Alexander, Proc. Natl. Acad. Sci. U. S. A. **110**(35), 14174 (2013)
232. G.P. Alexander, B.G.g. Chen, E.A. Matsumoto, R.D. Kamien. Rev. Mod. Phys. **84**(2), 497 (2012)
233. S.M. Hashemi, U. Jagodic, M.R. Mozaffari, M.R. Ejtehadi, I. Muševič, M. Ravnik, Nature Communications, **8**, 14026 (2017). doi:10.1038/ncomms14026
234. L. Tortora, O.D. Lavrentovich, Proc. Natl. Acad. Sci. U. S. A. **108**(13), 5163 (2011). doi:10.1073/pnas.1100087108
235. J. Jeong, Z.S. Davidson, P.J. Collings, T.C. Lubensky, A.G. Yodh, Proc. Natl. Acad. Sci. U. S. A. **111**(5), 1742 (2014). doi:10.1073/pnas.1315121111
236. E. Dubois-Violette, O. Parodi, J. Phys. Colloq. **30**(C4), C4 (1969)
237. G. Volovik, O. Lavrentovich, Zh Eksp, Teor. Fiz. **85**, 1997 (1983)
238. I.H. Lin, D.S. Miller, P.J. Bertics, C.J. Murphy, J.J. de Pablo, N.L. Abbott, Science **332**(6035), 1297 (2011)
239. J.K. Gupta, J.S. Zimmerman, J.J. de Pablo, F. Caruso, N.L. Abbott, Langmuir **25**(16), 9016 (2009)
240. Y. Bai, N.L. Abbott, Langmuir **27**(10), 5719 (2011)
241. M. Humar, M. Ravnik, S. Pajk, I. Muševič, Nature Photon. **3**(10), 595 (2009)
242. M. Humar, I. Muševič, Opt. Express **19**(21), 19836 (2011)
243. I. Muševič, M. Humar, Proc. SPIE **7955**, 795509 (2011)
244. M. Press, A. Arrott, Phys. Rev. Lett. **33**(7), 403 (1974)
245. M. Press, A. Arrott, J. Phys. Colloq. **36**(C1), C1 (1975)
246. A. Koval'chuk, M. Kurik, O. Lavrentovich, V. Sergan, Zh Eksp, Teor. Fiz. **94**, 350 (1988)
247. O. Lavrentovich, V. Sergan, Il Nuovo Cimento D **12**(9), 1219 (1990)
248. J. Doane, N. Vaz, B.G. Wu, S. Žumer, Appl. Phys. Lett. **48**(4), 269 (1986)
249. S. Žumer, S. Kralj, J. Bezič, Mol. Cryst. Liq. Cryst. **212**, 163 (1992)
250. P.P. Crooker, H.S. Kitzerow, F. Xu, in *Proceedings of SPIE*, vol. 2175 (International Society for Optics and Photonics, 1994), pp. 173–182
251. F. Xu, H.S. Kitzerow, P. Crooker, Phys. Rev. A **46**(10), 6535 (1992)
252. F. Xu, H.S. Kitzerow, P. Crooker, Phys. Rev. E **49**(4), 3061 (1994)
253. E.M. de Groot, G.G. Fuller, Liq. Cryst. **23**(1), 113 (1997)
254. J.K. Whitmer, X. Wang, F. Mondiot, D.S. Miller, N.L. Abbott, J.J. de Pablo, Phys. Rev. Lett. **111**(22), 227801 (2013)
255. C. Robinson, J. Ward, Nature **180**(4596), 1183 (1957)
256. C. Robinson, J.C. Ward, R.B. Beevers, Disc. Faraday Soc. **25**, 29–42 (1958)
257. M. Kurik, O. Lavrentovich, Mol. Cryst. Liq. Cryst. **72**(7–8), 239 (1982)
258. M. Kurik, O. Lavrentovich, J. Exp. Theor. Phys. Lett. **33**(10), 528 (1981)
259. D. Seč, T. Porenta, M. Ravnik, S. Žumer, Soft Matter **8**(48), 11982 (2012)
260. J. Bajc, J. Bezič, S. Žumer, Phys. Rev. E **51**(3), 2176 (1995)
261. F. Xu, P. Crooker, Phys. Rev. E **56**(6), 6853 (1997)
262. R. Swisher, H. Huo, P. Crooker, Liq. Cryst. **26**(1), 57 (1999)
263. J. Bezič, S. Žumer, Liq. Cryst. **11**(4), 593 (1992)
264. Y. Zhou, E. Bukusoglu, J.A. Martínez-González, M. Rahimi, T.F. Roberts, R. Zhang, X. Wang, N.L. Abbott, J.J. de Pablo, ACS Nano **10**(7), 6484 (2016). doi:10.1021/acsnano.6b01088
265. D. Seč, S. Čopar, S. Žumer, Nat. Commun. **5**, 3057 (2014)
266. S. Candau, P. Le Roy, F. Debeauvais, Mol. Cryst. Liq. Cryst. **23**(3–4), 283 (1973)
267. M. Kurik, O. Lavrentovich, J. Exp. Theor. Phys. Lett. **35**(9), 444 (1982)
268. T. Orlova, S.J. Asshoff, T. Yamaguchi, N. Katsonis, E. Brasselet, Nat. Commun. **6**, 7603 (2015)
269. J.K. Guo, J.K. Song, Opt. Express **24**(7), 7381 (2016)
270. E. Pairam, J. Vallamkondu, V. Koning, B.C. van Zuiden, P.W. Ellis, M.A. Bates, V. Vitelli, A. Fernández-Nieves, Proc. Natl. Acad. Sci. U. S. A. **110**(23), 9295 (2013)
271. V. Koning, B.C. van Zuiden, R.D. Kamien, V. Vitelli, Soft Matter **10**(23), 4192 (2014)
272. M.G. Campbell, M. Tasinkevych, I.I. Smalyukh, Phys. Rev. Lett. **112**(19), 197801 (2014)

273. M. Tasinkevych, M.G. Campbell, I.I. Smalyukh, Proc. Natl. Acad. Sci. U. S. A. **111**(46), 16268 (2014)
274. D.R. Nelson, Nano Lett. **2**(10), 1125 (2002)
275. K. Ho, C.T. Chan, C.M. Soukoulis, Phys. Rev. Lett. **65**(25), 3152 (1990)
276. T. Lubensky, J. Prost, J. Phys. II **2**(3), 371 (1992)
277. T. Lopez-Leon, A. Fernández-Nieves, Phys. Rev. E **79**(2 Pt 1), 021707 (2009)
278. T. Lopez-Leon, V. Koning, K. Devaiah, V. Vitelli, A. Fernández-Nieves, Nat. Phys. **7**(5), 391 (2011)
279. T. Lopez-Leon, A. Fernández-Nieves, Colloid Polym. Sci. **289**(4), 345 (2011)
280. V. Koning, T. Lopez-Leon, A. Fernández-Nieves, V. Vitelli, Soft Matter **9**(20), 4993 (2013)
281. M.A. Gharbi, D. Seč, T. Lopez-Leon, M. Nobili, M. Ravnik, S. Žumer, C. Blanc, Soft Matter **9**(29), 6911 (2013)
282. H.L. Liang, R. Zentel, P. Rudquist, J. Lagerwall, Soft Matter **8**(20), 5443 (2012)
283. D. Jesenek, S. Kralj, R. Rosso, E.G. Virga, Soft Matter **11**(12), 2434 (2015)
284. L. Mesarec, W. Góźdź, A. Iglič, S. Kralj, Sci. Rep. **6**, 27117 (2016)
285. L. Mirantsev, E. de Oliveira, I. de Oliveira, M. Lyra, Liq. Cryst. Rev. **4**(1), 35 (2016)
286. F. Serra, Liq. Cryst. (2016). doi:10.1080/02678292.2016.1209698
287. M. Kawachi, O. Kogure, Y. Kato, Jpn. J. Appl. Phys. **13**(9), 1457 (1974)
288. W.E. Haas, J.E. Adams, Appl. Phys. Lett. **25**(5), 263 (1974)
289. Y. Bouligand, J. Phys. **41**(11), 1307 (1980)
290. S. Pirkl, P. Ribiere, P. Oswald, Liq. Cryst. **13**(3), 413 (1993)
291. T.H.R. Skyrme, in *Proceedings of the Royal Society A*, vol. 260 (The Royal Society, 1961), pp. 127–138
292. T.H.R. Skyrme, Nucl. Phys. **31**, 556 (1962)
293. A. Belavin, A. Polyakov, J. Exp. Theor. Phys. Lett. **22**, 245 (1975)
294. I. Kézsmárki, S. Bordács, P. Milde, E. Neuber, L. Eng, J. White, H.M. Ronnow, C. Dewhurst, M. Mochizuki, K. Yanai, H. Nakamura, D. Ehlers, V. Tsurkan, A. Loidl, Nat. Mater. **14**(11), 1116 (2015)
295. A. Bogdanov, U. Rößler, A. Shestakov, Phys. Rev. E **67**(1), 016602 (2003)
296. A. Leonov, I. Dragunov, U. Rößler, A. Bogdanov, Phys. Rev. E **90**(4), 042502 (2014)
297. J.i. Fukuda, S. Žumer, Nat. Commun. **2**, 246 (2011). doi:10.1038/ncomms1250
298. R. Rajaraman, *Solitons and Instantons* (North-Holland, Amsterdam, 1982)
299. N. Manton, P. Sutcliffe, *Topological Solitons* (Cambridge University Press, Cambridge, 2004)
300. P.J. Ackerman, R.P. Trivedi, B. Senyuk, J. van de Lagemaat, I.I. Smalyukh, Phys. Rev. E **90**(1), 012505 (2014)
301. I. Smalyukh, O. Lavrentovich, Phys. Rev. E **66**(051703) (2002)
302. X. Hu, P. Jiang, C. Ding, H. Yang, Q. Gong, Nature Photon. **2**(3), 185 (2008)
303. H.G. Park, C.J. Barrelet, Y. Wu, B. Tian, F. Qian, C.M. Lieber, Nature Photon. **2**(10), 622 (2008)
304. O. Benson, Nature **480**(7376), 193 (2011)
305. B. Ellis, M.A. Mayer, G. Shambat, T. Sarmiento, J. Harris, E.E. Haller, J. Vučković, Nature Photon. **5**(5), 297 (2011)
306. D. Liang, J.E. Bowers, Nature Photon. **4**(8), 511 (2010)
307. H. Coles, S. Morris, Nature Photon. **4**(10), 676 (2010)
308. V. Kopp, B. Fan, H. Vithana, A. Genack, Opt. Lett. **23**(21), 1707 (1998)
309. B. Taheri, A. Munoz, P. Palffy-Muhoray, R. Twieg, Mol. Cryst. Liq. Cryst. **358**(1), 73 (2001)
310. H. Finkelmann, S.T. Kim, A. Munoz, P. Palffy-Muhoray, B. Taheri, Adv. Mater. **13**(14), 1069 (2001)
311. T. Matsui, R. Ozaki, K. Funamoto, M. Ozaki, K. Yoshino, Appl. Phys. Lett. **81**(20), 3741 (2002)
312. W. Cao, A. Muñoz, P. Palffy-Muhoray, B. Taheri, Nat. Mater. **1**(2), 111 (2002)
313. M. Kasano, M. Ozaki, K. Yoshino, D. Ganzke, W. Haase, Appl. Phys. Lett. **82**(23), 4026 (2003)

314. A. Chanishvili, G. Chilaya, G. Petriashvili, R. Barberi, R. Bartolino, G. Cipparrone, A. Mazzulla, L. Oriol, Appl. Phys. Lett. **83**(26), 5353 (2003)
315. S. Furumi, S. Yokoyama, A. Otomo, S. Mashiko, Appl. Phys. Lett. **82**(1), 16 (2003)
316. S. Furumi, S. Yokoyama, A. Otomo, S. Mashiko, Appl. Phys. Lett. **84**(14), 2491 (2004)
317. A. Chanishvili, G. Chilaya, G. Petriashvili, R. Barberi, M.P. De Santo, M.A. Matranga, F. Ciuchi, Appl. Phys. Lett. **88**(10), 101105 (2006)
318. S. Morris, A. Ford, C. Gillespie, M. Pivnenko, O. Hadeler, H. Coles, J. Soc. Inf. Disp. **14**(6), 565 (2006)
319. T. Ohta, M.H. Song, Y. Tsunoda, T. Nagata, K.C. Shin, F. Araoka, Y. Takanishi, K. Ishkawa, J. Watanabe, S. Nishimura, T. Toyooka, H. Takezoe, Jpn. J. Appl. Phys. **43**(9R), 6142 (2004)
320. M. Ozaki, M. Kasano, D. Ganzke, W. Haase, K. Yoshino, Adv. Mater. **14**(4), 306 (2002)
321. A.D. Ford, S.M. Morris, M.N. Pivnenko, H.J. Coles, Proc. SPIE **5289**, 213 (2004)
322. P.V. Shibaev, P. Rivera, D. Teter, S. Marsico, M. Sanzari, V. Ramakrishnan, E. Hanelt, Opt. Express **16**(5), 2965 (2008)
323. P. Shibaev, R. Sanford, D. Chiappetta, V. Milner, A. Genack, A. Bobrovsky, Opt. Express **13**(7), 2358 (2005)
324. J. Schmidtke, W. Stille, H. Finkelmann, S.T. Kim, Adv. Mater. **14**(10), 746 (2002)
325. P. Shibaev, V. Kopp, A. Genack, E. Hanelt, Liq. Cryst. **30**(12), 1391 (2003)
326. P. Shibaev, J. Madsen, A. Genack, Chem. Mater. **16**(8), 1397 (2004)
327. P. Shibaev, K. Tang, A. Genack, V. Kopp, M. Green, Macromolecules **35**(8), 3022 (2002)
328. P.V. Shibaev, L. Newman, Liq. Cryst. **40**(3), 428 (2013)
329. S.K. Wei, S.H. Chen, K. Dolgaleva, S.G. Lukishova, R.W. Boyd, Appl. Phys. Lett. **94**(4), 041111 (2009)
330. S. Yokoyama, S. Mashiko, H. Kikuchi, K. Uchida, T. Nagamura, Adv. Mater. **18**(1), 48 (2006)
331. G. Toquer, T. Phou, S. Monge, A. Grimaldi, M. Nobili, C. Blanc, J. Phys. Chem. B **112**(14), 4157 (2008)
332. M. Humar, I. Muševič, Opt. Express **18**(26), 26995 (2010)
333. I. Muševič, M. Škarabot, M. Humar, J. Phys.: Condens. Matter **23**(28), 284112 (2011)
334. I. Muševič, Philos. Trans. R. Soc. A **371**(1988), 20120266 (2013)
335. I. Muševič, Liq. Cryst. **41**(3), 418 (2014)
336. I. Muševič, H. Peng, M. Nikkhou, M. Humar, Proc. SPIE **8960**, 896016 (2014)
337. I. Muševič, Liq. Cryst. Rev. **4**(1), 1 (2016)
338. A. Ashkin, J. Dziedzic, Phys. Rev. Lett. **38**(23), 1351 (1977)
339. K.J. Vahala, Nature **424**(6950), 839 (2003)
340. S.X. Qian, J.B. Snow, H.M. Tzeng, R.K. Chang, Science **231**(4737), 486 (1986)
341. M. Saito, H. Shimatani, H. Naruhashi, Opt. Express **16**(16), 11915 (2008)
342. A. Kiraz, A. Kurt, M.A. Dündar, A.L. Demirel, Appl. Phys. Lett. **89**(8), 081118 (2006)
343. B. Maune, R. Lawson, C. Gunn, A. Scherer, L. Dalton, Appl. Phys. Lett. **83**(23), 4689 (2003)
344. T.J. Wang, C.H. Chu, C.Y. Lin, Opt. Lett. **32**(19), 2777 (2007)
345. C.H. Dong, L. He, Y.F. Xiao, V. Gaddam, S. Ozdemir, Z.F. Han, G.C. Guo, L. Yang, Appl. Phys. Lett. **94**(23), 231119 (2009)
346. R.J. Carlton, J.T. Hunter, D.S. Miller, R. Abbasi, P.C. Mushenheim, L.N. Tan, N.L. Abbott, Liq. Cryst. Rev. **1**(1), 29 (2013)
347. U. Manna, Y.M. Zayas-Gonzalez, R.J. Carlton, F. Caruso, N.L. Abbott, D.M. Lynn, Angew. Chem. **52**(52), 14011 (2013)
348. G. Tkachenko, E. Brasselet, Phys. Rev. Lett. **111**(3), 033605 (2013)
349. M. Donato, J. Hernandez, A. Mazzulla, C. Provenzano, R. Saija, R. Sayed, S. Vasi, A. Magazzù, P. Pagliusi, R. Bartolino et al., Nat. Commun. **5**, 3656 (2014). doi:10.1038/ncomms4656
350. P. Hands, D. Gardiner, S. Morris, C. Mowatt, T. Wilkinson, H. Coles, Appl. Phys. Lett. **98**(14), 141102 (2011)
351. J.L. Zhu, S.B. Ni, C.P. Chen, D.Q. Wu, X.L. Song, C.Y. Chen, J.G. Lu, Y. Su, H.P.D. Shieh, Appl. Phys. Lett. **104**(9), 091116 (2014)
352. J.L. Zhu, W.H. Li, Y. Sun, J.G. Lu, X.L. Song, C.Y. Chen, Z. Zhang, Y. Su, Appl. Phys. Lett. **106**(19), 191903 (2015)

353. Y. Geng, D. Seč, P.L. Almeida, O.D. Lavrentovich, S. Žumer, M.H. Godinho, Soft Matter **9**(33), 7928 (2013)
354. K. Peddireddy, P. Kumar, S. Thutupalli, S. Herminghaus, C. Bahr, Langmuir **28**(34), 12426 (2012)
355. K. Peddireddy, V. Jampani, S. Thutupalli, S. Herminghaus, C. Bahr, I. Muševič, Opt. Express **21**(25), 30233 (2013)

Index

I. Muševič, *Liquid Crystal Colloids*, Soft and Biological Matter,
DOI 10.1007/978-3-319-54916-3

D